Taxonomy of Porifera

From the N.E. Atlantic and Mediterranean Sea

NATO ASI Series

Advanced Science Institutes Series

A series presenting the results of activities sponsored by the NATO Science Committee, which aims at the dissemination of advanced scientific and technological knowledge, with a view to strengthening links between scientific communities.

The Series is published by an international board of publishers in conjunction with the NATO Scientific Affairs Division

A Life Sciences B Physics	Plenum Publishing Corporation London and New York
C Mathematical and Physical Sciences	D. Reidel Publishing Company Dordrecht, Boston, Lancaster and Tokyo
D Behavioural and Social Sciences E Applied Sciences	Martinus Nijhoff Publishers Boston, The Hague, Dordrecht and Lancaster
F Computer and Systems Sciences G Ecological Sciences H Cell Biology	Springer-Verlag Berlin Heidelberg New York London Paris Tokyo

Series G: Ecological Sciences Vol. 13

Taxonomy of Porifera

From the N.E. Atlantic and Mediterranean Sea

Edited by

Jean Vacelet and Nicole Boury-Esnault

Centre d'Océanologie de Marseille, Station Marine d'Endoume
Rue de la Batterie des Lions, 13007 Marseille, France

Springer-Verlag
Berlin Heidelberg New York London Paris Tokyo
Published in cooperation with NATO Scientific Affairs Divison

Proceedings of the NATO Advanced Research Workshop on Taxonomy of Porifera from the N.E. Atlantic and Mediterranean Sea held at Marseille, France, September 22–27, 1986

ISBN 3-540-16091-4 Springer-Verlag Berlin Heidelberg New York
ISBN 0-387-16091-4 Springer-Verlag New York Berlin Heidelberg

Library of Congress Cataloging in Publication Data. NATO Advanced Research Workshop on Taxonomy of Porifera from the N.E. Atlantic and Mediterranean Sea (1986 : Marseille, France) Taxonomy of Porifera, from the N.E. Atlantic and Mediterranean Sea. (NATO ASI series. Series G, Ecological sciences ; vol. 13) "Proceedings of the NATO Advanced Research Workshop on Taxonomy of Porifera from the N.E. Atlantic and Mediterranean Sea, held at Marseille, France, September 22–27, 1986."–Copr. p. 1. Sponges–Mediterranea Sea–Classification–Congresses. 2. Sponges–North Atlantic Ocean–Classification–Congresses. I. Vacelet, Jean. II. Boury-Esnault, Nicole. III. Title. IV. Series: NATO ASI series. Series G, Ecological sciences ; no. 13. QL372.4.M47N38 1986 593.4'09163'1 87-13129
ISBN 0-387-16091-4 (U.S.)

Printed in Germany

Printing: Druckhaus Beltz, Hemsbach; Bookbinding: J. Schäffer GmbH & Co. KG, Grünstadt
2131/3140-543210

PREFACE

Although sponges are one of the major components of littoral ecosystems, their systematics obviously lags behind that of the majority of other groups of marine invertebrates. Also the evolutionary trends in these most primitive of the Metazoans are poorly understood. The confused state of their taxonomy and the difficulties with their identification prevent their use in ecological studies. The emphasis of interest presently seems to have shifted towards chemistry, but many of the hundreds of interesting chemicals recently described in sponges probably have been found in mistakenly or imprecisely identified animals. This situation will become worse in the next few years, when a number of experts will retire and will not be replaced by new ones, owing to the difficulties that the science of systematics is facing in most countries — as exemplified during this meeting by an informal discussion on the endangered future of European Museums.

Under these circumstances, it is the responsability of present sponge taxonomists to leave a less confused state and tools such as simple fauna or computerized data bases to generations following. These tools do not exist even in one of the best studied areas, the North-Eastern Atlantic (including the Mediterranean Sea). A small group of European sponge taxonomists met in 1983 at Sherkin Island Marine Station (Ireland), in order to initiate the "first action". A few scientific papers, edited by W. C. Jones, have been published in the Journal of Sherkin Island and a decision was made to continue the effort. The present Advanced Research Workshop, financed by NATO Scientific Division and held, in September 1986, in Marseille, is the first step of this decision.

The workshop had three broad objectives : (1) To make an attempt to reach a consensus on the major unsolved problems. (2) To establish or reinforce collaborative works. For instance, a small group of eight participants will study the constitution of a common computerized data base, which is needed drastically, but which can only be established by a joint effort. (3) To introduce or develop new methods and new concepts in the taxonomy of these animals. The introduction of cladistic concepts and methods, the use of field characteristics observed *in situ* on large populations, enzyme studies, computerized analysis of spicule forms and characteristics, chemical data on secondary metabolites, cytological data at the ultrastructural level, were ideas emphasized during the meeting. They appear most promising both in the distinction and identification of the species and in the definition of higher taxa. However, it must be noted that sponge taxonomy will continue to rely on conventional skeletal features for a long time, although we are still extremely ignorant of the importance of their seasonal and ecological variations.

Most sincere thanks are due to the NATO Scientific Division and to all the people who contributed to the success of this workshop. Lauren Gollahon, who corrected the English of non-anglophone contributors, is greatly acknowledged. A special debt of gratitude is due to Monique Verdenal for her much appreciated work in retyping and preparing all the manucripts in their camera-ready form.

J. Vacelet
Marseille, March 1987

CONTENTS

CALCAREOUS SPONGES COLLECTED BY N.O. THALASSA ON THE CONTINENTAL MARGIN OF THE BAY OF BISCAYE : I. CALCINEA.

R. BOROJEVIC* & N. BOURY-ESNAULT**

* Museum national d'Histoire naturelle, Laboratoire de Biologie des Invertébrés marins UA-699, 57 rue Cuvier 75005-Paris, France.
** Centre d'Océanologie de Marseille, UA-41, Station marine d'Endoume, rue de la Batterie des Lions, 13007-Marseille, France.

SYNOPSIS

Ten species of calcinean sponges have been collected by the N.O. Thalassa on the continental margin of the Bay of Biscaye, from 322 to 760 m. Four of these species, two genera and one family are new to science. They represent missing links in the evolution of the Calcinea.

INTRODUCTION

The collections of calcinean calcareous sponges made by the N.O. Thalassa on the continental margin of the Bay of Biscaye between 322 m and 760 m on hard substrata have shown that these sponges are quite more abundant in the bathyal bottom than previous observations have led us to believe. With the exception of a few citations as those made for *Guancha blanca*, *Ascaltis lamarcki* and *Clathrina primordialis* by Poléjaeff (1883) and Hansen (1885), calcinean calcareous sponges are generally considered to be typically littoral and have never been collected on the continental margin.

The area studied is located to the West of Brittany on the Chapelle Bank by 47°58'N and 7°50'W between 322 m and 354 m depth and in the vicinity of Galicia, 44°1'N - 7°1'W between 490 m and 630 m depth and 44° 11'N - 8°40'W between 410 m and 760 m depth.

LIST OF SPECIES

Family CLATHRINIDAE Minchin
Genus *Clathrina* Gray, 1867 emend.
Clathrina ascandroides Borojevic,1971
Clathrina biscayae n.sp.

NATO ASI Series, Vol. G13
Taxonomy of Porifera
Edited by J. Vacelet and N. Boury-Esnault

Clathrina contorta (Bowerbank,1866)
Clathrina olynthus n.sp.
Clathrina reticulum (Schmidt,1862)

Genus *Guancha* Miklucho Maclay,1868 emend.
Guancha blanca Miklucho Maclay,1868
Guancha lacunosa (Johnston,1882)

Family LEUCASCIDAE Dendy
Genus *Ascaltis* Haeckel 1872 emend.
Ascaltis lamarcki Haeckel, 1872

Family LEUCALTIDAE Dendy & Row
Genus *Leuclathrina* n.g.
Leuclathrina asconoides n.sp.

Family LEVINELLIDAE Borojevic & Boury-Esnault, 1986
Genus *Levinella* Borojevic & Boury-Esnault, 1986
Levinella thalassae Borojevic & Boury-Esnault, 1986

DESCRIPTION OF SPECIES

Family CLATHRINIDAE Minchin

Calcinea with continuous choanoderm covering the whole internal cavity of the sponge; no common cortex to the cormus.

Genus *Clathrina* Gray, 1867 emend

Clathrinidae with a smooth choanoderm, or rarely raised up into conuli by the apical actines of tetractines, but never forming continuous folds. Cormus composed of anastomosing tubes never organized radially.

Clathrina ascandroides Borojevic,1971

Stations. T 450, T 451, T 477, T 503, U 843, U 847; depth 340-560 m.

Description.

The collection contains several specimens which all have the shape of solitary tubes reaching 1 cm in height and 1.5 mm in diameter. Their osculum is naked, their very thin walls are supported by a delicate skeleton composed of relatively large spicules. The wall of the tubes is perforated by regularly disposed ostioles. The choanoderm consists of a regular layer of choanocytes which does not form folds inside the choanocoele.

The skeleton is composed of triactines and tetractines. The tetractines are often noticeably larger than the triactines. The orientation of the spicules is in general irregular, but occasionally, mainly in the region near the osculum, it becomes regular with one of the rays orientated in the basipetal direction with the unpaired angle towards the osculum. In these regions, the unpaired actine can become slightly longer than the paired rays.

Spicules (fig. 1).

Triactines : equiangular, equiradial; conical and sharp-pointed actines, 142-164/13 µm.

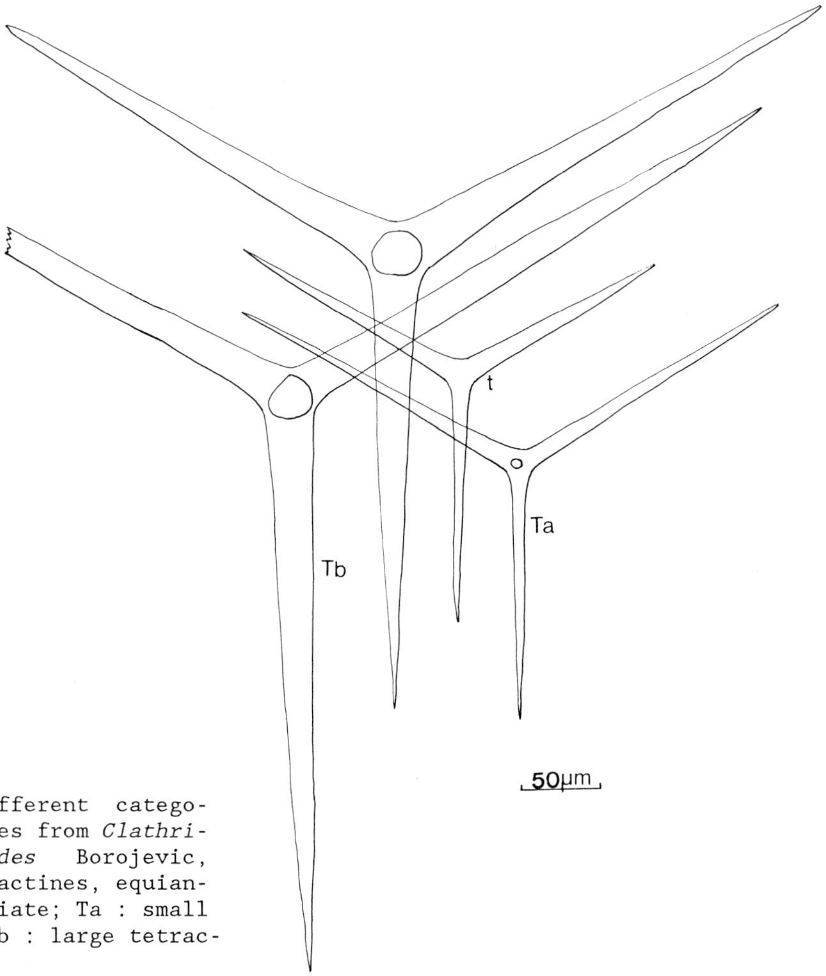

Figure 1 - Different categories of spicules from *Clathrina ascandroides* Borojevic, 1971. t : triactines, equiangular, equiradiate; Ta : small tetractines; Tb : large tetractines.

Tetractines a : small tetractines similar to triactines with a short apical actine. Actines of the basal system 161-200/11.7-13 µm; apical actine 52/5.2 µm.

b : large tetractines, straight apical actine or very slightly curved in its distal part. Actines of the basal system 299-364/26.6-31.2 µm; apical actine 62.4-114.4/15.6-20.8 µm.

Discussion.

The specimens of *Clathrina ascandroides* correspond exactly with those observed on the brazilian coasts (Borojevic,1971; Borojevic & Peixinho, 1976). We have already drawn attention to Topsent's (1892) record from the Azores of *Leucosolenia gegenbauri*, suggesting the strong possibility that it is *Clathrina ascandroides*. This species seems to have an amphi-atlantic distribution.

Distribution.

Brazil(75 m), Azores.

Clathrina biscayae n.sp.

Stations. T 437, T 450, T 453, T 475, T 476, T 477, T 787, U 847, U 852, U 878; depth 322-645 m.

Description.

The collection contains numerous small sized specimens of this sponge. They always have a clathroid body (fig.2), encrusting or more frequently globular, attached to the substrata by several peduncles. Their tubes are thin and regularly anastomose. Several young specimens are included whose body shape resembles a ramifying olynthus roughly representing the upper part of a mature clathroid body in its apical part. The osculum is always naked.

The skeleton is dense (fig.3) and composed of triactines to which tetractines are occasionally added. The presence of tetractines is not constant, but in some specimens they can be present in relatively large numbers. In bigger sponges, the skeleton is pluristratified and a differentiation of triactines of a larger size is observed at the external surface of the cormus.

In the clathroid body, the spicules are more often disordered, but in

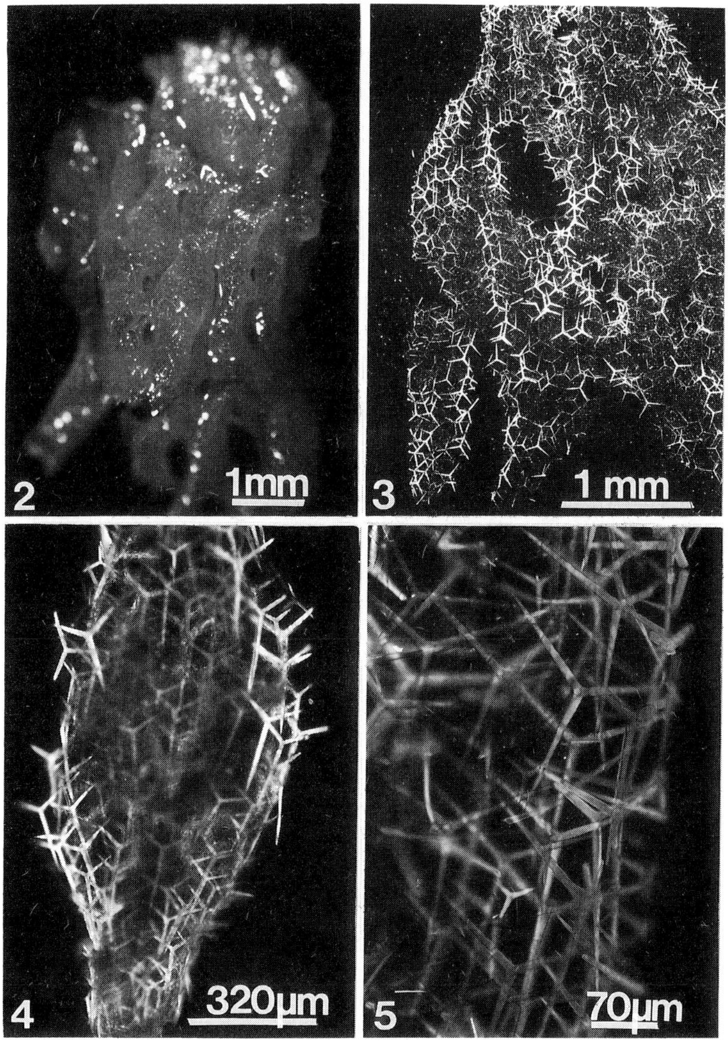

Clathrina biscayae n.sp. : figures 2/ Individual with regularly anastomosed tubes. x 10. 3/ Dense skeleton. Micrograph in interferential phase contrast. x20. 4/ Spicules are orientated in a parallel manner in young sponges. x63. Micrograph in interferential phase contrast. 5/ Pseudosagittal shape of the triactines with unpaired ray basipetal. x140. Micrograph in interferential phase contrast.

the peduncle, the oscular tube and the whole body in the young sponges, the spicules can be orientated in a parallel manner (fig.4). In this case, they assume a clearly pseudosagittal shape with the unpaired ray basipetal. Nevertheless, they always stay equiangular (fig.5).

Spicules (fig. 6).

Triactines of one size in young specimens or 2 sizes in adult specimens. Strong, conical and sharp-pointed rays. **1**- Paired actines 130-164/20.8 µm; unpaired actine 161-182/21 µm. **2**-Paired actines 143-223/10.8-18.2 µm; unpaired actine 143-270.4/10.8-21 µm.

Pseudosagittal triactines. Paired actines 114-161/9-13 µm; unpaired actine 174-22/9-13 µm.

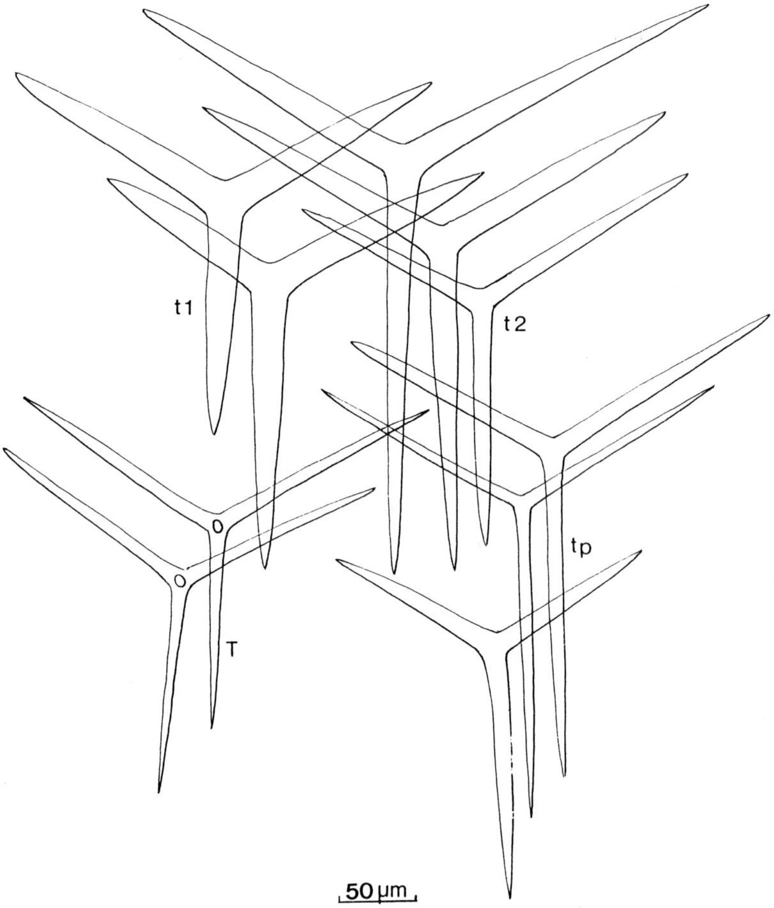

Figure 6 - Different categories of spicules from *Clathrina biscayae* n.sp. t1 : strong triactines ; t2 : triactines ; tp : pseudosagittal triactines ; T : tetractines.

Tetractines : rare, always small with an apical ray thinner and shorter than basal rays. Paired actines 156/10.8 µm; unpaired actine 143/10.8 µm; apical actine 20-26/5.2 µm.

Discussion.

In practice it is extremely difficult to classify with any degree of certainty the *Clathrina*, the skeleton of which is composed of tri- and tetractines. Haeckel (1872) had already underlined this difficulty by creating "generic varieties" to indicate, for example, the occasional presence of tetractines in the genus such as *Ascetta* characterized by the presence of triactines only. He recognized differences in the shape of spicule rays in *Clathrina* species having only triactines, and separated *C. primordialis* from *C. coriacea* on the basis of this distinction. This criterion has proved to be very difficult to apply and Topsent (1936) in his revision decided to unite all such species together under the name *C. coriacea*. We have already indicated (Borojevic & Peixinho, 1976) that it is possible to distinguish the *C. coriacea* from the Channel and *C. primordialis* from the tropical Atlantic. Nevertheless it would be essential to undertake a revision of the specimens of this group of *Clathrina* from different regions, using for instance numerical analysis of the shape, the size and the distribution of their spicules before making a final decision. *Clathrina biscayae* is characterized by the presence of strong triactines of two sizes, by the presence of pseudosagittal triactines and by the occasional presence of tetractines.

The holotype has been deposited in the Museum national d'Histoire naturelle of Paris under the number LBIM.C.1985.3, and the type-locality is station U 842 (44°11'3N/8°41'2W by 500-520 m depth).

Clathrina contorta (Bowerbank,1866)

Stations. T 450, T 475, T 477, U 854; depth 340-360 m.

Description.

The collection contains several globular specimens with a clathroid body composed of small regularly anastomosing tubes. The central tube which functions as an osculum, is distincly larger than the rest and stands one or several millimiters above the body. The large diactines are always laid down in the wall of the external tubes of the cormus in a characteristic way. They are absent inside the clathroid body.

<u>Spicules</u> (fig. 7).

Triactines : equiangular, equiradiate. Actines 94-140/10.4-13 µm.

Tetractines : with an apical actine thinner and in general shorter than the basal actines. Actines of the basal system 104-130/10.4 µm ; apical actine 50/10.4 µm.

Diactines : very large with a central point generally marked by a slight thickening, 503-762/44 µm.

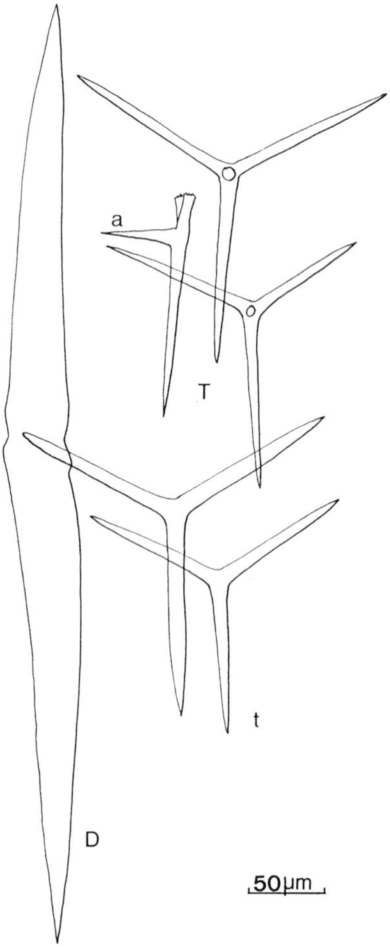

Figure 7 - Different categories of spicules from *Clathrina contorta* (Bowerbank, 1866). D : diactines ; T : tetractines ; a : apical actine of a tetractine ; t : triactines.

<u>Distribution</u>.

Arctic, East-Atlantic, Mediterranean sea; depth from the littoral zone to 95 m.

Clathrina olynthus n.sp.

Stations. T 450, T 474, T 476, T 477, T 478, T 503, T 506, U 842; depth 340-620 m.

Description.

The collection contains numerous specimens of this sponge. Typically they have the shape of an olynthus which can reach 8 mm in height and 2 mm in diameter at the base (fig.8). They can form several stolons or a small body of anastomosing tubes encrusting on the surface of the substrata. These sponges are often hispid especially at the base and on the stolons.

The skeleton of *C. olynthus* is very variable. The presence of large tetractines is constant. They are always directed in a parallel manner with their unpaired ray basipetal; their apical actine is very strong and bent in the direction of the osculum (fig.9). Small triactines are present in some specimens, rare or absent in others. In some samples, a collar composed of triactines and tetractines surrounds the osculum. It is not covered by the choanocytes. The very large diactines lie obliquely in the wall of the sponge. The choanocyte layer is thin, and riddled by large pores; it never forms folds inside the choanocoele.

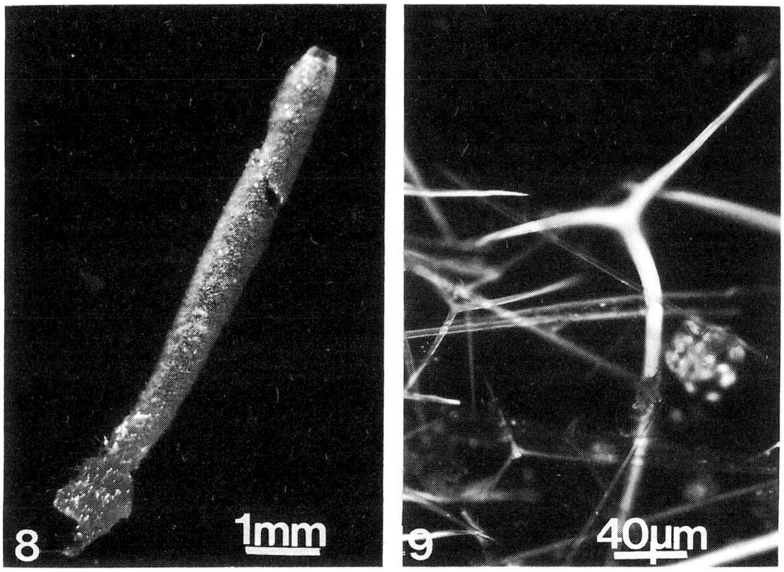

Clathrina olynthus n.sp. : figures 8/ Typical shape. x10. 9/ Large tetractines with unpaired ray basipetal. x250. Micrograph in interferential phase contrast.

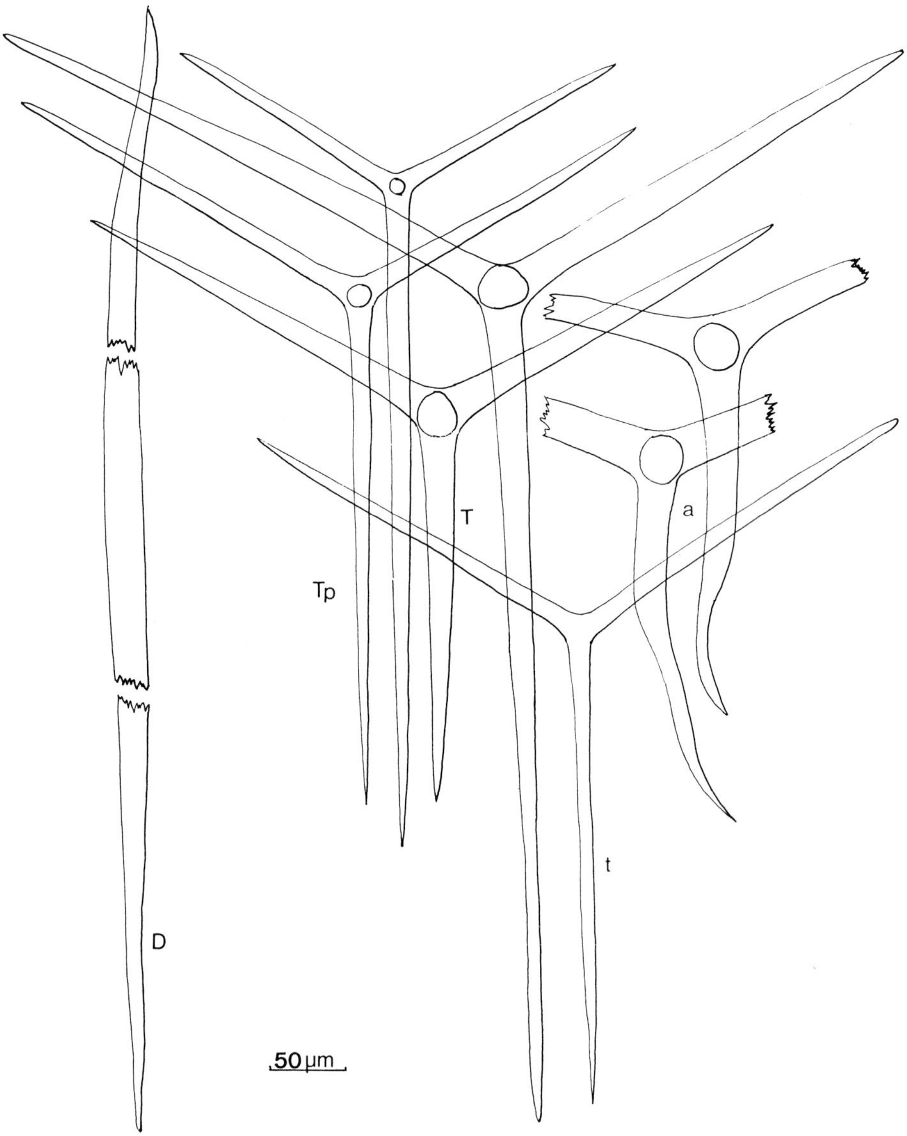

Figure 10 - Different categories of spicules from *Clathrina olynthus* n.sp ; D : diactines ; T : equiangular, equiradiate tetractines ; Tp : pseudosagittal tetractines; a : apical actine of the tetractines; t : triactines.

Spicules (fig. 10).

Tetractines : equiangular, equiradiate or with a larger basipetal actine. Apical ray bent with a tapering head. Paired actines 252.2-364/13-23.4 µm; unpaired actine 260-546/13-23.4 µm; apical actine 247/13.2 µm.

Triactines : when present, always noticeably smaller than tetractines; equiangular, equiradiate or with a basipetal actine slightly larger. Paired actines 247/11.7 µm; unpaired actine 312/14.3 µm.

Diactines : straight or slightly curved with a terminal differentiation sometimes in the shape of a spear-head; 1560/26 µm.

Discussion.

All the specimens that are identified as *C. olynthus* are characterized by a remarkable constancy in their growth form: **a large olynthus**, which is seldom observed in the genus *Clathrina*. This species has, nevertheless, a variability of spiculation which is as marked between the specimens collected at one site as between the different regions of the same sponge. By its spicular composition, it is close to *C. atlantica* (Thacker,1908) described from Cape Verde Islands and redescribed from the coasts of Brazil (Borojevic & Peixinho, 1976 p.994). It is distinguished by the shape of the cormus, by the very large size of diactines and by the shape and extension of the apical actine of tetractines. *C. olynthus* is also without any doubt, close to *C. irregularis* (Jenkin,1908) and *C. stolonifera* (Dendy,1891). The three species are characterized by forming individual tubes, linked at their base by stolons. *C. irregularis* has apparently a confused skeleton of tetractines while in *C. olynthus* the spicules are strictly parallel in the wall of the sponge. On the other hand, the tetractines often present in *C. olynthus* always seem to be absent in *C. irregularis*. The sponge described by Dickinson (1945) as *Leucosolenia irregularis* Jenkin, 1908 is not recognizable from its description; but if figure 193 is an accurate representation of its spiculation, it is certainly a different species.

C. stolonifera (Dendy,1891) is characterized by giant tetractines with apical actines clearly larger than basal actines which makes it quite distinct from our species. Its diactines are small, giving to the surface a furry aspect and not just roughly hispid as in *C. olynthus*. *Ascandra falcata* (Haeckel,1872) resembles *C. olynthus* by the presence of both large tetractines and diactines at its surface and by the presence of well individualized tubes at the superior part of its cormus. *Ascandra falcata* is nevertheless well characterized by the formation of folds of the choanoderm in its choanocoele. The specimens which have been identified as *Clathrina ascandroides* can be easily confused with certain specimens of *C. olynthus*. In typical cases, the presence of diactines and tetractines with a slender apical actine in *C. olynthus* allows easy recognition of this species. Nevertheless in some specimens where the diactines are rare, the distinction is more difficult.

The holotype has been deposited in the Museum national d'Histoire naturelle of Paris under the number LBIM.C.1985.4. The type-locality is station T 478 (44°09'9N/8°45'9W by 513-550 m depth).

Clathrina reticulum (Schmidt,1862)

Stations. T 450, T 474, T 475, T 476, T 477, T 503, T 506, T 512; depth 340-360 m.

Description.

In the collection are included several samples in the shape of small pedunculated sponges with a clathroid globular body and a single oscular tube. The peduncle is not differentiated as it is composed of functionnal tubes. The oscular tube further extends a large choanocytary central tube which concentrates the exhalant current. The sponge is hispid due to diactines which are distributed regularly on its surface.

Spicules (Fig.11).

Triactines : equiangular and equiradiate. Actines 109.2-122.2/9.1 µm.

Tetractines of variable size with apical actine thinner and generally larger than the basal actines; they are always much more numerous than triactines. Basal actines 83.2-166.4/10.4-13 µm; apical actine 111.8-175/5.2-7.8 µm.

Diactines slightly bent with a spear-head termination differentiation, 624-1011/22-36.4 µm.

Distribution.

Arctic, Atlantic, Mediterranean sea; depth 5-171 m.

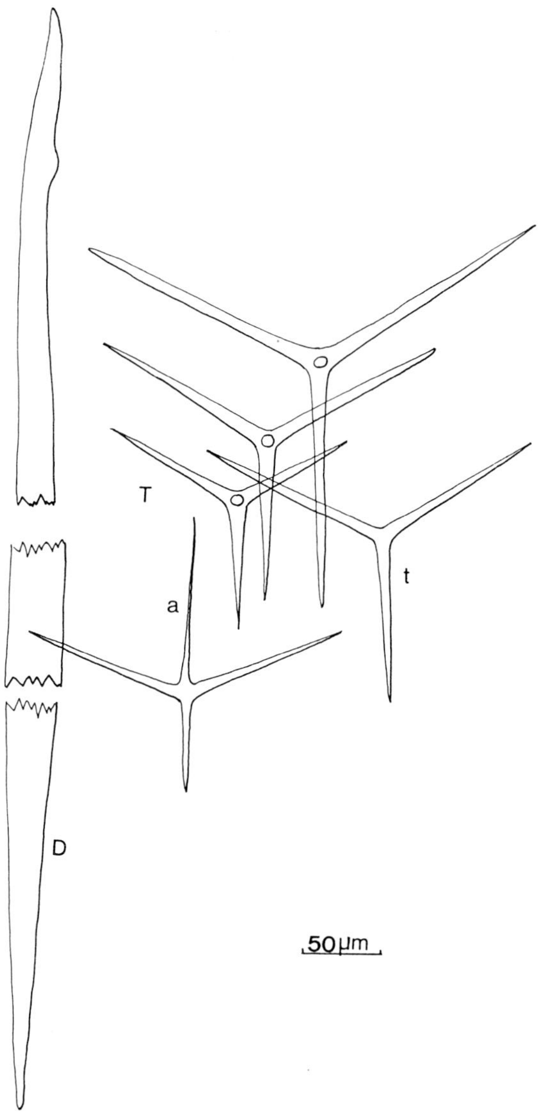

Figure 11 - Different categories of spicules from *Clathrina reticulum* (Schmidt, 1862). D : diactines ; T : tetractines ; a : apicale actine of a tetractine ; t : triactine.

Genus *Guancha* Miklucho Maclay,1868

Clathrinidae with a cormus composed of a peduncle and a clathroid body. Spicules : regular and parasagittal, or only parasagittal, orientated in a parallel manner in the walls of the sponge, with a basipetal unpaired actine.

Guancha blanca Miklucho Maclay,1868

Stations. T 437, T 450, T 474, T 475, T 476, T 477, T 503, T 512, U 842, U 847, U 852; depth 322-645 m.

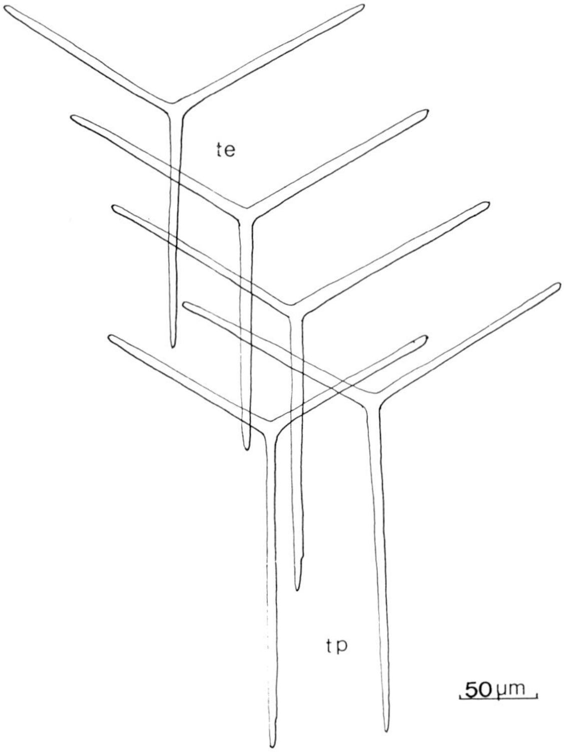

Figure 12 - Different categories of spicules from *Guancha blanca* Miklucho Maclay, 1868. te : equiangular, equiradiate triactines - tp : parasagittal triactines.

Description.

The collection contains numerous specimens of this sponge, which is abundant on corals and rocks at different depths. A large number of the specimens collected in August are young, whereas in October the young forms have disappeared. This sponge is considered to be an annual in the littoral zone, therefore it is possible that this synchrony in its development indicates a recent growth stage in summer.

The specimens are pedunculated with either a well-developed clathroid body, or in young specimens, showing some bush-like apical ramifications. The peduncle, always hollow, is formed by tubes with a well developed choanosome.

Spicules (fig. 12).

Triactines. In the apical part of the sponge, the triactines are equiangular and equiradiate; actines 130-169/5-7.8 µm. They are parasagittal in the basal part and the peduncle, with a characteristic basipetal orientation of the unpaired actine; paired actine 130-143/5-7.8 µm, unpaired actine 208-216/7.8 µm.

Distribution.

Arctic, Atlantic, Mediterranean sea, Japan?, California?; littoral to 800 m.

Guancha lacunosa (Johnston,1882)

Stations. T 450, T 451, T 453, T 478, T 503; depth 340-550 m.

Description.

With *Guancha blanca*, this is the most typical calcareous sponge of the continental margin in the Bay of Biscaye, where it is found at all depths on hard substrata.

All specimens are characteristically globular with a clathroid body and a long peduncle, always solid but without any choanosome.

Spicules (fig. 13).

Regular triactines of the clathroid body. Actines 83.2-124.8/6.5-7.8 µm.

Parasagittal triactines of the body. Paired actines 80-84/7.8 µm; unpaired actines 203/7.8 µm.

Parasagittal triactines of the peduncle with paired actines bent in a characteristic manner. Paired actines 33-39/7.8 µm; unpaired actine 364-832/18-21 µm.

Diactines with a central angle. 182-507/10.4-13 µm.

__Distribution__

Arctic, Atlantic, Mediterranean sea; depth from littoral to 200 m.

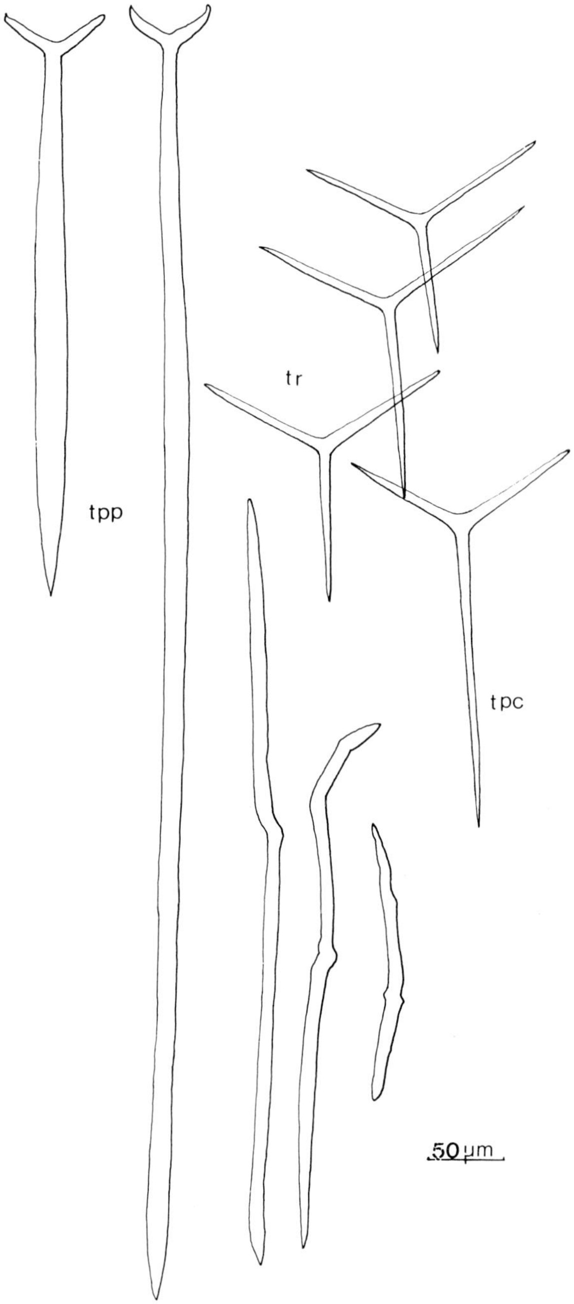

Figure 13 - Different categories of spicules from *Guancha lacunosa* (Johnston, 1842). tr: regular triactines; tpc: parasagittal triactines of the body ; tpp : parasagittal triactines f the peduncle ; d : diactines.

Family LEUCASCIDAE Dendy

Calcinea with a cormus surrounded by a well-developed cortex. Choanocyte chamber tube-shaped often very ramified and anastomosing. Choanoskeleton limited to the bent walls of choanocyte chambers.

Genus *Ascaltis* Haeckel,1872 emend.

Leucascidae with the inhalant aquiferous system formed by the lacunae delimited by the cortex and the walls of the choanocyte tubes; the exhalant aquiferous system is restricted to the osculum or to a secondary atrium formed by the cup-shaped growth of the sponge.

Ascaltis lamarcki Haeckel,1872

Stations. T 450, T 451, T 474, T 475, T 476, T 477, T 503, T 506, U 804, U 842, U 844, U 847, U 851, U 852, U 854; depth 340-645 m.

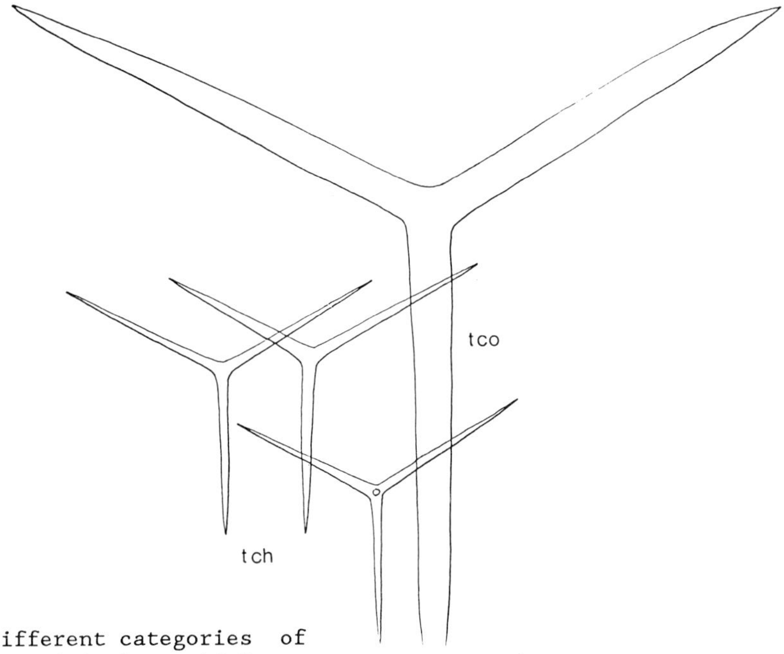

Figure 14 - Different categories of spicules from *Ascaltis Lamarcki* Haeckel, 1872. tco : cortical triactines; tch : triactines of the choanosome.

Description.

The collection has numerous specimens which all have the typical globular shape (1-7 mm diameter), pedunculated with a single osculum. The surface is perforated by the ostioles which open into the exhalant aquiferous system. The spicules, giant triactines, are easily visible at the surface of the sponge.

Spicules (fig. 14).

Cortical triactines : giant, equiangular, equiradiate, actines 338-377/26-31 µm.

Triactines of the choanosome : equiangular, equiradiate, actines 114.4-130/7.8-10.4 µm.

Occasionally cortical and choanosomal triactines can develop a rudimentary apical actine.

Discussion.

In comparison with the examples found in the littoral zone, the specimens of this collection are slender and small and their cortex is weakly developped. The organization of the sponge is, however, typical of the genus and allows it to be distinguished from *Clathrina*.

Distribution.

Arctic, Atlantic, Mediterranean sea, Antarctic? Australia? depth 60-290m.

Family LEUCALTIDAE Dendy & Row

Calcinea with wall composed of a particularly well-developed cortex supported by triactines and (or) tangential tetractines, and by a choanosome without a skeleton or with a weak and dispersed skeleton consisting of very small triactines and tetractines. Atrial skeleton badly developed or absent.

Genus *Leuclathrina* n.g.

Leucaltidae with leuconoid organization, the skeleton of which is limited exclusively to the cortex. The choanosome is completely devoid of a skeleton.

Leuclathrina asconoides n.sp.

Stations. T 476, U 842, U 844; depth 500-760 m.

Description.

The collection contains several samples of this very interesting sponge. They have a globular shape from 3 to 5 mm in diameter and bear one or two osculum situated at the end of a chimney several millimeters in height (fig. 15). The surface is smooth and the wall of the sponge is transparent and perforated by small pores. The body of the sponge is divided into two parts : 1/ the external wall which corresponds physiologically to the cortex; 2/ the choanosome.

The external wall is supported by a skeleton composed of triactines, the size of which is variable, but the shape is very constant (fig.16). The choanosome is completely devoid of a skeleton. Thus the preservation of the organization of the aquiferous system is precarious and large spaces can be observed between the choanosome and the external wall. Nevertheless, it is not certain that these are not artefacts. The aquiferous system seems to be lacunar and irregular, and the subspherical choanocyte-chambers are irregularly distributed.

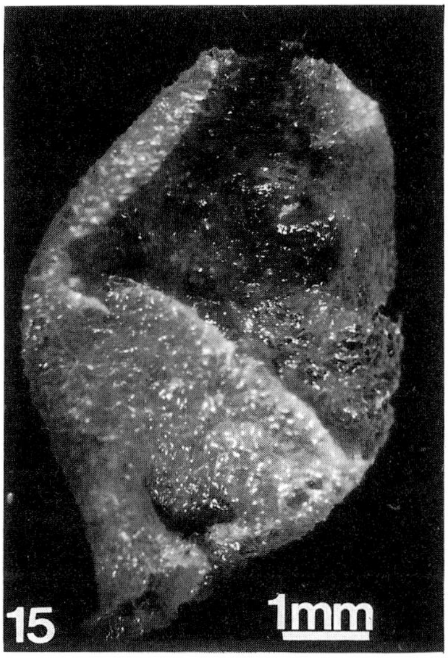

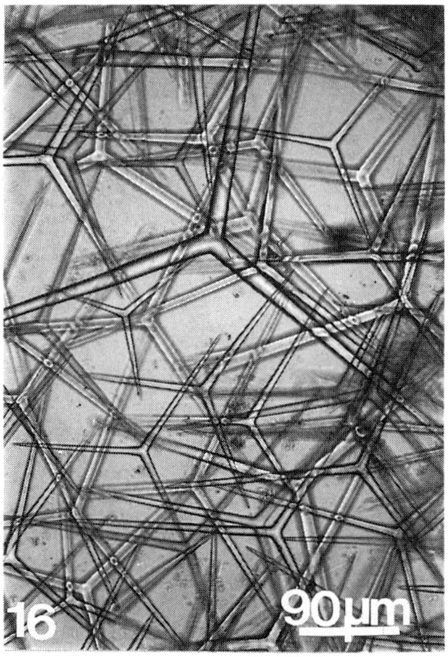

Leuclathrina asconoides n.g., n.sp. : figures 15/ Type-specimen. A piece of the wall has been cut and the central cavity is apparent. x 10. 16/ Aspect of the skeleton of the external wall. x 110. Micrograph in phase contrast.

<u>**Spicules**</u> (fig. 17).

Triactines : equiangular, equiradiate with cylindrical actines of variable sizes, actines 156-598/13-36 µm.

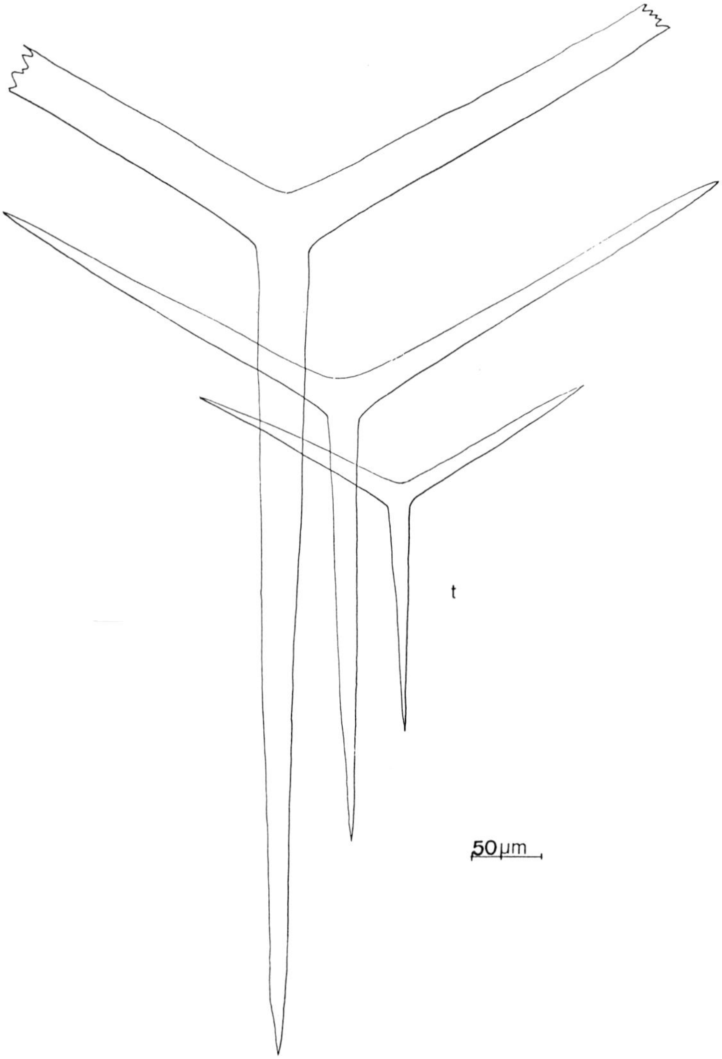

Figure 17 - Different categories of spicules from *Leuclathrina asconoides* n.g., n.sp. t : triactines of variable size.

<u>Discussion</u>.

Leuclathrina asconoides is to our knowledge the only calcareous sponge with an heterocoel organization in which the choanosome is entirely devoid of a skeleton.

Among the Calcinea, several species show a very developed cortical skeleton and a reduced choanoskeleton. In the case of *Ascandra minchini* Borojevic, 1966, the choanosome, formed by the corrugation of the choanoderm inside the asconoid tube, is sustained solely by the apical actines of the tetractines of the extended wall. The organization of the choanosome in *Leucettusa simplicissima* Burton, 1932, recalls that of *A. minchini* : it is composed of radiating tubes uniquely supported by the apical actines of the cortical tetractines. Very unusual and very small spicules occur on the atrial rim of this sponge, delimiting the exhalant system. Therefore, one could postulate that the simple *Leucettusa* are derived from a form similar to *A. minchini*, where the organization of the choanosome in radial tubes would be the consequence of the corrugation of the choanoderm inside the asconoid tube, directed and supported by the apical actines of cortical tetractines. The leuconoid system of the organization of the choanosome is observed only in austral evolved *Leucettusa*. The development by progressive stages from an organization of elongated radial tubes to one with subspherical chambers can be observed in this genus and sometimes in species such as *Leucettusa vera* (Poléjaeff, 1883).

Although the organization of *Leuclathrina asconoides* is analogous to that of *A. minchini* and *L. simplicissima*, the leuconoid organization of its choanosome seems to be primary, and could be more or less derived from the organization observed in *A. minchini*. This could be due to the fact that its skeleton does not possess tetractines and thus the corrugation of the choanoderm cannot be induced by the radial disposition of the apical actines of the cortical tetractines. Thus, it is impossible to assign the sponge described here among the primary *Leucettusa*, and therefore we have created a new genus, *Leuclathrina*, which must be placed in the family Leucaltidae, close to the most primitive *Leucettusa*. It is probable that the sponges classified in the genus *Leucettusa*, having a lacunar organization and a reduced skeleton, such as *L. corticata* Haeckel, 1872 and *L. dictyogaster* Row & Hozawa, 1931, have been derived from a form similar to *Leuclathrina asconoides*.

The holotype has been deposited in the Museum national d'Histoire naturelle of Paris under the number LBIM.C.1985.5, the type-locality is station T 476, (44°11'2N/8°41'3W by 400 m depth).

Family LEVINELLIDAE Borojevic & Boury-Esnault, 1986

Calcinea with a cormus composed of a central tube, which can be ramified, and diverticles isolated or in clusters. The skeleton of the central and radial tubes is composed of equiangular spicules. The skeleton of the diverticles is composed of equiangular or parasagittal spicules, always clearly distinct from the skeleton of the central tube. Choanoderm lining all the central cavity or limited to the diverticular region.

Genus *Levinella* Borojevic & Boury-Esnault, 1986

Levinellidae with a central tube without ramification. The choanoderm covers all the internal cavities of the sponge.

Levinella thalassae Borojevic & Boury-Esnault, 1986

Stations. T 476, T 478, T 503, U 842; depth 490-620 m.

Description.

Levinella thalassae has been described in detail recently (Borojevic & Boury-Esnault, 1986). Numerous specimens of this interesting sponge have been collected. They have the shape of elongated sacs covered by diverticles. The skeleton of the central tube is clearly differentiated from the diverticular skeleton and is composed of charasteristic tetractines. The skeleton of the diverticles is composed of triactines, tetractines and occasionally diactines.

Spicules (fig. 18).

Tetractines of the central tube. Paired actines 195-226/7.8-10.4 µm ; unpaired actine 221-260/7.8-10.4 µm; apical actine 369-421/10.4µm.

Tetractines of the diverticles. Paired actines 78-125/10-13 µm; unpaired actine 62.4-125/10-13 µm; apical actine 13-21/7.8-10 µm.

Triactines of the diverticles : rare. Paired actines 81-86/13 µm; unpaired actine 77-81/13 µm.

Diactines very rare 468-1250/10-47 µm.

<u>Distribution</u>.

Bay of Biscaye; depth 490-620 m.

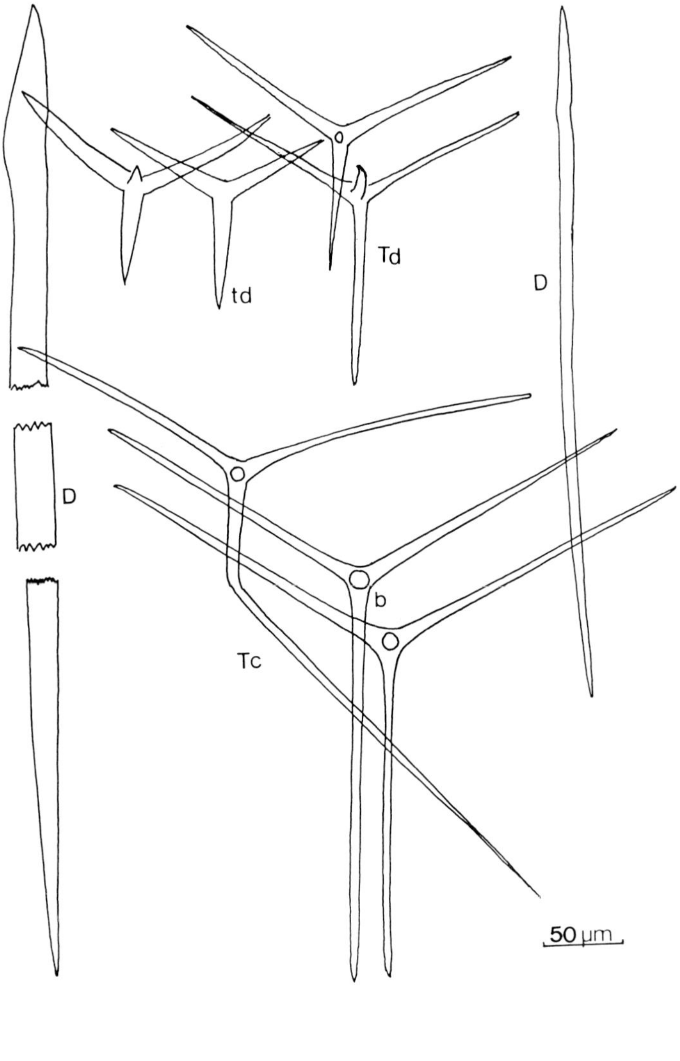

Figure 18 - Different categories of spicules from *Levinella thalassae* Borojevic & Boury-Esnault, 1986. D : diactines ; td : triactines of diverticules ; Td : tetractines of diverticules ; Tc : tetractines of the central tube ; b : basal system of tetractines of the central tube.

CONCLUSION

All the stations on the Chapelle Bank, and west and east of Galicia have shown a remarkable homogeneity in their faunistic composition, which remains stable from one year to another. The most representative and the richest stations are those where the hard substrata (rocks and corals) are most abundant. The richest station is station T 450 of the Chapelle Bank where eight of the ten described species in this work have been found fixed on coral branches at 340 m. In two other stations off west Galicia, T 476 and T 477, and one off east Galicia T 503, on bottoms of rocks and pebbles, seven species have been collected.

The most frequent species is *Ascaltis lamarcki* which has been found at 15 stations out of 21 in which Calcinea occurred. *Guancha blanca* is present at eleven stations, *Clathrina biscayae* at nine, *Clathrina olynthus* and *Clathrina reticulum* at eight.

This fauna of calcareous sponges is largely representative of the Calcinea, since four of the five known families and five of the twelve known genera are present.

It is worthwhile drawing attention to the high proportion of new species that this fauna contains (4/10), and also the two new genera and one new family. This shows the interest in exploring the continental margin of the Atlantic. *Leuclathrina asconoides* n.g., n.sp., is the first Leucaltidae described from the Atlantic.

This fauna of the continental margin shows a remarkable affinity with that of the littoral zone of the Atlantic-Mediterranean region where the species *C. contorta*, *C. reticulum*, *G. blanca*, *G. lacunosa* and *A. lamarcki* are present up to the subtidal zone and are characteristic of the fauna of walls and rocks not colonized by algae.

The study of the Calcinea on the continental margin has not only shown that littoral species can live in the bathyal zone, but also that new forms representing particular evolutionary steps can be discovered. This is of tremendous interest because they represent missing links in the evolution of Calcinea. As a consequence, this continental margin may be a conservative environment for representatives of steps in the evolutionary lines and/or a place where speciation occurs in an active manner.

LIST OF SPECIES BY STATIONS

W Brittany-Chapelle Bank - august 1967.

T 437: 47°57'0 N/07°49'0 W - 322 m - dead and living coral.
Clathrina biscayae; Guancha blanca.

T 450: 47°58'3 N/7°50'0 W - 340 m - coral.
Clathrina ascandroides; Clathrina biscayae; Clathrina reticulum; Guancha blanca; Guancha lacunosa; Ascaltis lamarcki.

T 451: 47°57'5 N/7°50'7 W - 358 m - coral very abundant.
Clathrina ascandroides; Guancha lacunosa; Ascaltis lamarcki.

T 453: 47°57'3 N/7°51'0 W - 344-354 m - pebble, few coral.
Clathrina biscayae; Guancha lacunosa.

W Galicia-august 1967.

T 474: 44°11'0 N/8°41'3 W - 519 m - rocks and stones.
Clathrina olynthus; Clathrina reticulum; Guancha blanca; Ascaltis lamarcki.

T 475: 44°10'8 N/8°41'3 W - 400 m - rock.
Clathrina biscayae; Clathrina reticulum; Guancha blanca; Ascaltis lamarcki.

T 476: 44°11'2 N/8°40'9 W - 620 m - rocks and stones.
Clathrina biscayae; Clathrina olynthus; Clathrina reticulum; Guancha blanca; Ascaltis lamarcki; Leuclathrina asconoides; Levinella thalassae.

T 477: 44°11'1 N/8°42'0 W - 500 m - rocks and stones.
Clathrina ascandroides; Clathrina biscayae; Clathrina contorta; Clathrina olynthus; Clathrina reticulum; Guancha blanca; Ascaltis lamarcki.

T 478: 44°09'9 N/8°45'9 W - 513-550 m - 1 large boulder.
Clathrina olynthus; Guancha lacunosa; Levinella thalassae.

E Galicia- august 1967.

T 503: 44°00'7 N/7°06'9 W - 490 m - rock (1 large boulder).
Clathrina ascandroides; Clathrina olynthus; Clathrina reticulum; Guancha blanca; Guancha lacunosa; Ascaltis lamarcki; Levinella thalassae.

T 506: 44°01'7 N/7°00'8 W - 490 m - large stones and pebbles.
Clathrina olynthus; Clathrina reticulum; Ascaltis lamarcki.

T 512: 44°01'6 N/7°01'9 W - 510-630 m - stones.
Clathrina reticulum; Guancha blanca.

W Galicia- october 1968.

U 804: 44°11'7 N/8°41'9 W - 455-500 m - stones and boulders.
Ascaltis lamarcki.

U 842: 44°11'3 N/8°41'2 W - 500-520 m - stones.
Clathrina olynthus; Guancha blanca; Ascaltis lamarcki; Leuclathrina asconoides; Levinella thalassae.

U 843: 44°11'4 N/8°41'1 W - 540-560 m - boulders 20-30 cm.
Clathrina ascandroides.

U 844: 44°12'1 N/8°42'1 N- 695-760 m - dead and living coral.
Ascaltis lamarcki; Leuclathrina asconoides.

U 847: 44°10'9 N/8°34'1 W - 500 m - stones and boulders (1-20 cm).
Clathrina ascandroides; Clathrina biscayae; Guancha blanca; Ascaltis lamarcki.

U 851: 44°12'0 N/8°31'4 W - 520-530 m - stones and boulders.
Ascaltis lamarcki.

U 852: 44°12'0 N/8°34'0 W - 615-645 m - stones and boulders.
Clathrina biscayae; Guancha blanca; Ascaltis lamarcki.

U 854: 44°10'0 N/8°22'3 W - 410-640 m - boulders.
Clathrina contorta; Ascaltis lamarcki.

ACKNOWLEDGMENTS

We express our appreciation to Marie-Rose CAUSI for her help with photographic plates and we especially thank Shirley STONE for her helpful assistance for the improvement of the English.

This work was made in the associate laboratory CNRS UA 41, and received the support of Greco Ecoprophyce 130068.

REFERENCES

Borojevic R., 1966a - Eponges calcaires des côtes de France II.- Le genre *Ascandra* Haeckel emend. *Archs Zool. exp. gén.*, 107(2) : 357-368.

Borojevic R., 1966b - Eponges calcaires des côtes de France III.- Discussion sur la taxonomie des éponges calcaires: *Aphroceras ensata* (Bowerbank) et *Ute gladiata* n.sp. *Archs Zool. exp. gén.*, 107(4) : 703-724.

Borojevic R., 1968 - Eponges calcaires des côtes de France IV.- Le genre *Ascaltis* Haeckel emend. *Archs Zool. exp. gén.*, 109(2) : 193-210.

Borojevic R., 1971 - Eponges calcaires des côtes sud-est du Brésil, épibiontes sur *Laminaria brasiliensis* et *Sargassum cymorum*. *Rev. Brasil. Biol.*, 31(4) : 525-530.

Borojevic R. & Boury-Esnault N., 1986 - Une nouvelle voie d'évolution chez les éponges calcaires Calcinea: description des genres *Burtonulla* et *Levinella*. *Bull. Mus. natn. Hist.nat.*, 3 : 443-455.

Borojevic R. & Peixinho S., 1976 - Eponges calcaires du nord-nord-est du Brésil. *Bull. Mus. natn. Hist. nat.*, 402 : 987-1036.

Dendy A. & Row H., 1913 - The classification and phylogeny of the calcareous sponges, with a reference list of all the described species, systematically arranged. *Proc.Zool.Soc.*, London, 704-813.

Dickinson M.G., 1945 - Sponges of the Gulf of California. *Allan Hancock Pacific Expeditions*, 11(1) : 1-55.

Haeckel E., 1872 - Die Kalkschwämme, eine monographie. Berlin.

Hansen G.A., 1885 - Spongiadae. *Norw. North-Atlantic Expedition 1876-1878*, 1-25.

Poléjaeff N., 1883 - Report on the Calcarea dredged by H.M.S. Challenger during the years 1873-76. *Zool. Chall. Exp.*, 24 : 1-76.

Topsent E., 1892 - Contribution à l'étude des Spongiaires de l'Atlantique nord. *Résults. Camp. scient. Albert Ier Monaco*, 2 : 1-165.

Topsent E., 1936.- Etude sur les *Leucosolenia*. *Bull. Inst. océanogr. Monaco*, 711 : 1-47.

THE *POLYMASTIA* SPECIES (DEMOSPONGES, HADROMERIDA) OF THE ATLANTIC AREA.

N. BOURY-ESNAULT

Centre d'Océanologie de Marseille, UA 41, Station marine d'Endoume, Rue de la Batterie des Lions, 13007-Marseille, France.

SYNOPSIS

A revision of the genus *Polymastia* has been undertaken within the Atlantic area. The importance is stressed for a very precise description of the skeleton. Thirteen *Polymastia* species are recognized and redescribed. A table with discriminating characters for the 13 species is included.

INTRODUCTION

The order Hadromerida is characterized by a skeleton composed of monactinal spicules which are tylostyles or subtylostyles organized in a radial pattern. Microscleres are present in some families but absent in others.

There are eight families within the order Hadromerida. Five of them have microscleres (Clionidae, Spirastrellidae, Timeidae, Placospongiidae, Stylocordylidae), two (Polymastiidae and Suberitidae) are without microscleres and the last one (Tethyidae) has a genus with microscleres, *Tethya*, and a genus without microscleres, *Aaptos*. The systematic of the families without microscleres is still not clear and the border between the families and the genera seems to be very arbitrary. Also, the families Polymastiidae, Suberitidae and the genus *Aaptos* (Tethyidae) badly needs revision.

At first glance, the family Polymastiidae appears to be well defined. All the authors emphasize the presence of erect oscular and pore-bearing papillae, and give this as a characteristic of the family. This character, in fact, seems to be strong in the genus *Polymastia*, but not for other genera of the family such as *Quasillina*, *Ridleia*, *Tentorium* etc... On the other hand a species like *Aaptos papillata*, which is for the moment classified within the Tethyidae, has papillae.

NATO ASI Series, Vol. G13
Taxonomy of Porifera
Edited by J. Vacelet and N. Boury-Esnault

In order to redefine these families and genera, it is necessary to begin by a revision of the *Polymastia* species, insisting on the characters described in a previous work (Boury-Esnault, 1974). The importance of a precise description have been stressed for the skeleton, the repartition of the aquiferous cavities in the papillae and also the necessity of transverse sections of the papillae, to understand their architecture. It also appears necessary to make many measurements of the spicules in determining the presence of different size categories. Finally, a precise description of the architecture of the cortex, with sections perpendicular to the surface, has to be done.

The genus *Polymastia* was described in 1864 by Bowerbank (p. 177). He gave *Spongia mammillaris* Müller, 1806 as the type-species. His definition reads : - "Skeleton. Basal mass. Central portion consisting of a plexus contorted anastomosing fasciculi resolving themselves near the surface into short straight bundles disposed at nearly right angles to the surface. Oscula congregated, elevated on numerous long fistulae. Fistulae composed of numerous parallel fasciculi, radiating from the base to the apex of each in straight or slightly spiral lines".

In 1866, Bowerbank added six other species to the genus : *P. ornata, P. bulbosa, P. robusta, P. brevis, P. spinula, P. radiosa* and in 1874 *P. conigera*.

Ridley & Dendy (1887, p. 210) added more precise data: "...megasclera tylostyli or styli. Sponges usually attached and without any supporting fringe of spicules". They found two new species of Polymastia in the Atlantic : *P. corticata* and *P. agglutinans*.

Topsent (1900, p.131) gave a similar definition: - " Polymastiidae massives, sessiles, avec des papilles de nombre et de longueur variables. Mégasclères, tylostyles et styles. Charpente disposée en lignes rayonnant vers la surface. Ecorce épaisse pleine de spicules de plus petite taille rangés verticalement".

Dendy (1921, p. 150) precisely described the structure of the cortex : - " The cortical skeleton consists typically of a deeper layer of large tangentially placed spicules, often arranged in bundles or fibres and a superficial layer of small, radially arranged spicules".

In this work the 13 known species of Atlantic *Polymastia* are redescribed in order to view more accurately their specific characters and to define, more precisely, the genus. The same will be done with the other

Map 1 - Distribution of the species of *Polymastia* in the North Atlantic Ocean. ■ *P. mammillaris*, ▼ *P. agglutinans*, ● *P. inflata*, ▲ *P. robusta*, ☆ *P. tenax*, △ *P. conigera*, ○ *P. polytylota*, Φ *P. uberrima*, Ψ *P. grimaldi*, Θ *P. spinula*, Δ *P. corticata*, Χ *P. infrapilosa*.

genera and species of Polymastiids in a future work in order to determine the limits of each genera.

MATERIAL AND METHODS

The type-specimens of Bowerbank, Ridley & Dendy, Vosmaer, and Koltun have been studied in the British Museum of Natural History.

The type-specimens of Topsent from the Musée Océanographique de Monaco, have been reexamined whereas the type-specimens of Lévi, Cabioch and Vacelet, in the Museum national d'Histoire naturelle of Paris, underwent reexamination.

A schizotype of *Polymastia gleneni* has been given by Descatoire from Concarneau and specimens of *P. conigera* and *P. inflata* were lent by Cabioch (Roscoff). All the collections of the Museum National d'Histoire Naturelle of Paris (Prof. Lévi) and of the Station Marine d'Endoume (J. Vacelet) have also been studied.

The following descriptions of the different species give a short account of the external morphology of the specimens and whenever possible, an estimation of the number of papillae, the precise description of the skeleton, the shape and the size of the spicules and their localization. When the original description is sufficient, it is only referred to. The distribution of each species in the North-Atlantic area will also be given.

DESCRIPTION OF THE DIFFERENT SPECIES

Polymastia mammillaris (Müller, 1806)

This species is the type-species of the genus *Polymastia*.

<u>**External characters**</u>.

Polymastia mammillaris is an encrusting species, pale yellow *in vivo*. It may cover large areas of the substrata (several dm2). The surface of the sponge is often covered by a layer of sediment and the specimen can only be located by the papillae projecting above the sediment. The presence of 30-50 inhalant papillae and 1 exhalant papilla has been estimated for a surface of 10cm2.

Skeleton.

Choanosome (fig. 1a) : the choanosomal skeleton of *P. mammillaris* is composed of fascicles of large tylostyles perpendicular to the surface which they passed through. Between these bundles free intermediary tylostyles are interlaced.

Ectosome (fig. 1a) : the skeleton of the ectosome is composed of a tangential network of intermediary tylostyles on which rests a layer of small tylostyles disposed perpendicularly to the surface like a palissade. The thickness is about 500 µm.

Papillae (fig. 1b) : the bundles of the principal tylostyles constitute the axial framework. The number of bundles varies from 6 to 10 in the inhalant papillae and from 10 to 20 in the exhalant papillae. They are located at the periphery of the papillae. Each bundle contains 20 to 70 spicules. They are the extension of the bundles from the choanosome. From the inside to the outside the skeleton of the papilla is composed of bundles of tylostyles, and a layer of intermediary tylostyles, located in a perpen-

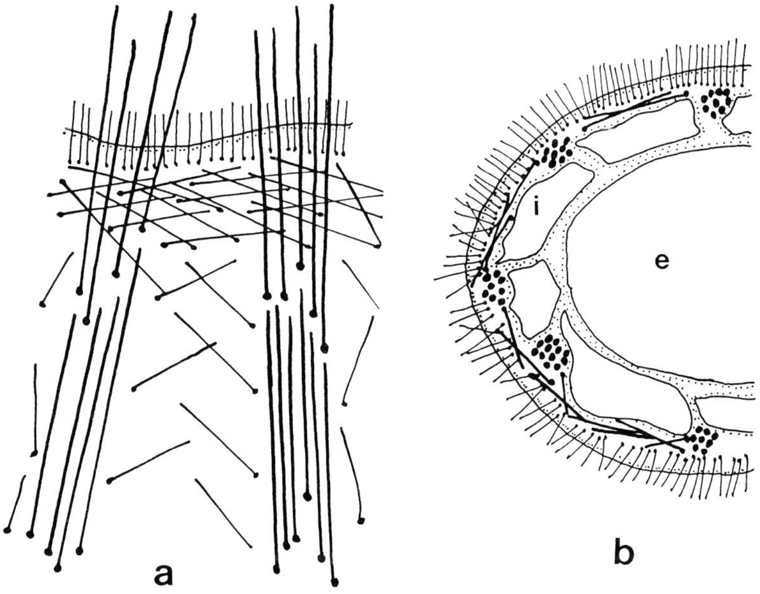

Figure 1 - *Polymastia mammillaris* (Müller, 1806). Specimen from The Channel. a/ Longitudinal section of the main body. b/ Transverse section of an exhalant papilla. e : exhalant canal, i : inhalant canal.

dicular plan to the bundles and tangential to the surface. They are similar to those of the ectosome and are continuous with them. The tangential tylostyles are distinguished from principal tylostyles by their location and by their dimensions. At the surface a palissade of ectosomal tylostyles is observed to be continuous with that of the main body.

A detailed description of the aquiferous cavities was given by Boury-Esnault (1974). It is only necessary to point out here that the cellular laminae which, in the exhalant papillae, separate the exhalant canal from the inhalant canals, are free of spicules.

Spicules.

Measurements are made on 2500 spicules belonging to the three types of tylostyles (principal, intermediary and ectosomal). The minimum and maximum sizes obtained are 50 and 1700 µm respectively. A trimodal distribution is obtained by a histogram of frequencies. The extreme values for the first population, which corresponds to the ectosomal tylostyles of the palissade, are 50 and 250 µm and the mean is 136 µm ± 1. The extreme values for the second population, which corresponds to the tangential tylostyles, are 250 and 725 µm and the third, which corresponds to the principal tylostyles, are 725 and 1700 µm. The means are respectively, 540 µm ± 1 and 930 µm ± 1. The difference between these three means is highly significant. Three different populations of tylostyles are present that were not indicated in the previous descriptions of the species (Bowerbank, 1866; Topsent, 1900).

Distribution.

This species is distributed in the Atlantic from the Arctic region (East Greenland) to the North Sea, The Channel, Bay of Biscaye, Coasts of Galicia, Portugal and Mediterranean. It was signalized twice on the West Atlantic coast, Newfoundland and Jan Mayen, but was very deep (map 1).

Burton reported *P. mammillaris* from South Africa, and Dendy from Sandy Island, but these records have to be confirmed.

From the North Sea to the Mediterranean the species is known from the littoral zone to 600 m deep. In the Arctic area, it was found deeper than 2500 m around Jan Mayen and 1267 m near Newfoundland.

Polymastia agglutinans Ridley & Dendy, 1886

Polymastia agglutinans was very well described by Ridley & Dendy, 1887 (p. 212, pl. XLI fig. 6, pl. XLII figs 1, 2, 2a, 2b, 3).

It is a small species from 1 to 3 cm in diameter with a few papillae that may reach 2 cm in length. The colour *in vivo* is yellow to orange. The characteristic of this species is the incorporation, at the surface of the cortex, of numerous fragments of shells, grains of sand etc... The presence of these foreign objects modifies the architecture of the skeleton of the cortex. In particular, the palissade of ectosomal tylostyles can be observed only between the incorporated fragments (fig. 2a).

The papillae have the same structure as those of *P. mammillaris* (fig. 2b). The surface of the exhalant canal is echinated by ectosomal tylostyles.

As with *P. mammillaris* there is three kinds of spicules. Principal styles which constitute the bundles of the choanosomal skeleton and of the papillae, the "intermediate forms" of Ridley and Dendy which constitute the tangential layer of intermediary tylostyles and the small tylostyles of the palissade.

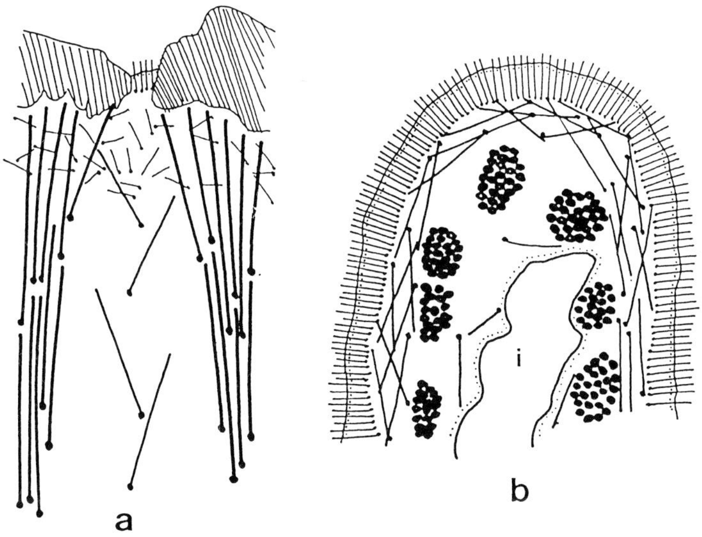

Figure 2 - *Polymastia agglutinans* Ridley & Dendy, 1886. According to their figure 6, plate XLI. a/ Longitudinal section of the main body. b/ Transverse section of an inhalant papilla. i : inhalant canal.

Distribution.

Described from the Azores Islands by Ridley & Dendy, this species is known from 59°3N-4°08W to 38°38'N-28°28'W and 9°20'N-14°15'W, in depths between 15 to 800 m. Burton (1956) found it on the coasts of the Guinea Gulf, Lévi (1960) in the Seminole Bank near Dakar (Senegal coast) and Cabioch (1968) in The Channel (map 1).

Remarks.

There is in the Antarctic a very similar species, *P. isidis* Thiele, 1905 which also incorporates foreign bodies in its cortex, but the structure of the papillae is quite different.

Polymastia azorica Lévi & Vacelet, 1957

The description of this species does not allow it to be recognized. The type-specimen seems to have disappeared.

Polymastia conigera Bowerbank, 1874

Bowerbank, 1874 p. 192, pl. LXXII. There is no recent description of the Atlantic *P. conigera*. The type-specimen from Bowerbank, which appears to be a fragment, and a specimen from Cabioch collected in The Channel (N-W Batz), are the only specimens available.

External characters.

Encrusting like *P. mammillaris, P. conigera* seems to be a small species. The specimen from Cabioch (fig. 3) is 3.5 cm long, 1.5 cm large and

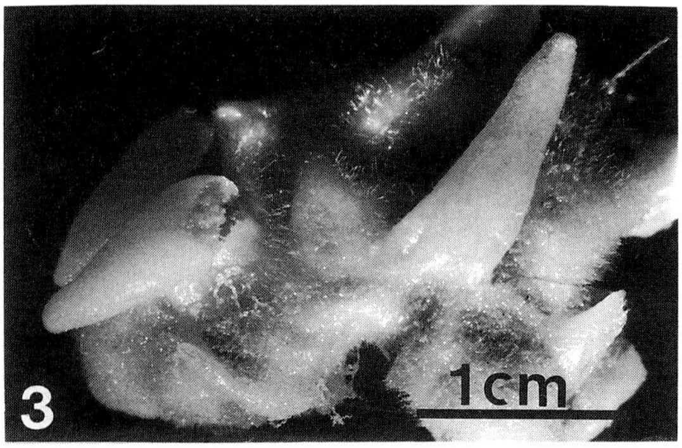

Figure 3 - *Polymastia conigera* Bowerbank, 1874. Specimen from The Channel (Cabioch's collection).

0.5 cm thick. There are eight very conical papillae about 2 cm long. The colour in alcohol is cream. The surface of the body is very hispid.

Skeleton.

Choanosome (fig. 4a) : the choanosomal skeleton is composed by bundles of principal tylostyles of 300 µm in diameter. The ends of these bundles pass through the surface. Intermediary tylostyles are loose between the bundles.

Ectosome (fig. 4a) : the cortex is 290 µm thick. Its skeleton is composed of a palissade of ectosomal tylostyles, the basis of which is embedded in a layer of tangential tylostyles 200 µm thick.

Papillae (fig. 4b) : all the papillae are simultaneously inhalant and exhalant. The exhalant canal is central and the inhalant canals are distributed around it. The thickness of the wall is approximately 600 µm. The longitudinal bundles of tylostyles are located within the wall between the inhalant canals. They are linked by a dense, irregular network of intermediary tylostyles, which echinate the surface of the canals. Near the surface, the same arrangement as in the ectosome is found, i.e. a tangential layer of intermediary tylostyles and a palissade of small tylostyles.

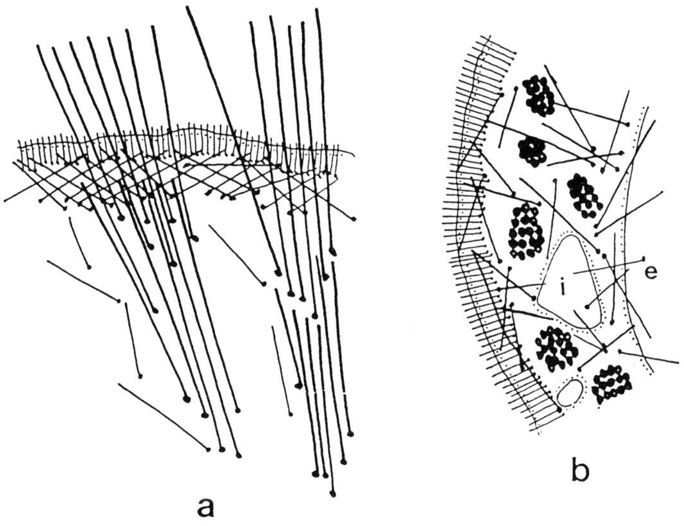

Figure 4 - *Polymastia conigera* Bowerbank, 1874. Type-specimen. a/ Longitudinal section of the main body. b/ Transverse section of an exhalant papilla. e : exhalant canal, i : inhalant canal.

Spicules.

The principal and intermediary tylostyles are polytylote. The central canal of the spicule is straight without any dilatation at the level of the swelling.

Principal tylostyles : 1196-1533 / 16-30 µm.

Intermediary tylostyles : 516-780 / 10-16 µm.

Ectosomal tylostyles : 110-130 / 3-5 µm, with an assymetric and tapering point.

Distribution

Described by Bowerbank from the Shetland, it was found in The Channel near Roscoff at depths from 15-85m (Cabioch, 1968) (map 1).

Dendy (1921) and Bergquist (1968) described this species from the Indian Ocean and New Zealand. I am not convinced that their *Polymastia conigera*, as the shape of the sponge and the size of the spicules differ from those of the type.

Polymastia inflata Cabioch, 1968

Polymastia inflata was well described under the name *P. bulbosa* Sarà & Siribelli by Vacelet, 1961 (p.31, fig.1).

External characters.

It is a small encrusting species about 2 cm2 with one to three papillae, about 1.5 cm in length. The colour *in vivo* is pale yellow.

Skeleton.

Choanosome (fig. 5a) : the choanosomal skeleton, as usual, consists of bundles of tylostyles approximately 250 µm in diameter. Between these bundles intermediary and ectosomal tylostyles are loosely dispersed.

Ectosome (fig. 5a) : the cortex has a width of circa 600 µm. It is made by a dense layer of tangential, inflated tylostyles which embeds more or less completely the palissade of small cortical tylostyles.

Papillae (fig. 5b) : the papillae are simultaneously inhalant and exhalant. The exhalant canal is surrounded by the inhalant ones. The bundles of tylostyles are at the periphery. Just above this there is the layer of tangential fusiform tylostyles in which are embedded the ectosomal tylostyles. Free fusiform tylostyles lay between the canals in the cellular laminae.

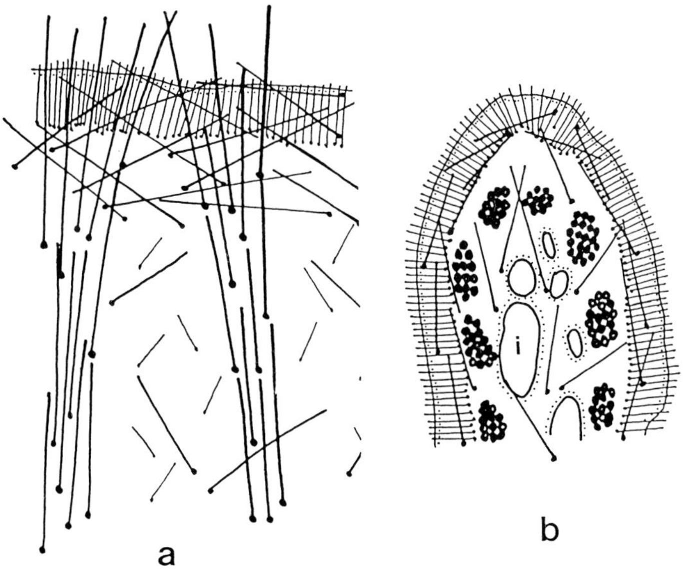

Figure 5 - *Polymastia inflata* Cabioch, 1968. Type-specimen. a/ Longitudinal section in the main body. b/ Transverse section of an inhalant papilla, i : inhalant canal.

Spicules.

The three types of tylostyles are well distinguishable here.

The principal tylostyles are in fact fusiform strongyloxes, with a badly-rounded head : 620-1250 / 11-18 µm.

The intermediary ones are very fusiform with a well-rounded head : 250-660/11-26 µm. But just below the head, the width may be 4-6 times narrower.

The ectosomal tylostyles have a well-rounded head, and are slightly fusiform : 90-290/3-5 µm.

Distribution.

This species has been discovered in Mediterranean Sea on the Riviera Ligure by Sarà & Siribelli (1960) under the name *P. bulbosa*. This species was known from Corsica in 63-73 m depth (Vacelet,1961), from The Channel near Roscoff at 15 m deep (Cabioch, 1968), from Galicia (North of Spain) in the mediolittoral zone (Rodriguez-Babio, 1978). It is also known from the north coasts of Guinea Gulf near Abidjan by 100-150m depth (Vacelet, personal communication) (map 1).

Polymastia infrapilosa Topsent, 1927

Polymastia infrapilosa was well described by Topsent, 1928 (p. 147, pl.II 25,26, pl.VI 3).

External characters.

This pale yellow species has a massive, somewhat spherical shape. It is fixed on the substrata. The surface is smooth on the papillae and the upper face, but hispid at the base near the level of the fixation zone.

Skeleton.

Choanosome (fig. 6a) : the choanosomal skeleton is made of bundles of tylostyles of about 350 µm of diameter. These bundles do not pass through the upper face. They stop at the basis of the tangential layer, but at the level of the fixation zone the bundles of tylostyles go through the whole cortex which they strongly echinate. In the choanosome between the bundles, ectosomal tylostyles are arranged in groups of 3 to 5 spicules.

Ectosome (fig. 6a) : the cortex is about 900 µm thick at the upper face and 1250 µm at the base. It is composed at the upper face by a layer of tangential intermediary tylostyles approximately 600 µm thick and a palissade of ectosomal tylostyles. At the base, the cortex is reinforced by the termination of the bundles of the choanosomal skeleton which pass through the surface by approximately 500 µm.

Papillae (fig. 6b) : the papillae are conical and vary in length from 0.5 to 1 cm. The exhalant canal is central and the inhalant canals are peripherical. The structure from inside to outside is formed by : the exhalant canal, the surface of which is echinated by intermediary tylostyles, bundles of principal tylostyles, a layer of tangential tylostyles and the superficial layer of small tylostyles.

Spicules.

They are fusiform tylostyles. The principal ones vary between tylostyles and strongyloxea.

Principal tylostyles : 1700-1900/23 µm.

Intermediary tylostyles : 300-700 µm.

Ectosomal tylostyles : 150-240/7-8 µm.

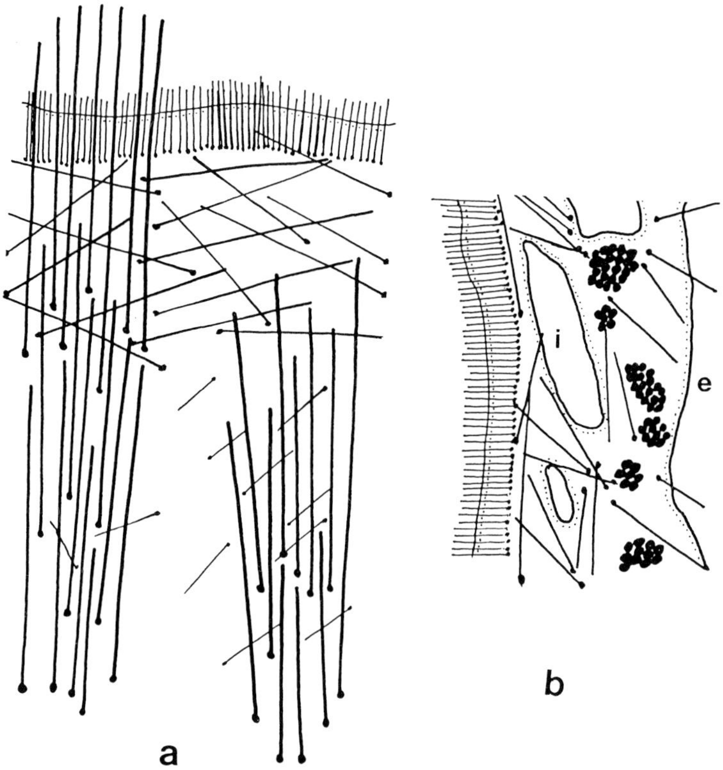

Figure 6 - *Polymastia infrapilosa* Topsent, 1927. Type-specimen. a/ Longitudinal section in the main body. b/ Transverse section of an exhalant papilla, e : exhalant canal, i : inhalant canal.

Distribution.

Found N-E of Halifax by Topsent, 1927a, this species was never collected again (map 1).

Remarks.

Here appears the problem of the limits between *Trichostemma* and *Polymastia*. *Polymastia infrapilosa* may be considered within the genus *Polymastia* as the species similar to that from which the *Trichostemma* evolved. It is always fixed, and it has the fringe of long spicules around the base but this fringe is not made by a special type of spicule. In the choanosome, the presence of small ectosomal tylostyles in groups of 3 to 5 is observed as in *Trichostemma*. But in my opinion it has to be still considered as a true *Polymastia* because of the structure of the papillae and of its irregular shape.

Polymastia grimaldi (Topsent, 1913)

Trichostemma grimaldi Topsent, 1913 p. 21, pl I, fig. 4

This species, very well recognizable, is an other step in the evolutionary line between *Polymastia* and *Trichostemma*.

External characters.

This species is disc-shaped and fixed only by certain points to the substrata. At the limit between the upper face and the lower face a somewhat long fringe of bristles may be observed. The papillae are very numerous and located on the upper face. The upper face is hispid and the lower face smooth. A specimen of Topsent of about 2 dm2 had almost 300 papillae. The colour *in vivo* is not known; it is whitish in alcohol.

Skeleton.

Choanosome (fig. 7a) : bundles of tylostyles, 100-200 µm in diameter, echinate the surface and pass through the palissade of small tylostyles. In the choanosome between the bundles, free small tylostyles are scattered.

Ectosome (fig. 7a) : the upper cortex about 650 µm thick is composed by three layers : a layer of tangential intermediary tylostyles, 250 µm thick; a collagenous layer, 150 µm thick; on this lays the cortical layer of ectosomal tylostyles. These three layers are traversed by the bundles of principal tylostyles. At the junction between the upper and the lower faces there is a fringe of very long, thin bristles. On the lower face the two external layers of the cortex disappear and only the layer of tangential intermediary tylostyles remains. The bundles of principal tylostyles run parallel to the surface.

Papillae (fig. 7b) : exhalant papillae and inhalant papillae are distinct but built on the same scheme. The only difference is the thickness of their wall. From the center to the surface they are made of a central canal (inhalant or exhalant), bundles of longitudinal tylostyles, a layer of tangential tylostyles and a cortical layer of small tylostyles.

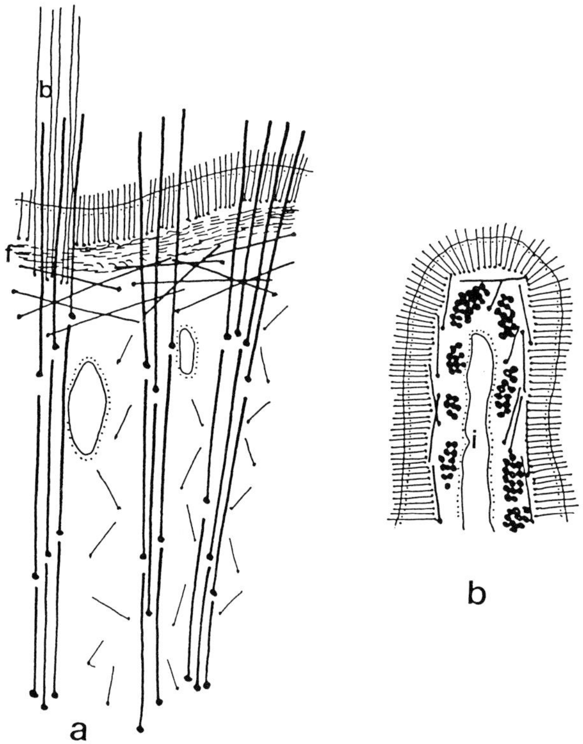

Figure 7 - *Polymastia grimaldi* Topsent, 1913. Type-specimen. a/ Longitudinal section of the main body. b/ Transverse section of an inhalant papilla, i : inhalant canal, f : collagenous layer, b : bristles.

Spicules.

There are four different kinds of spicules. The principal and the intermediary are fusiform strongyloxea or fusiform tylostyles.

Principal : 1450-2275/21-26 µm.

Intermediary : 210-676/10.4-21 µm.

Ectosomal tylostyles : 172-286/5.2-7.8 µm.

The bristles of the edge are long and thin styles ; the smallest are 2 mm long and their maximum length is greater than 4 mm/10 µm thick.

Distribution.

This species is quite characteristic from the boreal region of the

Atlantic. It is known from 65°21'N to 72°5'N / 10° 42' W - 37°57' E (map 1). The bathymetric range is from 70 m to 650 m.

Remarks.

There is a large amount of confusions in the literature between *P. mammillaris*, *P. penicillus* (Montagu, 1818) and *P. grimaldi*. Vosmaer (1885) described as *P. mammillaris* an obvious specimen of *P. grimaldi*. *Polymastia penicillus* is normally considered as a synonym of *P. mammillaris*. Therefore a part of the sponges described as *P. penicillus* by Vosmaer (1882), Hansen (1885), Levinsen (1886), Fristedt (1887) are also *P. grimaldi*. The differences between *P. grimaldi* and *P. mammillaris* are absolutely clear and no confusion is possible between the two species. I do not agree with Koltun (1966) who considers "*grimaldi*" as a subspecies of *mammillaris*. According to Koltun (1966) *P. mammillaris* var. *hyperborea* Hentschel is also a synonym of *P. grimaldi*.

Polymastia grimaldi was described in 1892 as a *Trichostemma*, and Topsent prefered to consider it as a *Polymastia* in 1927b. The same problem appears for this species as for *P. infrapilosa*, although here bristles appear on the outside edge. *Polymastia grimaldi* may be considered as a step on the evolutionary line which starts at *Polymastia* advancing to *Trichostemma*.

Polymastia robusta (Bowerbank, 1861)

Bowerbank, 1866, p. 62-64; 1882, p. 31.

External characters.

Polymastia robusta is a very common species of the N-E Atlantic. Its color *in vivo* is orange and its shape is spherical. Its mean volume is 40 cm3. One inhalant papilla and one exhalant papilla corresponds to a volume of 1 cm3. The surface is smooth, without any epibiosis and/or sediments. The inhalant apertures are very obvious *in situ* and are scattered on the whole surface of the sponge, including the papillae. The oscules open at the top of almost all the papillae.

Skeleton.

Choanosome (fig. 8a) : the choanosomal skeleton consists of bundles of tylostyles, 350 µm in diameter. These bundles open out when they reach the surface, which they never pass through. Intermediary tylostyles are scattered between the bundles.

Ectosome (fig. 8a) : the ectosome is about 700 µm thick and is composed of two layers: a tangential layer (450 µm) composed of intermediary tylostyles and a ectosomal layer (250 µm) of small tylostyles perpendicular to the surface.

Papillae (fig. 8b) : they are conical and have a length of 2 to 8 mm. On a perpendicular section of a papilla, there are always several inhalant canals, varying in number from 3 to 12, which are distributed around the exhalant canal in the exhalant papilla. The skeleton is composed of bundles of principal tylostyles which are the axial framework. The number of bundles varies from 10 to 30 and each bundle contains 5 to 20 spicules. These bundles are located around the papillae and in the cellular laminae which separate the different canals. They are linked by a net of intermediary tylostyles located in a perpendicular plan. At the surface of the papillae, there is a layer of small tylostyles similar to that of the ectosome.

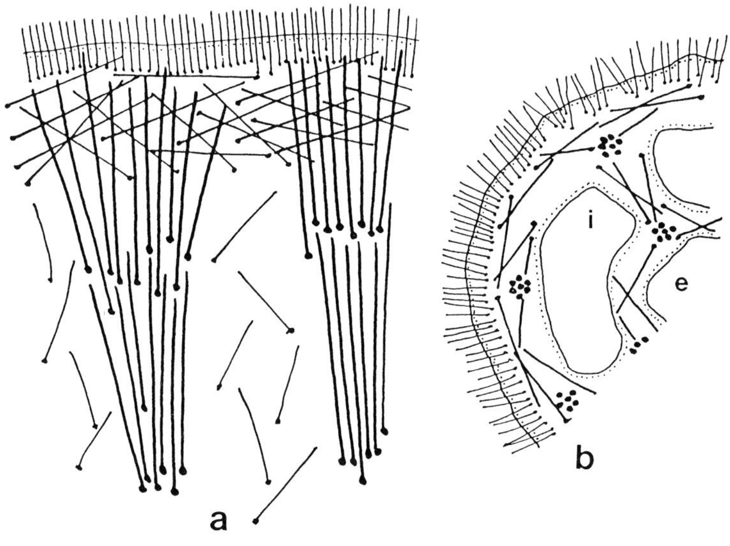

Figure 8 - *Polymastia robusta* (Bowerbank, 1861). Specimen from The Channel. a/ Longitudinal section of the main body. b/ Transverse section of an exhalant papilla, i : inhalant canal, e : exhalant canal.

<u>Spicules</u>.

Principal and intermediary tylostyles : in contrast to *P. mammillaris*, it was impossible to separate them into two size classes (Boury-Esnault, 1974). They are distinguishable only by their location in the skeleton : 250-700/8-15 µm (mean of 400 µm from 500 spicules measurements).

The ectosomal tylostyles are 90-230 µm in length with a mean 150 µm (measurements on 500 spicules)

Distribution

Known from the Artic zone of the Atlantic 72°37'N-20°00'E, the species was collected on the two sides of the Atlantic, east until the Mediterranean and Adriatic coasts and west until the Canada and Newfoundland coasts (map 1). The bathymetric distribution is from 10 to 1267 m in depth. The citations of Procter (1933) and Wells *et al.* (1960) are not *P. robusta.*

Remarks.

In 1959 Burton reintroduced the name "*boletiformis*" for *P. robusta* without giving any justification. In fact Topsent (1933) indicated in his revision of the Lamarck's sponge from the Museum of Paris that a specimen of Lamarck labelled *Alcyonium boletiforme* is a *P. robusta.* As the name was not used since Lamarck and the description of Lamarck inexistant, Topsent prefered to keep the name given by Bowerbank as it is based on a good description; he considered *Alcyonium boletiforme* Lamarck as a *nomen nudum.* Between the description of Lamarck and the reintroduction of "*boletiforme*" by Burton there are 144 years without any use of "*boletiforme*". In the literature there are four citations as *P. boletiforme* starting with Burton, 1959 versus 38 citations of *P. robusta* since Bowerbank, 1861.

Polymastia spinula Bowerbank, 1866

Bowerbank, 1866, p. 66-68; 1882, p.32.

This species considered as a synonym of *P. mammillaris* by Topsent (1900), seems to be a good species. I have reexamined the type specimen from Bowerbank and many different features distinguish the two species.

External characters.

Bowerbank (1866, fig. 10) gave the general description of this little sponge characterized externally by the length of the papillae which overtook 5 cm for a body of about 1 cm2. Their point is sharp. The colour of the dry specimen is cream white. The specimen of Topsent (1892) which is in Monaco Museum has the same traits. It is a small sponge with one very long papilla sharply ended.

Skeleton.

Choanosome (fig. 9a) : the choanosomal skeleton consists of bundles of tylostyles of 250 µm in diameter. These bundles end arched just below the cortical layer of ectosomal tylostyles. Between the bundles few intermediary spicules are scattered.

Ectosome (fig. 9a) : the ectosome is 250 µm thick. It is composed of intermediary tangential tylostyles, and a layer of small tylostyles arranged in palissade.

Papillae (fig. 9b) : only the inhalant papillae were observed. In contrast with the other species of the genus, the papilla of *P. spinula* is very simple. The axial skeleton is composed by the bundles of tylostyles slightly protuberant in the single canal of the papilla, a thin layer of tangential tylostyles and the palissade of ectosomal tylostyles. In fact, the wall of the papilla has the same thickness as the palissade, i.e. about 150 µm.

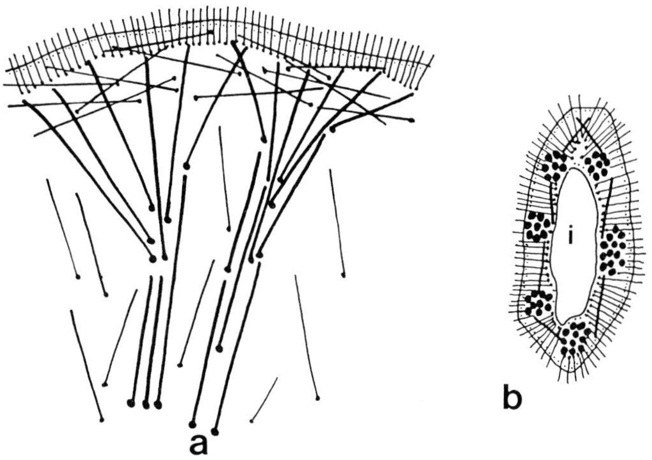

Figure 9 - *Polymastia spinula* Bowerbank, 1866. Type-specimen. a/ Longitudinal section of the main body. b/ Transverse section of an inhalant papilla, i : inhalant canal.

Spicules.

All the spicules of *P. spinula* are bent or slightly flexuous.

The principal ones are fusiform strongyloxea, bent or slightly flexuous : 724.5-926.8/10.4-15.6 µm.

The intermediary ones are also bent; but they are fusiform tylostyles with a slightly marked head : 332.8-478.4/5.2-10.4 µm.

The ectosomal ones are tylostyles with a well-marked head; they are bent once or twice : 124.8-156/5.2 µm.

Distribution.

Described from the Shetland by Bowerbank, this species was found again only by Topsent in the Bay of Biscaye 63 m deep (map 1).

Polymastia uberrima (Schmidt, 1870)

Polymastia uberrima was badly described by Schmidt, 1870. The type species has not been seen, but only the specimen described as *P. uberrima* by Topsent, 1913.

non *P. uberrima* Thiele, 1903; non *P. uberrima* Hentschel, 1929.

External characters.

It is a spherical sponge with about 15 papillae at the top in the Topsent's specimen. It is fixed on pebbles and fragments of bryozoans. All the papillae end by an osculum.

Skeleton.

Choanosome (fig. 10a) : the bundles of principal spicules are 200-250 µm in diameter. They reach the basis of the palissade of tylostyles without passing through. Between these bundles small fascicles of intermediary spicules are dispersed.

Ectosome (fig. 10a) : it is a very thick cortex of about 2.2 mm. It is composed of two layers: a layer of tangential tylostyles 1520 µm thick, and a layer of tufts of small tylostyles which constitute an irregular palissade of about 760 µm thick.

Papillae (fig. 10b) : the papillae are very wide. There is a central exhalant canal and the inhalant canals are at the periphery. The structure from inside to outside is respectively : the central canal, the surface of which is echinated by intermediary tylostyles; the bundles of spicules which are the axial framework of the papilla and which are linked by intermediary tylostyles; the inhalant canals which open between these bundles; at the surface, the tufts of small tylostyles similar to those of the cortex.

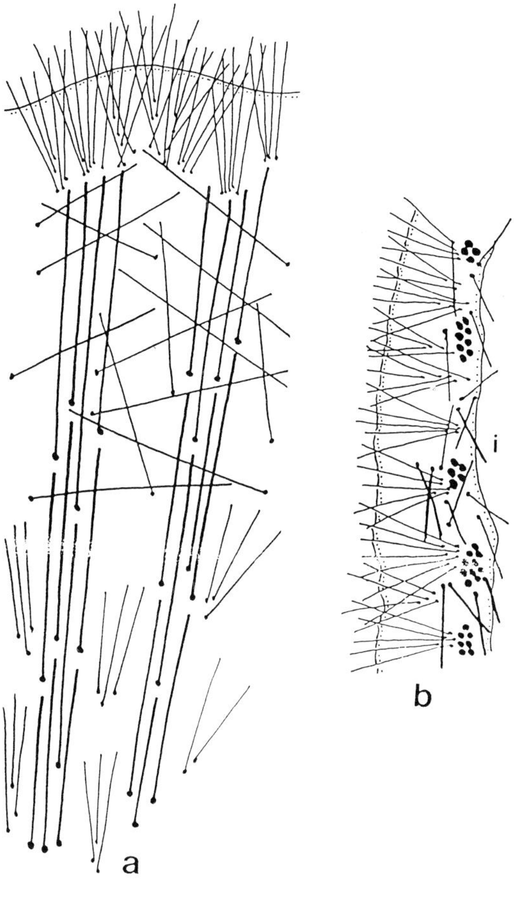

Figure 10 - *Polymastia uberrima* (Schmidt, 1870). Specimen of Topsent, 1913. a/ Longitudinal section in the main body. b/ Transverse section of an inhalant papilla, i : inhalant papilla.

Spicules.

Principal straight, slightly fusiform strongyloxea: 1315-1770/17-25 µm.

Intermediary straight tylostyles, with a poorly marked head: 556-754/8-10 µm.

Ectosomal slightly bent tylostyles : 307-350/5 µm.

Distribution.

Described from Iceland by Schmidt, the species has always been collected in cold water. It was found along the coasts of Norway (Topsent, 1913), and in Greenland (Lundbeck, 1909) (map 1). The bathymetric distribution is from 90 to 645 m.

Remarks.

Polymastia uberrima is a valid species, but it was confused with *Weberella bursa* by Hentschel (1929), and Burton (1959) synonymised the two species. The confusion was heightened when Koltun (1966) separated the specimens described by the different authors under the name "*uberrima*", into two groups. One group with only Schmidt's citation is *P. uberrima sensu stricto* and the other one with all the others citations is erected as a new species : *P. thielei*. But Koltun did not give any argument for this separation or why he decided that Lundbeck's and Topsent's specimens have to belong to this new species.

In fact, in his description of *P. uberrima*, Thiele (1903) mentionned groups of small tylostyles arranged like trichodragmata between the choanosomal bundles. The creation of a new species for the sponge of Thiele seems consistent. But these trichodragmata are not present in the Topsent's specimen.

Many of the citations of *P. uberrima* are very brief without giving any details on the skeleton, thus it is impossible to decide without a careful reexamination of the specimens which are related to *P. uberrima* and which to *P. thielei*.

Polymastia corticata Ridley & Dendy, 1886

This species was well described by Ridley & Dendy, 1887 (p.211 pl. XLII figs 4, 5, 5a, 5b; pl. XLIV fig. 3). It is characterized by its dense and leathery cortex.

External characters.

The type specimen is a spherical sponge (12 mm in diameter and 25 mm in height) with inhalant and exhalant papillae. These papillae are approximately 8 mm high with a diameter of 3 mm. There are about 100 inhalant papillae and 4 to 5 exhalant papillae on the type-specimen.

<u>Skeleton</u>.

Choanosome (fig. 11a) : it is composed of bundles of principal spicules 150-250 µm in diameter. They reach the basis of the cortex. Free intermediary tylostyles are scattered between the bundles.

Ectosome (fig. 11a) : the cortex is very thick (1.5 to 2 mm) with two very distinct layers. The outermost portion consists of a layer of small tylostyles which are approximately 250 µm thick. A layer measuring about

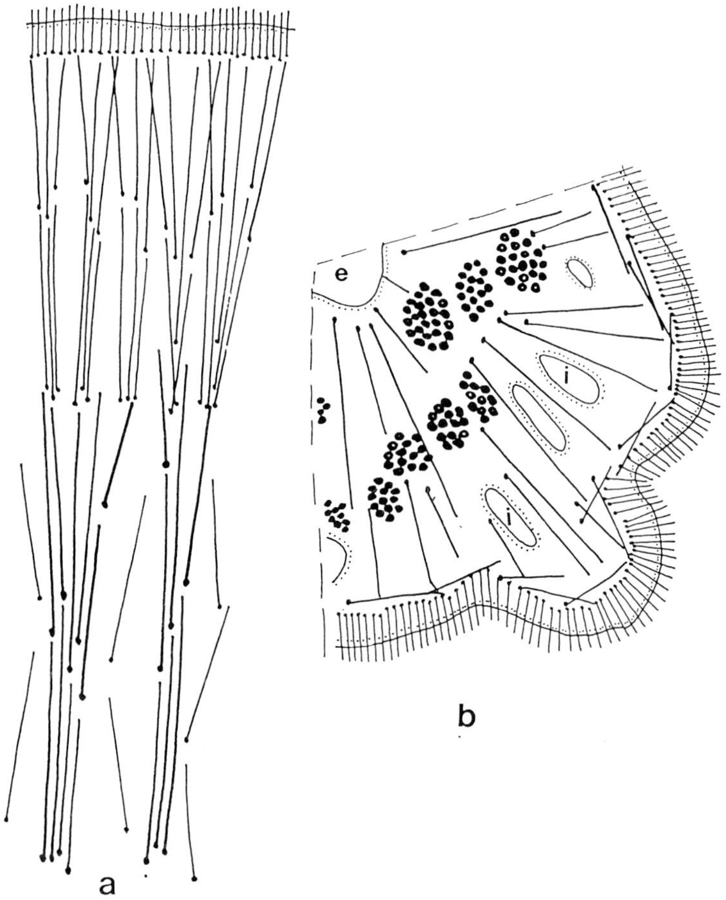

Figure 11 - *Polymastia corticata* Ridley & Dendy, 1886. a/ Longitudinal section of the main body. b/ Transverse section of an exhalant papilla, e : exhalant canal, i : inhalant canal.

1500 µm, of intermediary styles more or less vertically placed is located immediately below.

Papillae (fig. 11b) : inside to outside : the exhalant canal; the axial framework of principal spicules; the inhalant canals; intermediary spicules more or less radially placed; a thin tangential layer of intermediary spicules and the external layer of small tylostyles.

Spicules.

On a preparation of the type specimen, I have observed three categories of spicules. The principal and intermediary are fusiform styles with a slender head, which are distinguishable by the size.

Principal styles : 1634-2106/17-34 µm; **intermediary styles** : 674-893/10.4-15.6 µm. **Ectosomal tylostyles** : 166-255/5.2-10.4 µm.

Distribution.

Found by Ridley & Dendy off the coast of Brazil (Bahia), it was collected a second time by Topsent (1904) east of São Miguel (Azores) (map 1). The bathymetric range is from 200 to 1385 m.

Remarks.

The sponges collected by Topsent (1892) are not *P. corticata*.

Polymastia polytylota Vacelet, 1969

Vacelet, 1969 p.172, fig.7. This very small species is a one-papilla *Polymastia*. Described by Vacelet (1969) from Corsica it is well characterized.

External characters.

The sponge is fixed on shells or fragments by the whole base. This base is a somewhat disc-shaped cushion of 5-8 mm in diameter. The surface is slightly hispid. In the center there is one single papilla, 3-5 mm in height, the surface of which is smooth.

Skeleton.

Choanosome (fig. 12a) : the choanosomal skeleton consists of bundles of principal tylostyles 100-150 µm in diameter. They reach the basis of the cortex without passing through. Intermediary and ectosomal tylostyles are scattered between these bundles.

Ectosome (fig. 12a) : the ectosome is 350 µm thick and consists of two layers. The internal is a tangential layer of intermediary tylostyles about 230 µm thick. The external layer is a palissade of small tylostyles.

Papilla (fig. 12b) : the papilla is inhalant and exhalant. It has the same architecture than the exhalant papillae of *P. robusta* : a central exhalant canal is surrounded by the bundles of tylostyles which constitute the axial framework. The tangential tylostyles are present in the cellular laminae which separate the canals. The papilla is surrounded by an external layer of ectosomal tylostyles.

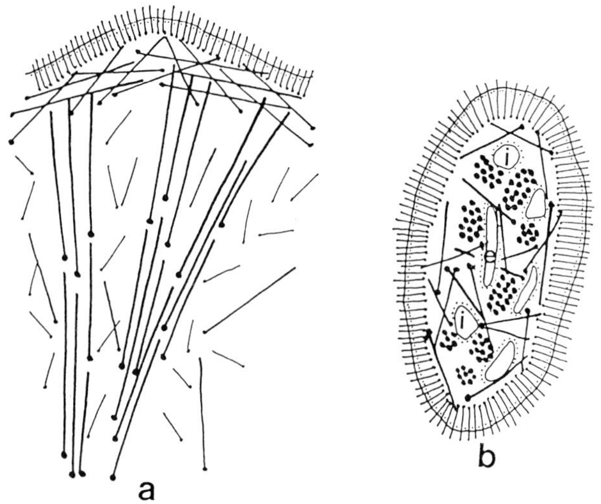

Figure 12 - *Polymastia polytylota* Vacelet, 1969. Type-specimen. a/ Longitudinal section of the main body. b/ Transverse section of the papilla, e : exhalant canal, i : inhalant canal.

<u>Spicules</u>.

The **principal tylostyles** are straight and fusiform and show polytylotism : 650-990/10-12.5 µm.

The **intermediary tylostyles** are straight, fusiform and often have a vesicle in the head : 210-490/7-10 µm.

The **ectosomal tylostyles** are slightly bent : 70-180/2-5 µm.

<u>Distribution</u>.

Always found off the coasts of Corsica (Vacelet, 1969; Pulitzer-Finali, 1983) (map 1). The bathymetric range is from 117 to 270 µm.

Polymastia littoralis Stephens, 1915

It was not possible to reobserve the type-specimen but the accurate description by Stephens (1915, p.436, pl XXXVIII fig.4, pl XL fig.3) enables the schemes of organisation of the skeleton to be constructed.

This species is characterized by its cortex (3 mm), which is composed of two layers. Groups of small tylostyles are scattered between the choanosomal bundles (fig. 13a).

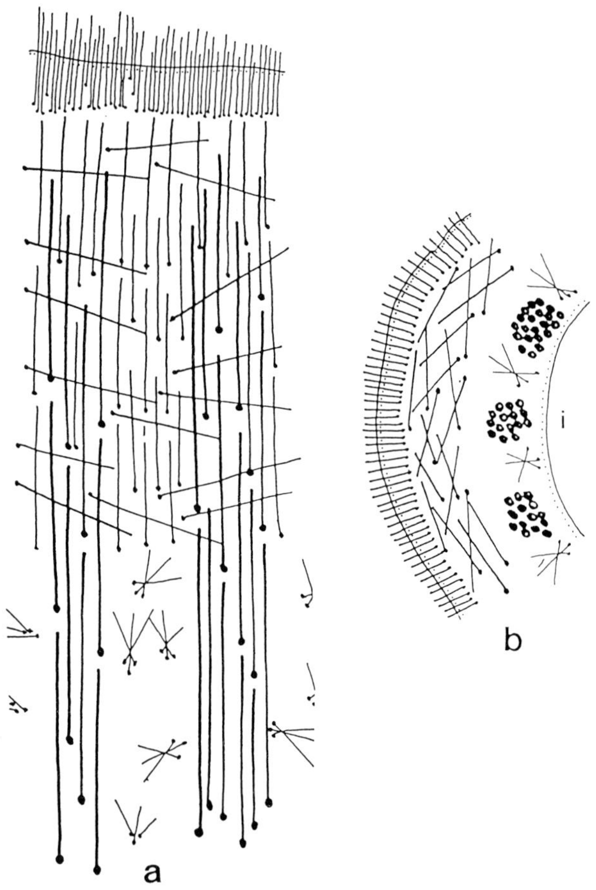

Figure 13 - *Polymastia littoralis* Stephens, 1915. According to the original description. a/ Longitudinal section of the main body. b/ Transverse section of a papilla, i : inhalant canal.

In the papillae towards the center strong fibres of large styli run longitudinally, with scattered groups of small tylostyles. At the surface there is a palissade of ectosomal tylostyles and beneath them is a thick layer of tangential tylostyles (fig. 13b).

Distribution.

Described by Stephens from S-E Atlantic (Saldanha Bay), the species was also found by Burton (1936) in S-Africa (Clitts Point). It is a littoral species.

Polymastia tenax Pulitzer-Finali, 1986

Insufficiently described by Pulitzer-Finali, 1986 p. 89, figs 19-20; the specimens at our disposal allow for a new description.

External characters.

This is the largest known species of *Polymastia* in the Atlantic. It is an encrusting species which can cover about 1 m2 with a thickness of 3 cm (fig. 14). Its upper surface is covered by numerous mammiform processes, the appearance of which is very similar to the papillae of *Cliona viridis* (fig.15). There is one exhalant papilla for 15 to 20 inhalant papillae. The size of the inhalant papillae is quite variable as they are often fused together. Their diameter observed on underwater photograph varies from 5 to 21 mm. These papillae are roundish and have a maximum height of 4 mm. The exhalant papillae are larger; their height varies from 15 to 25 mm and the

Polymastia tenax Pulitzer-Finali, 1986. Underwater photograph (J. Vacelet). Fig. 14/ Specimen from Bas Baudry (Martinique). Fig. 15/ Detail of a specimen from Cap Enragé (Martinique).

diameter at their basis varies from 13 to 38 mm. The osculum at the summit of these exhalant papillae have a diameter which varies from 5 to 19 mm. The surface of the papillae is smooth. It is hispid in the depressions between adjacent papillae. The inhalant apertures seem limited to the surface of the inhalant and exhalant papillae. The colour *in vivo* is dull yellow for the choanosome and whitish to light brown for the cortex.

Skeleton.

Choanosome (fig. 16a) : the choanosomal skeleton consists of bundles of tylostyles which are about 350 µm in diameter. At the level of the inhalant papillae, the bundles penetrate only to the layer of intermediary tylostyles of the cortex whereas between the papillae, the bundles reach and pass through the surface. Between the bundles, cortical and intermediary tylostyles are scattered.

Ectosome (figs 16a, 17d) : the ectosome is relatively thick (1 to 3 mm). It is composed of two layers which contain spicules perpendicular to the surface. The superficial layer is made of ectosomal tylostyles arranged in palissade and the inner is formed by intermediary fusiform tylostyles

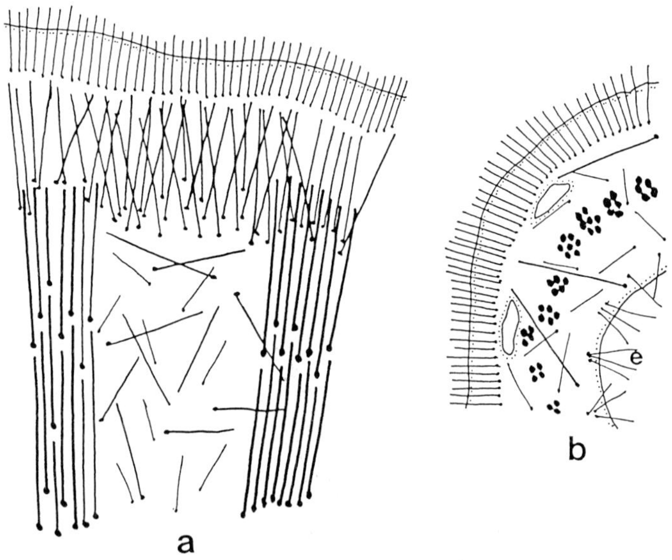

Figure 16 - *Polymastia tenax* Pulitzer-Finali, 1986 - Specimen from Martinique. a/ Longitudinal section of the main body. b/ Transverse section of an exhalant papilla, e : exhalant canal.

which are more randomly arranged. The inner layer is twice as thick as the external layer.

Papillae (fig. 16b) : the inhalant papillae have the same organization as the cortex. All the inhalant ostioles open in small canalicules which converge in the direction of the central canal (fig. 17d). The exhalant papillae vary in some respects (fig. 17a). The surface of the exhalant canal is folded and tufts of ectosomal tylostyles echinate these folds (fig. 17b). In the wall of the papillae, the bundles of principal tylostyles composed the axial framework. The palissade of ectosomal tylostyles is present at the

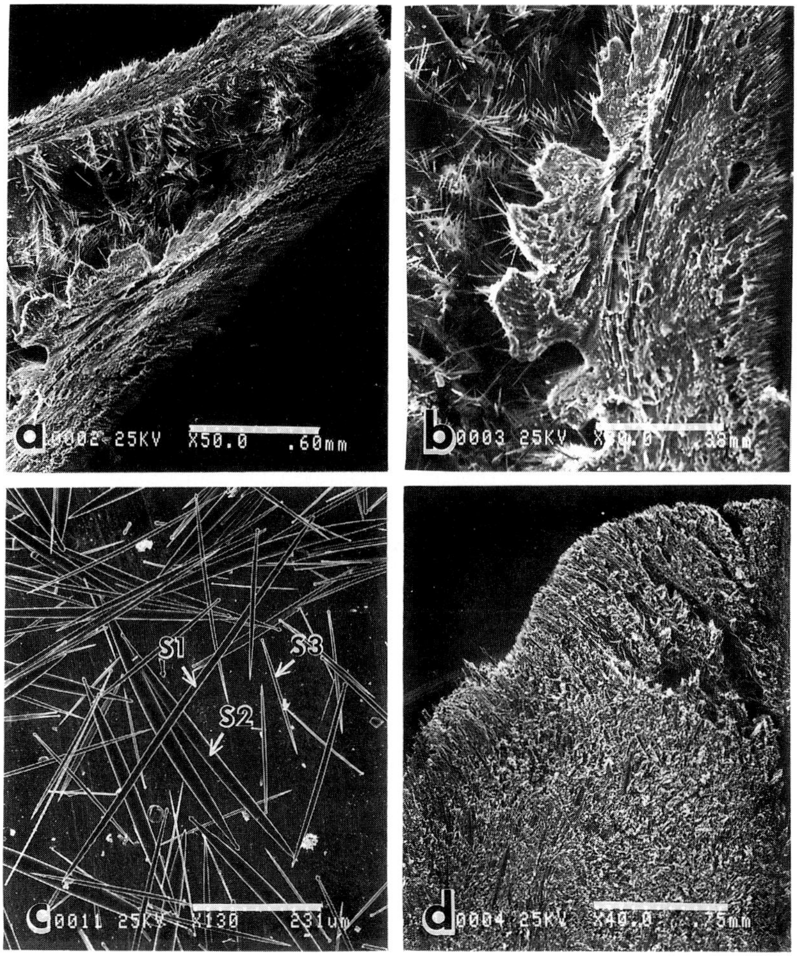

Figure 17 - *Polymastia tenax* Pulitzer-Finali, 1986 - a/ Longitudinal section in an exhalant papilla. Scanning electron micrograph. b/ Detail of the wall of the exhalant papilla. The exhalant canal is echinated by ectosomal tylostyles. c/ Three types of tylostyles, S1 : principal tylostyles, S2 : intermediary tylostyles, S3 : ectosomal tylostyles. d/ Longitudinal section in an inhalant papilla and the ectosome.

surface. Between the bundles numerous ectosomal tylostyles and a few intermediary ones are scattered. The intermediary layer of fusiform tylostyles is absent.

Spicules(fig. 17c).

Principal tylostyles : straight and fusiform : 806-1100/13-26 µm.

Intermediary tylostyles : straight, fusiform and stout. The end is sharp or rounded and the head is very distinguishable : 436-666/18-47 µm at the thickest point, and 2.6-10.4 µm just below the head.

Ectosomal tylostyles : straight, head distinctive : 114-327/2.6-10.4 µm

Distribution.

Dominican Republic : 19.4.1964, Sosua, 7 m, shadowed cavity (Pulitzer-Finali, 1986).

Martinique : 23.12.1983, Bas Baudry, overhanging wall, 45 m; 24.12.1983, Diamant, overhanging cliff, 35 m; 5.1.84, Cap Enragé, basis of a vertical slope, 12 m.

Anguilha : 1.5.1986, Dog Island, cave, 10 m.

St Barthélémy: 19.4.1986, Pointe Nègre, vertical cliff, 5 m; 20.4.1986, Ile Fourche, below rocks, 6 m.

All these specimens were collected by J. Vacelet during the missions Corantille II December 1983-January 1984 and Ecorecif April 1986.

This species appears characteristic of shaded habitats between 5-45 m and it was collected only in the Caribbean.

Remarks.

The most characteristic feature of this species is the fusiform shape of the intermediary tylostyles. Two species have been described with this trait before the recent description of *P. tenax* (Pulitzer-Finali, 1986): *P. littoralis* Stephens, 1915 from South Africa and *P. maeandria* Wilson, 1904 from Galapagos. Although *P. tenax* was poorly described, the size and the shape of the spicules, and its geographical distribution allow the assumption that the specimens described here are identical to the specimen described by Pulitzer-Finali.

Polymastia littoralis differs from *P. tenax* by the structure of the choanosomal skeleton consisting of stout bundles (500-1000 µm) and in small tylostyles arranged in tufts visible between these bundles. It differs also by the structure of the papillae where the intermediary layer is quite distinct. The length of the three kinds of spicules is similar, but the intermediary tylostyles are thinner and are not as fusiform as those of *P. tenax*.

Polymastia maeandria is closer to *P. tenax*. The principal differences are:

- the shape and size of the specimens and their papillae,
- the structure of the cortex, the intermediary layer of *P. maeandria* being tangential,
- the shape of principal spicules which are styles in *P. maeandria* and true tylostyles in *P. tenax*,
- the shape of the ends of the intermediary tylostyles, which are always sharp in *P. maeandria* and very often rounded in *P. tenax*,
- the size of the spicules, the principal and the intermediary being smaller and the ectosomal ones being larger.

The structure of the papillae and of the choanosomal skeleton and the general shape of the spicules is similar.

These two species are very close and appear to be sibling species. It is suspected that here there are two populations of a species which have evolved separately since the closing of the Isthmus of Panama, in such a way that to-day, two species can be recognized.

Polymastia nigra Alcolado, 1984

Insufficiently described by Alcolado, 1984 p. 12, figs 6C, 8C; the type-specimen needs to be reexamined.

Polymastia sola Pulitzer-Finali, 1983

Insufficiently described by Pulitzer-Finali, 1983 p. 489, figs 20, 21; the type-specimen needs to be reexamined.

DISCUSSION

In the literature, 34 species have been described in the genus *Polymastia* (table 1) in the Atlantic area. Among them seven were transferred in other genera, eight declared as synonymous of other species and four are unrecognizable. For the moment, 13 good species of *Polymastia* are recognizable in this area.

Table 1 - List of the species of *Polymastia* from the Atlantic area in the literature.

P. agglutinans Ridley & Dendy,1886	valid species
P. azorica Lévi & Vacelet, 1957	unrecognizable, type-specimen not available
P. boletiforme Lamarck, 1815	synonym of *P. robusta*
P. brevis (Bowerbank, 1861)	transferred to *Quasillina* by Norman, 1869
P. bulbosa Bowerbank, 1866	synonym of *P. robusta*
P. bulbosa Sarà & Siribelli, 1961	synonym of *P. inflata*
P. conigera Bowerbank, 1874	valid species
P. corticata Ridley & Dendy, 1887	valid species
P. disclera Lévi, 1964	description needs more data
P. gleneni Descatoire, 1966	synonym of *Aaptos papillata*
P. grimaldi (Topsent, 1913)	valid species
P. inflata Cabioch, 1968	valid species
P. infrapilosa Topsent, 1927	valid species
P. littoralis Stephens, 1915	valid species
P. mammillaris (Müller, 1806)	valid species
P. mespilus Schmidt, 1873	unrecognizable
P. nigra Alcolado, 1984	insufficiently described
P. ornata Bowerbank, 1866	synonym of *P. robusta*
P. paupera Fristedt, 1887	Suberitid
P. penicillus (Montagu, 1818)	some records synonym of *P. mammillaris* some records synonym of *P. grimaldi*
P. polytylota Vacelet, 1969	valid species
P. radiosa Bowerbank, 1866	synonym *P. robusta*
P. robusta (Bowerbank, 1861)	valid species
P. sola Pulitzer-Finali, 1983	insufficiently described
P. spinula Bowerbank, 1866	valid species
P. stipitata Carter, 1876	transferred to *Stylocordyla*
P. tenax Pulitzer-Finali, 1986	insufficiently described,valid species
P. uberrima (Schmidt, 1870)	valid species some record *P. thielei* Koltun, 1964.
P. varia Verril, 1907	unrecognizable

Some descriptive characters appears to be constant in the 13 described species:

1) Presence in all cases of erect oscular and pore-bearing papillae. In all cases they are inhalant and inhalant/exhalant papillae. The number of papillae is quite variable, from one in *P. polytylota* to hundreds in *P. grimaldi* or *P. tenax*.

2) The choanosomal skeleton is always composed of radiate bundles of spicules. These bundles may either reach the surface that they echinate, or stop below the palissade layer or just below the cortex. In this last case, the cortex can be easily detached, as in *P. corticata*. Between the bundles there are always free spicules scattered, but they can be intermediary, intermediary and ectosomal or ectosomal spicules.

3) The cortex always has two layers, with the exception of *P. grimaldi* where there is a third one between the upper layer and the intermediary layers. The upper layer is always a palissade of small tylostyles. In general their size varies from 90 to 130 µm. *Polymastia uberrima* has small tylostyles slightly larger (300-350 µm). In *P. agglutinans* the palissade is modified by the presence of foreign bodies included in the upper face of the cortex. The layer of intermediary spicules is in most cases tangential, but is perpendicular to the surface in *P. corticata* and *P. tenax*. In most cases, these intermediary spicules are true tylostyles, except in *P. grimaldi* where they are strongyloxea and in *P. corticata* where they are styles.

4) The shape of the principal spicules is much more variable. They are true tylostyles in *P. mammillaris*, *P. infrapilosa*, *P. robusta* and *P. tenax*, polytylote tylostyles in *P. polytylota* and *P. conigera*, styles in *P. agglutinans*, *P. infrapilosa*, *P. corticata* and *P. littoralis*; strongyloxea in *P. inflata*, *P. infrapilosa*, *P. spinula*, *P. uberrima* and *P. grimaldi*.

5) In one case, *P. grimaldi*, there is a fourth category of spicules which forms a fringe of bristles on the external edge of the sponges.

In the table 2, the diagnostic characters which distinguish the 13 species of *Polymastia* of the Atlantic area are given.

Table 2 - Discriminating characters for the 13 species of *Polymastia* from the Atlantic area. 1) *Polymastia mammillaris*, 2) *P. agglutinans*, 3) *P. conigera*, 4) *P. inflata*, 5) *P. infrapilosa*, 6) *P. grimaldi*, 7) *P. robusta*, 8) *P. spinula*, 9) *P. uberrima*, 10) *P. corticata*, 11) *P. polytylota*, 12) *P. littoralis*, 13) *P. tenax*. t: tangential; P: perpendicular; I: intermediary spicule; E: ectosomal spicule; T: Tylostyle; St: Style; S: Strongyloxea; p: polytylote; f: fusiform; b: bent; s: straight.

	1	2	3	4	5	6	7	8	9	10	11	12	13
CORTEX													
-Thickness	500	700	300	600	900-1250	650	700	250	2200	2000	350	3000	1000-3000
-Layers	2	2	2	2	2	3	2	2	2	2	2	2	2
-Foreign bodies	-	+	-	-	-	-	-	-	-	-	-	-	-
-Orientation internal layer	t	t	t	t	t	t	t	t	t	P	t	t	P
CHOANOSOME													
-Nature free spicules	I	I	I	I+E	E	E	I	I	I	I	I+E	E	I+E
PAPILLA													
-Number/specimen	>10	<10	<10	<10	>10	>100	>10	>10	>10	>100	1	>10	>10
SPICULES													
-In cellular lamina	-	+	+	+	+	+	+	?	+	+	+	+	+
-Echinating exhalant canal	-	E	I	-	I	-	-	-	I	-	-	-	E
-Number of types	3	3	3	3	3	4	2	3	3	3	3	3	3
-Type of principal	T	St	Tp	Sf	S-Tf	S-Tf	T	Sfb	Sfs	Stf	Tpfs	Stf	Tf
-Type of intermediary	T	St	Tp	Tf	Tf	S-Tf	= T	Tfb	Ts	Stf	Tfs	Tfb	Tf
-Type of ectosomal	T	T	T	T	T	T	T	Tb	Tb	T	Tb	T	T

For the moment, the following definition may be given for the genus *Polymastia* :

Polymastids encrusting or spherical always with papillae. The skeleton is composed of radial bundles of principal spicules between which free spicules are scattered. There is always a cortex of at least two layers. The upper layer is always a palissade of small tylostyles. The lower layer is made of intermediary spicules tangential or perpendicular to the surface. The principal spicules may be tylostyles, styles or strongyloxea, the intermediary is most often tylostyles and the ectosomal spicules are always tylostyles. The architecture of the papillae reflects the architecture of the body.

ACKNOWLEDGEMENTS

I would like to acknowledge Miss Shirley Stone from the British Museum of Natural History, Dr Christian Carpine from the Monaco Museum, Dr Annie Descatoire-Girard from the Laboratoire de Biologie marine of Concarneau, Dr Louis Cabioch from the Station biologique in Roscoff, Professor Claude Lévi from the Museum National d'Histoire Naturelle of Paris and Dr Jean Vacelet from the Station marine d'Endoume for providing type-specimens and collection specimens necessary for this revision.

I would also like to express my gratitude to Chantal Bezac and Marie-Rose Causi for their technical assistance.

REFERENCES

Alcolado P., 1984 - Nuevas especies de esponjas encontradas en Cuba. *Poeyana*, 271, 1-22.

Bergquist P., 1968 - The Marine Fauna of New Zealand : Porifera, Demospongiae, Part 1. (Tetractinomorpha and Lithistida. *Mem. N. Zeal. Oc. Inst.*, 188, 1-106.

Boury-Esnault N., 1974 - Structure et ultrastructure des papilles d'éponges du genre Polymastia Bowerbank. *Archs Zool. exp. gén.*, 115(1), 141-165.

Bowerbank J.S., 1864 - A Monograph of the British Spongiadae. Vol. I, 1-290. Ray Society London

Bowerbank J.S., 1866 - A Monograph of the British Spongiadae. Vol. II, 1-388. Ray Society London.

Bowerbank J.S., 1874 - A Monograph of the British Spongiadae. Vol. III, 1-360. Ray Society London.

Bowerbank J.S., 1882 - A Monograph of the British Spongiadae. Vol. IV, 1-250. Ray Society London.

Burton M., 1936 - Notes on Sponges from S.Africa, with descriptions of a new species. *Ann. Mag. nat. Hist.*, 141-147.

Burton M., 1959 - Sponges. *John Murray Expedition 1933-34 Science Report*, 10(5), 151-281.

Cabioch L., 1968 - Contribution à la connaissance de la faune des Spongiaires de la Manche occidentale. Démosponges de la région de Roscoff. *Cah. Biol. Mar.*, 9, 211-246.

Dendy A., 1921 - Report on the Sigmatotetraxonida collected by H.M.S. "Sealark" in the Indian Ocean. *Trans. Linn. Soc. London*, 18, 1-164.

Fristedt K., 1887 - Sponges from the Atlantic and Arctic Oceans and the Behring Sea described by ..., *Vega-Expeditionens Vetensk. Iakttagelser (Nordensköld)*, 4, 401-471.

Hansen G.A., 1885 - Den Norske Nordhavs-Expedition 1876-1878. Zoologi, Spongiadae, Christiana. 25pp.

Hentschel E., 1929 - Die Kiesel- und Hornschwämme des Nördlichen Eismeers. *Fauna Arctica*, 5(4), 859-1042.

Koënnecker G., 1973 - Littoral and benthic investigations on the west coast of Ireland. I. The Sponge fauna of Kilkieran Bay and adjacent areas. *Proc. r. Irish Acad.*, 73b, (26), 451-472.

Koltun V.M., 1966 - Les éponges Tetraxonida des mers du Nord et de la région extrême-orientale d'URSS. Opredel. Faune S.S.S.R., 90, 1-111. (en russe).

Lévi C., 1960 - Spongiaires des côtes occidentales africaines. *Bull. Inst. fr. Afr. noire*, sér.A, 22(3), 743-769.

Lévi C. & Vacelet J., 1957 - Eponges récoltées dans l'Atlantique Nord-oriental par le "Président Théodore-Tissier" 1955-56. *Revue Trav. Inst. Pêch. marit.*, 22, 225-246.

Levinsen G.M.R., 1886 - Kara-Havets Svampe (Porifera). Ved...in : *Dijmpha-Togtets zool. bot. Udbytte*, 341-372.

Lundbeck W., 1909 - The Porifera of East Greenland. *Meddelelser om Gröland*, 29, 423-464.

Procter W., 1933 - Marine Fauna. *Biol. Survey Mount Desert Region*, 5, 78-115.

Pulitzer-Finali G.,1983 - A collection of Mediterranean Demospongiae (Porifera) with, in appendix, a list of the Demospongiae hitherto recorded from the Mediterranean Sea. *Ann. Museo Civ. Stor. nat.* Genova, 34, 445-621.

Pulitzer-Finali G., 1986 - A collection of West Indian Demospongiae (Porifera). In appendix, a list of the Demospongiae hitherto recorded from the West Indies. *Ann. Mus. Civ. Stor. nat.* Genova, 86, 65-216.

Ridley S.O. & Dendy A., 1886 - Preliminary Report on the Monaxonida collected by H.M.S. "Challenger". *Ann. Mag. nat. Hist.*, 5(18), 325-351.

Ridley S.O. & Dendy A., 1887 - Report on the Monaxonida collected by H.M.S. Challenger during the years 1873-1876. *Rep. Scient. Results Challenger Zoology*, 20, 1-275.

Rodriguez-Babio C. & Lorenzo-Gondar E., 1978 - Fauna marina de Galicia II. Contribución al conocimiento de los poriferos del litoral gallego. *Monografias de la Universidad de Santiago de Compostela*, 42, 1-68.

Sarà M. & Siribelli M., 1960 - La fauna di Poriferi delle "secche" del Golfo di Napoli. I. La "secca" della Gaiola. *Ann. Ist. Mus. Zool. Univ.* Napoli, 12, 1-93.

Schmidt O., 1870 - Grundzüge einer Spongien-Fauna des atlantischen Gebietes. Leipzig, 1-88.

Stephens J., 1915 - Atlantic sponges collected by the Scottish National Antarctic Expedition. *Trans. Roy. Soc.* Edinburgh, 1(2), 423-467.

Thiele J., 1903 - Beischreibung einiger unzureichend bekannten monaxonen Spongien. *Arch. Naturgesch.* 66, 374-398.

Thiele J., 1905 - Die Kiesel- und Hornschwämme des Sammlung Plate. *Zool. Jahrb. Suppl.* 6, 407-496.

Topsent E. 1892 - Contribution à l'étude des Spongiaires de l'Atlantique nord. *Rés. Camp. scient. Prince Albert I Monaco*, 17, 645-722.

Topsent E., 1900 - Etude monographique des Spongiaires de France. III. Monaxonida (Hadromerina). *Archs Zool. exp. gén.*, 3(8), 1-331.

Topsent E., 1904 - Spongiaires des Açores. *Résult. Camp. scient. Prince Albert I Monaco*, 25, 1-280.

Topsent E. 1913 - Spongiaires provenant des campagnes scientifiques de la Princesse Alice dans les Mers du Nord (1898-1899, 1906-1907). *Résult. Camp. scient. Prince Albert I Monaco*, 45, 1-67.

Topsent E., 1927a - Diagnoses d'éponges nouvelles recueillies par le Prince Albert I de Monaco. *Bull. Inst. Ocanogr., Monaco*, 50, 1-19.

Topsent E., 1927b - Eponges du Scoresby Sund (Groënland oriental). *Ann. Inst. ocanogr.*, 4(6), 253-260.

Topsent E., 1928 - Spongiaires de l'Atlantique et de la Méditerranée provenant des croisières du Prince Albert I de Monaco. *Rés. Camp. Scient. Prince Albert Ier Monaco*, 25, 1-263.

Topsent E., 1933 - Eponges de Lamarck conservées au Museum de Paris. III - *Archs Mus. nat. Hist. nat.*, 6(10), 1-60.

Vacelet J., 1961 - Spongiaires (Démosponges) de la région de Bonifacio. (Corse). *Recl Trav. Stn Mar. Endoume*, 22(36), 21-45.

Vacelet J., 1969 - Eponges de la roche du large et de l'étage bathyal de Méditerranée. *Mém. Mus. Hist. Nat. Paris*, sér.A, 59(2), 145-219.

Vosmaer G.C.J., 1882 - Report on the Sponges dredged up in the Arctic Sea by the "Willem Barrents" in the years 1878 & 1879. Niederl. *Arch. Zool. Suppl.*, 1, 1-58.

Vosmaer G.C.J., 1885 - The Sponges of the "Willem Barents" Expedition 1880 & 1881. *Bijdr. Dierk.*, 12, 177-368.

Wells H.W., Wells M.J. & Gray E., 1960 - Marine Sponges of North Carolina. *J. Elisha Mitchell Sc. Soc.*, 76, 220-245.

Wilson H.V., 1904 - The Sponges. Reports on an exploration off the West coasts of Mexico, Central and South America and off the Galapogos Islands, incharge of Alexander Agassiz, by the U.S. Fish. Commission Steamer "Albatros", during 1891. *Mem. Mus. Comp. Zoöl. Harvard*, 30, 1-164.

DISTINCTIVE CHARACTERS WITHIN THE ORDER PETROSIDA (= NEPHELIOSPONGIDA)

R. DESQUEYROUX-FAUNDEZ

Muséum d'Histoire Naturelle
CH. 1211 Genève, Suisse.

SYNOPSIS

The morphological characters utilized to establish the classification and to define the families and genera of the Petrosida order are analysed.

An explanation of the value which the introduction of a new character - the silica level in relation to the dry weight of the sponge - has in classification and its relationship with traditional factors is given.

INTRODUCTION

The members of the order Petrosida are often associated with the order Haplosclerida. They were considered a family of this order until 1980, when Bergquist proposed this family to be considered as a new order. This decision is mainly based upon embryological and biochemical characters. However, some authors are not in agreement with this concept (i.e. de Weerdt, 1985).

In both Haplosclerida and Petrosida, the difficulties encountered in taxonomy are mainly due to the uniformity in the skeletal elements and to the lack of obvious discriminating characters. At first glance, the only difference between the two orders is the amount of silica as compared to the amount of spongin. The composition of the skeletal elements such as fibres or tracts and their arrangement in the skeleton need to be carefully studied in order to establish differences. Consequently, these two orders are badly known and the definitions of their families and genera need to be more precisely established.

NATO ASI Series, Vol. G13
Taxonomy of Porifera
Edited by J. Vacelet and N. Boury-Esnault

The definitions of the taxa are very similar for both orders ; Table 1 shows the main features used by authors to define families and genera.

In the present work, the characters which allow a distinction between the families and the genera of Petrosida are analysed in order to estimate their value and relevance in separating these taxa.

Table 1 - Morphological characters used to differentiate families and genera in Haplosclerida and Petrosida, according to the authors.

Authors and areas	Family characters	Genus characters
Griessinger (1971) Mediterranean Sea	-Reticulation of the surface. -Organisation of the skeleton.	-Spicules and skeleton types. -Anatomical and skeletal organisation.
Wiendenmayer (1977a & b) Bahamas	-General architecture of the skeleton. -Surface skeleton.	-Presence of microscleres.
Bergquist & Warne (1980) New Zealand	-Consistency. -Surface skeleton.	-Categories of spicules.
van Soest (1980) Caribbean Sea	-Surface skeleton associated with the structure of the main skeleton.	-Surface skeleton.
de Weerdt (1985) North-Eastern Atlantic	-Complexity of the surface skeleton as shown by the number of layers.	-Main skeleton.
Desqueyroux-Faundez (1984 & in press) Western Pacific	-Surface skeleton and cortical structure. -Global content in silica. -Conditions of isodiametry linked to the rigidity.	-Main skeleton and internal structure. -Size categories of spicules. -Microscleres, if present.

MORPHOLOGICAL CHARACTERS USED TO DIFFERENTIATE FAMILIES AND GENERA IN THE ORDER PETROSIDA.

The definition of the families and genera in this order relies on several characters which are usually present but whose variations are often difficult to clearly appreciate. This is due to the variability of the relative amounts of diverse elements which compose the skeleton. Therefore, the taxa are difficult to identify.

Families

1/ Character : surface skeleton — Among the characters used at the family level, one of the most significant is the surface skeleton. In the species of Petrosida, there is a strong tendency to develop a spicule crust, whose thickness and structure differ according to the family.

In the family Petrosiidae, the surface skeleton is made by very compact spicular tracts or lines (Desqueyroux-Faundez, in press). Free spicules are very abundant in some genera. In transverse sections, the skeleton of the surface displays the shape of a reticulum made of regular, isodiametric meshes. This aspect is due to the non differentiation of fibres or tracts into main secondary or transverse lines, a characteristic of the Petrosida. The variations of the surface skeleton have been analysed in two genera : *Xestospongia* and *Petrosia*. Both possess a cortex or surface layer, which varies due to differences in thickness and in the number of free spicules. It may be confused, hispid or clear. However a fundamental alveolar structure, which is characteristic of the family, is always apparent. The surface layer can be considerably thick, but its lamellar structure, consisting of spicular tracts lacking visible spongin, is always clearly evident. In certain species with a thick cortex, the ramifications of the longitudinal tracts which construct the surface skeleton can clearly be observed.

In the Oceanapiidae, the surface skeleton is composed of a confused network of isolated spicules grouped into one or two layers. More simply, it may be composed of an isodictyal network in which the extremities of the spicules are linked by spongin. The extremities of the longitudinal tracts divide in their terminal third. They then spread out tangentially and form a subjacent network which is isodiametric, alveolar, and also functions to reinforce the surface mesh.

2/ Character : variability of the global level of silica — It has been previously illustrated that in New Caledonian Haplosclerida (Desqueyroux-Faundez, 1984), the global level of silica, measured in percentage of the dry weight of the sponge, is a good character for the distinction of families. For instance, this level is very low in the Callyspongiidae (0.08 to 30%). In contrast, this character appears to be significantly high for New Caledonian Petrosida (Desqueyroux-Faundez, in press). In the Petrosiidae, the level of silica varies between 44.58 and 74.92% of the dry weight of the specimens analysed. In the Oceanapiidae, it is more variable, with limits between 12.81 and 54.90% (with the exception of *Oceanapia papula* Desqueyroux-Faundez).

Genera

1/ Organization of the main skeleton — The genera of the Petrosida have been defined mainly from the organization of the main skeleton. The characters which have been used for New Caledonian species are : the structure and arrangement of the fibres ; their differentiation or non-differentiation in primary, secondary and transverse fibres and tracts ; the distinction between branching or non-branching fibres ; the distinction between compact fibres forming a clear network and less compact fibres forming an irregular network ; presence or absence of free spongin within the fibres. The shape and the disposition of the mesh of the main skeleton determine the internal structure. Thus, the genus *Petrosia* displays a compact, lamellar and isodiametric internal structure, whereas the genus *Xestospongia* has a loosely constructed lamellar, trabecular structure. The genus *Oceanapia* has a pulpy, lamellar concentric structure.

2/ Size of the spicules — Size categories of the spicules seem to be a good character for generic distinction. But up to now, observations have been few, and limits between the different categories are not very clear-cut. It appears, in New Caledonia, that *Petrosia* species possess larger spicules (>200 µm) than *Pellina* species(80-100µm), but the limit, between *Petrosia* and *Xestospongia*, for instance, is not clear.

3/ Presence of microscleres — Numerous authors have pointed out the existence of microscleres in the Petrosida as a character without taxonomic value at the genus level. This relies on the presumed, uncertain evolutionary origin of these spicules. In New Caledonian Petrosida, as in Haplosclerida, microscleres are useful to separate forms within a genus which otherwise appear to be very similar.

CONCLUSIONS

Based on these morphological characters, it is possible to give the following definitions of the families of the Petrosida :

Family Petrosiidae : Petrosida with a surface skeleton composed of dense spicule tracts or of spicules pressed closely together, grouped in a regular reticulum with isodiametric meshes (isodiametry = rigid skeleton). No differentiation between primary and secondary fibres. Global level of silica = 58.26% (mean from seven studied species in New Caledonian region).

Family Oceanapiidae : Petrosida with a surface skeleton composed of close spicules or of an isodictyal reticulum of non-jointed spicules, with spongin at the nodes of the reticulum, in one or several layers. Primary and secondary fibres building a reticulum witl a irregular, alveolar appearance, non-isodiametric. Global level of silica : 28.5% (mean from seven studied species in New Caledonian region).

In her recent taxonomic study of the North-Eastern Atlantic Haplosclerida (which in her opinion includes the Petrosida), de Weerdt (1985) considered the following morphological traits as good taxonomic characters :

- Surface skeleton, the only character whose presence is constant at the family level,
- Main skeleton, a discriminating character at the genus level,
- Consistency, reproductive period and larval features, which could allow the identification of certain families (Petrosiidae and Oceanapiidae), whereas the growth form gives taxonomic indications for certain genera (*Oceanapia*).

These characters are close to those that have been used for New Caledonia Haplosclerida and Petrosida. Thus, in future descriptions, it will be necessary to describe these characters meticulously, in order to obtain the unification of criteria on a world level.

In spite of the progress made to sponge systematics by the applications of cytology, chemistry and embryology, the conventionnal morphological characters still remain useful for the discrimination of taxa in the order Petrosida.

LITERATURE CITED

Bergquist P.R., 1980 - The ordinal and subclass classification of the Demospongiae (Porifera); appraisal of the present arrangement and proposal of a new order. *N.Z.J. zool.*, 7 : 1-6.

Bergquist P.R. & Warne K.P., 1980 - The marine fauna of New Zealand : Porifera, Demospongiae. Part 3 (Haplosclerida and Nepheliospongida). *Coll. Repr. N.Z. oceanogr. Inst. Mem.*, 87 : 4-43.

Burton M., 1932 - Sponges. *Discovery Rep., Cambridge*, 6 : 239-392.

Desqueyroux-Faundez R., 1984 - Description de la faune des Haplosclerida (Porifera) de la Nouvelle-Calédonie. 1. Niphatidae - Callyspongiidae. *Revue Suisse Zool.*, 91(3) : 765-827.

Desqueyroux-Faundez R., sous presse - Description de la faune des Petrosida de la Nouvelle-Calédonie. Petrosiidae - Oceanapiidae. *Revue Suisse Zool.*

Griessinger J.M., 1971 - Etude des Réniérides de Méditerranée. (Demosponges-Haplosclerides). *Bull. Mus. natn. Hist. nat. Paris* (sér.3), 3 : 1-181.

Laubenfels M.W. de, 1936 - A discussion of the sponge fauna of the Dry Tortugas in particular and the West Indies in general, with material for a revision of the families and orders of the Porifera. *Publ. Carnegie Inst. Wash. 476, Pap. Tortugas Lab.*, 30 : 1-225.

Lévi C., 1973 - Systématique de la classe de Demospongiaria (Démosponges) : pp 577-631 in *Traité de Zoologie. 3*. P.P. Grassé (ed.), Masson.

Soest R.M.W. van, 1980 - Marine sponges from Curaçao and other Caribbean localities. Part II. Haplosclerida. *Stud. Fauna Curaçao*, 62 : 1-132.

Weerdt W.H. de, 1985 - A systematic revision of the North-Eastern Atlantic shallow-water Haplosclerida (Porifera, Demospongiae). Part I. Introduction, Oceanapiidae and Petrosiidae. *Beaufortia*, 35(5) : 61-91.

Weerdt W.H. de & van Soest R.M.W., 1986 - Marine shallow-water Haplosclerida (Porifera) from the South-Eastern part of the North Atlantic Ocean. *Zool. Verh.*, Leiden, 225 : 1-49.

Wiedenmayer F., 1977a - A monograph of the shallow-water sponges of the Western Bahamas. *Experientia*, suppl. 28 : 287 p.

Wiedenmayer F., 1977b - The Nepheliospongiidae Clarke, 1900 (Demospongiae, upper Devonian to Recent) an ultraconservative, chiefly shallow-water sponge family. *Eclog. geol. Helv.*, 70(3) : 885-918.

ANISOCHELAE ANALYSIS AND TAXONOMY OF THE GENUS *MYCALE* GRAY (DEMOSPONGIAE)

D. DOUMENC & C. LEVI

Laboratoire de Biologie des invertébrés marins et Malacologie
Muséum National d'Histoire Naturelle
57, rue Cuvier, 75005 Paris, France.

SYNOPSIS

The spicular biometry and the anisochelae-1 shape analysis is used to improve the systematics of *Mycale*. A comparison is made on the length of : the megascleres ; the larger anisochelae ; the larger sigmata ; the toxa ; the raphides. Each full-face anisochela-1 drawing is characterized by eight parameters computed from six digitalized points. These parameters form the subject of a classical statistic analysis (mean, standard variation, parametric tests), and a ratio classification analysis (correlation, regression, non parametric tests). The ratio calculation has been classified to help future identifications. The full-face anisochelae-1 outlines studied by a computerized method allows a definition of five shape types. Some general remarks on the taxonomy of the genus *Mycale* are also given.

INTRODUCTION

Several difficulties become apparent for a practical taxonomy of sponges. The descriptions of species are quite scattered in many publications. They are written in several languages, not universally understood by the readers. Drawings and photographs, if they are included, are not always of the best quality.

An important part of taxonomical work on sponges consists of exeges to correct misinterpretations by previous taxonomists. Another problem is that the main collections are stored in Museums scattered around the world and type-specimens are not localized enough.

Although spicules are an excellent material for statistic analysis of subspecific and subgeneric variability, the biometric or statistic methods

NATO ASI Series, Vol. G13
Taxonomy of Porifera
Edited by J. Vacelet and N. Boury-Esnault

have rarely been applied. Other morphological characters are never quantified and states of characters are never clearly described. A huge work needs to be done to supply objective data for new taxonomists.

Many monospecific sponge genera are doubtful. On the other hand, in several cases, more than a hundred species were grouped in only one main genus and taxonomists tried to split it into subgenera or other subtaxa. For example, *Geodia* was split in new genera by Sollas, after the aquiferous system anatomy. Ijima divided *Hyalonema*, according to the morphological structure of pinulae and also *Hymedesmia* was split by Topsent after several spiculation characters. The genus *Mycale* Gray, in its modern acceptance, was subject to some splitting attempts by Topsent (1924) and Laubenfels (1936). Topsent was using the ectosomal characters, while Laubenfels discriminated subgenera after the combination of microscleres.

In 1913, Hentschel published a paper intitled : "Über einen fall von orthogenese bei der Spongien", collecting 112 biometric data groups relative to about one hundred species of *Mycale*. He arranged these data after increasing order of the length of megascleres and gave some conclusions :

- The length of the larger anisochelae is correlated with the length of megascleres (m).
- The number of microsclere types first increases with m, then remains constant.
- The frequence of rosettes increase with m.
- These correlations have no immediate interest for taxonomists but suggest some evolutionary trends into the genus *Mycale*.

Mycale species are able to produce eight categories of microscleres : three size classes of anisochelae (anisochelae-1, anisochelae-2, anisochelae-3), two size classes of sigmata, toxa, raphides and rarely palmate isochelae, equivalent of the smaller anisochelae.

In his classification of sponges, Laubenfels (1936) discerns for example, the genera *Carmia* with toxa, *Zygomycale* with palmate isochelae, *Mycalecarmia* without toxa or sigmata. Following this rule, it would be necessary, but absurd, to give a generic name to each spicular combination. It has been shown that microscleres are often inconstant, so that they can not be used for taxonomy. In *Mycale*, it is true that the microscleres density varies, according to the specimen or to the fragment used for spicule slides. So, descriptions of species need to be read carefully and it is necessary to remember that the described diversity of microscleres is not always the true diversity.

To improve the systematic of *Mycale*, we try here to cluster all species already described, according only to the shape of the larger anisochela (anisochela-1), the typical spicule of this genus. The shape variability, but the constant geometry of the anisochelae are obvious inside the genus, and sometimes inside the species (Plancher, DEA communication, 1985). In addition to this shape analysis of anisochelae-1 and following Hentschel, new correlations between spicular types (biometric data and shape clusters) have been searched in order to help future systematical works.

ANISOCHELAE-1 BIOMETRY AND SHAPE ANALYSIS

1. From classical data (mean size from literature).

The mean anisochela-1 size has been classified on Fig. 1. The analysis shows that from 290 data (from 7 µm to 134 µm), the distribution is

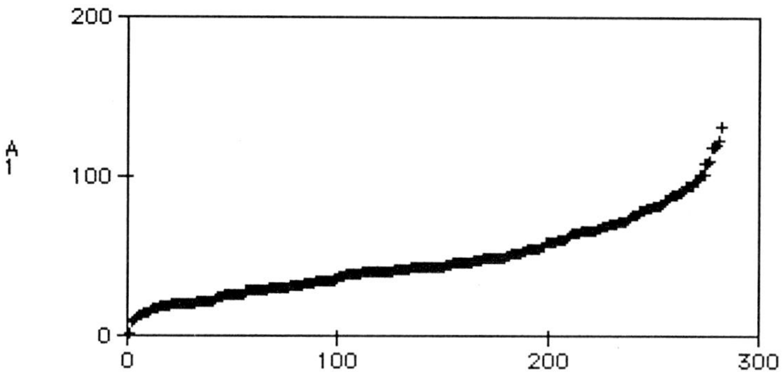

Figure 1 - Distribution of species according to their anisochela-1 length in um (y-axis).

continuous and slightly oblique and linear for nine tenths of the measurements (from 20 to 100 µm). The size frequency distribution of these data confirms this continuity. It is thus impossible to directly use these data for the biometric type anisochela-1 discrimination.

2. From new data measured using full-face anisochelae-1 drawings in the literature (Fig. 2)

Through the structural complexity of the anisochelae-1, eight parameters are necessary for characterization : the maximum length (D1) ; the two vertically opposite angles formed by the lateral teeth of the spicule (A1 & A2) ; the two maximum widths at the larger end (head) (D2) ; the two maximum widths at the smaller end (foot) of the spicule (D3) ; the head height (DU) ; the foot height (DV) ; the shaft between the head and the foot of the spicule (DW).

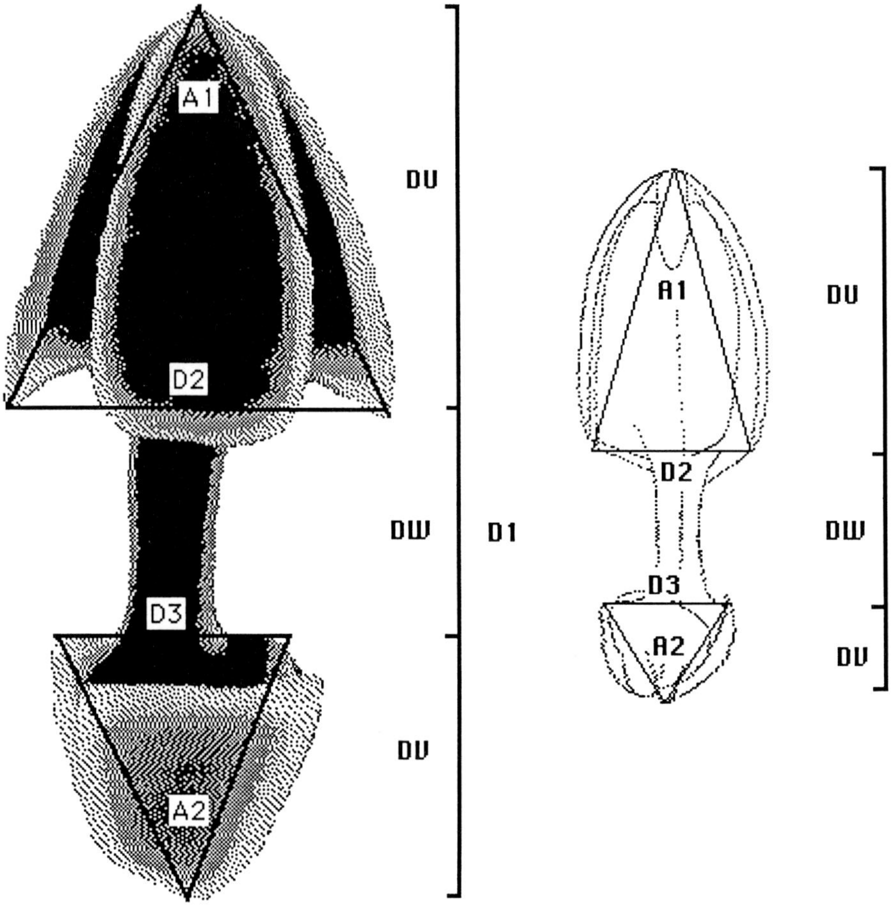

Figure 2 - Full-face anisochelae-1 digitalized points.

These eight parameters computed from six digitalized points taken from full-face anisochelae-1 drawings in the literature form the subject of a classical statistic analysis (mean, standard variation, parametric tests) for each of them. A ratio classification analysis is applied when two parameters are compared (i.e. correlations, regression, non parametric tests).

2.1. Statistical analysis.

The data (D1, D2, D3, DU, DV, DW, A1, A2) computed from the six points are distributed according to very different statistic laws :

- D1, D2, D3, have no modal peak in their distribution. D1 has a minimum value of 9 µm and a maximum value of 97 µm. The mean is 31.2 µm. D2 has a minimum value of 2 µm and a maximum value of 41 µm. The mean is 11.7 µm. D3 has a minimum value of 1 µm and a maximum value of 28 µm. The mean is 7.3 µm. Most of D1 and D2 values (40%) are included in one class (9 < D1 < 8.5 and 2 < D2 < 8.5) (Fig. 3).

- DU and DV have the same distribution with no modal peak. DU has a minimum value of 3 µm and a maximum value of 43 µm. The mean is 14.1 µm. DV has a minimum value of 1 µm and a maximum value of 28 µm with a mean of 6.4 µm. Most of DU and DV values are included in few classes (3 < DU < 18 and 1 < DV < 14) (Fig. 4).

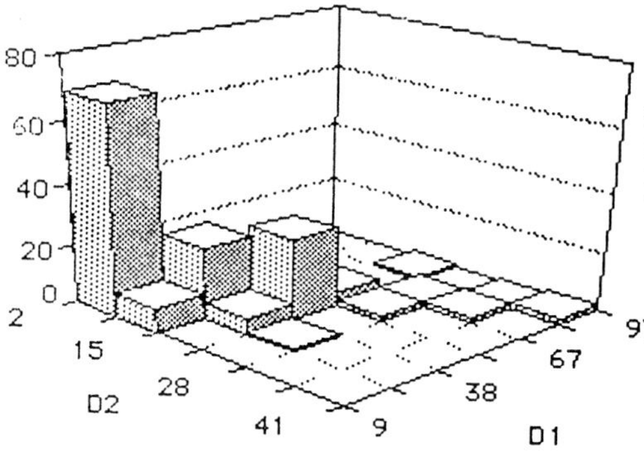

Fig.3 - D1, D2 frequency histogram.

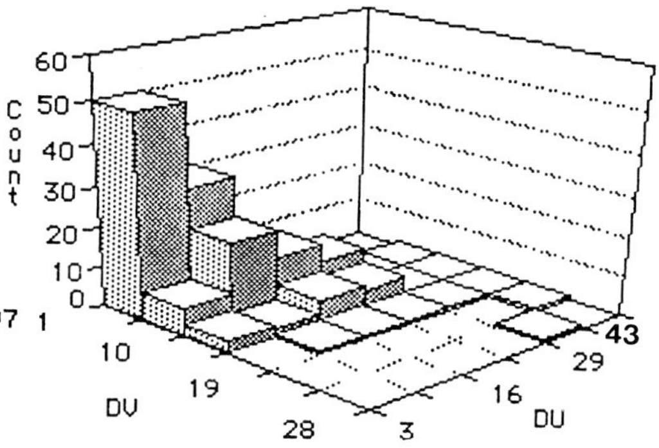

Fig.4 - DU, DV frequency histogram.

- DW is the shaft length of the anisochelae-1. Its minimum value is 2 µm and maximum value 52 µm. The mean is 11 µm and the distribution is close to D1 with 85% of the data being situated between 2 and 27 µm (Fig. 5).

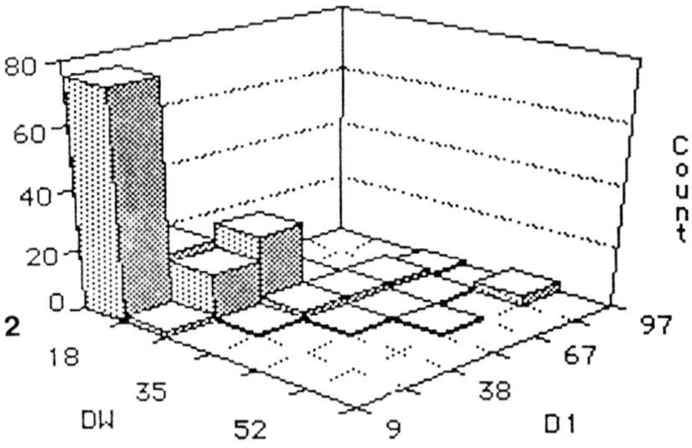

Fig. 5 - D1, DW frequency histogram.

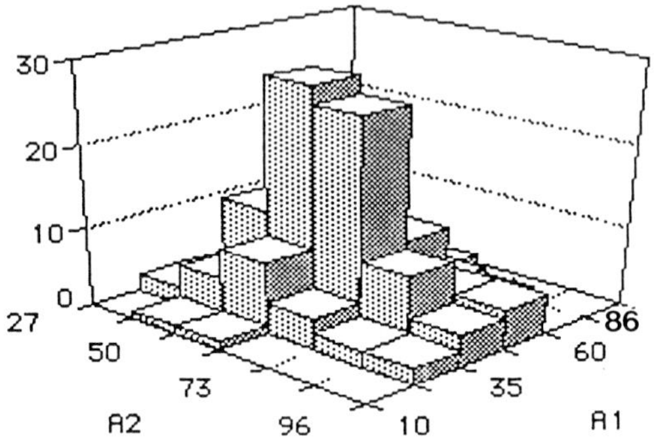

Fig. 6 - A1, A2 frequency histogram.

- A1 and A2 ; the angles formed by the alae of the anisochelae-1, have a modal peak in their distribution (Fig.6). The value of the Chi 2 in a contingency table indicates a good degree of homogeneity in A1 and A2 size. A1 has a minimum value of 10 and a maximum value of 86, but the mean is 45.2°. A2 has a minimum value of 27 and a maximum value of 96, and the mean is 61.4°. A1 and A2 are nearly constant (and independent from other parameters D1, D2, D3, DW, DU, and DV).

2.2. Ratio classification analysis.

The absolute values of computerized data, because of the homogeneity of their distribution, do not allow several geometrical types of anisochelae-1 to be defined. Therefore a relative approach by the ratio calculation of most representative proportions has been chosen. These ratios have been classified by increasing values to determine a possible discontinuity. Among the most representative ratios, DW/D1 and DU/DV (Fig.7) have been chosen. For DW/D1 and DU/DV, between their minimums (0.14 and 0.7) and their maximums (0.7 and 7) respectively, the distribution of the values is continuous. These classified lists, useful for the taxonomists, do not help in defining possible anisochelae-1 types.

Species	Author code	DW/D1	DU/DV	Species	Author code	DW/D1	DU/DV
toxifera	L63	0.14	2.00	cockburniana	H11	0.31	3.33
curvisigma	L63	0.15	2.00	trichela	L63	0.31	3.50
macrosigma	LI98	0.17	2.20	rotalis	T24	0.31	3.67
contarenii	SO77	0.17	2.75	cleistochela	VV71	0.31	2.40
sulevoidea	L61	0.17	4.00	contarenii	T24	0.31	3.13
cavernosa	BQ65	0.19	5.50	sulevoidea	H12	0.32	2.38
fascifibula	T04	0.19	1.00	profunda	K64	0.32	2.50
fibrosa	NBE82	0.20	5.17	syrinx	L60	0.32	2.50
porosa	RD86	0.20	3.00	tunicata	T24	0.32	3.08
imperfecta	BR06	0.20	7.00	cucumis	K	0.32	1.80
strelnicovi	RZ31	0.21	1.71	graveleyi	VV65	0.32	2.20
erythraeana	ROW11	0.21	1.67	bellabellensis	LB32	0.32	4.00
euplectellioides	ROW11	0.21	2.20	papillosa	K59	0.32	1.25
euplectelloides	ROW11	0.21	2.20	cleistochela	VV71	0.32	1.83
nuda	RD86	0.21	4.00	macilenta	L63	0.32	2.54
profunda	K64	0.23	2.38	fusiformis	L60	0.32	1.44
adhaerens	LA94	0.24	2.75	moluccensis	H11	0.32	2.00
massa	L57	0.24	2.13	similaris	T24	0.33	2.40
thaumatochela	LU05	0.25	1.78	fusca	RD86	0.33	1.50
hispida	K59	0.25	2.00	helios	LA94	0.33	1.50
lissochela	BQ65	0.25	2.00	lobata	K59	0.33	1.67
lissochela	BQ65	0.25	2.25	lingua	K59	0.33	1.70
angulosa	SO84	0.25	2.33	ovulum	LU05	0.33	1.83
mannarensis	TH68	0.25	3.00	sp2	VV71	0.33	1.88
meridionalis	L63	0.25	3.33	atlantica	ST17	0.33	2.00
antarctica	H14	0.26	2.88	contarenii	LB51	0.33	2.00
erythraeana	ROW11	0.26	1.50	retifera	K59	0.33	2.33
tylotornata	K64	0.26	2.50	angulosa	L59	0.33	2.50
fistulata	H11	0.26	2.13	topsenti	B59	0.33	2.67
grandis	H12	0.27	1.83	isochela	H11	0.33	2.80
macilenta	L63	0.27	4.30	paradoxa	LB35	0.33	3.14
diminuta	S78	0.27	3.00	senegalense	L52	0.33	3.67
psila	BK66	0.27	3.50	dentata	S58	0.33	4.00
longistyla	K59	0.28	1.75	rotalis	T24	0.33	4.20
adhaerens	LA94	0.28	2.25	adhaerens	HO81	0.33	4.33
curvisigma	L63	0.28	2.25	sp1	VV71	0.34	3.00
fibrosa	NBE82	0.28	2.80	raphidotoxa	H12	0.34	2.14
similaris	T24	0.28	2.56	intermedia	K59	0.35	1.43
acerata	K08	0.28	4.56	cucumis	K59	0.35	1.70
thaumatochela	K59	0.28	2.00	macginitiei	HO81	0.35	2.33
adhaerens	K59	0.28	3.25	syrinx	T24	0.35	2.80
retifera	T24	0.28	3.14	quadripartita	NBE73	0.35	1.67
ochotensis	K59	0.29	1.67	digitata	BQ67	0.35	4.50
richardsoni	BK66	0.29	1.75	massa	L60	0.36	2.33
bidentata	D05	0.29	2.33	cofundata	LB54	0.36	3.00
hispida	LA93	0.29	2.33	crassissima	D05	0.36	3.50
minima	DS82	0.29	2.50	lingua	K59	0.36	1.60
massa	L60	0.29	2.67	pectinicola	H11	0.36	1.29
fistulifera	ROW11	0.29	4.00	tenuis	S78	0.36	3.00
rossi	H14	0.29	4.14	loveni	K59	0.36	4.50
madraspatana	B37	0.29	3.24	grandis	VV65	0.37	1.69
orientalis	R81	0.29	2.25	profunda	K64	0.37	1.78
sp	TH73	0.30	1.67	contarenii	BI81	0.37	3.00
minima	DS82	0.30	2.50	parishii	L63	0.37	3.33
syrinx	L60	0.30	2.75	contarenii	BI81	0.37	5.00
rotalis	T24	0.30	3.38	obscura	H11	0.37	1.73
minima	T24	0.30	2.56	penicillium	LB88	0.37	2.89

Species	Author code	DW/D1	DU/DV	Species	Author code	DW/D1	DU/DV
grandis	VV65	0.37	2.57	massa	BA22	0.42	2.67
rotalis	T24	0.38	2.71	tylota	K64	0.43	1.50
massa	T24	0.38	2.53	phyllophila	H11	0.43	3.25
graveleyi	B37	0.38	2.50	sulcata	H11	0.43	0.96
crassissima	VV71	0.38	3.00	plumosa	D16	0.44	3.40
bidentata	B37	0.38	1.67	grandis	L58	0.44	2.44
plumosa	HO81	0.38	3.00	stecarmia	LB54	0.44	2.17
placoides	C74-76	0.39	2.00	grandis	TH73	0.44	2.67
plumosa	AR03	0.39	3.00	stegoderma	LB54	0.45	2.40
macginitiei	LB32	0.40	1.08	strongylophora	LB54	0.46	2.33
adhaerens	HO81	0.40	2.25	pellucida	R84	0.46	2.75
tunicata	BA22	0.40	2.25	sulcata	H11	0.48	0.83
adhaerens angulosa	H81	0.40	3.00	lindbergi	K59	0.49	1.13
helios?	F87	0.40	3.50	japonica	K59	0.50	1.00
lingua	T24	0.40	1.42	grandis	B37	0.51	2.50
tylota	K59	0.41	1.57	novaezelandiae	D24	0.53	0.90
tylota	K59	0.41	1.71	myriasclera	LL83	0.53	1.00
macilenta	H11	0.41	3.00	parasitica	H11	0.61	0.70
armata	LB54	0.41	2.17	tridens	H14	0.62	1.21
serrulata	S60	0.41	2.33	parasitica	VVL76	0.66	0.82
mammiformis	RD86	0.42	1.50	parasitica	C85	0.67	1.00
mytilorum	B37	0.42	3.00	carteri	DF24	0.70	1.00

Figure 7 - DW/D1 : length of the shaft (DW) on the maximum length (D1) of the anisochelae-1. DU/DV : length of the head (DU) on the length of the foot (DV) of the anisochelae-1. The ratio DW/D1 is classified by increasing values. (AR : Arnesen / B : Burton / BA : Babic / BI : Bibiloni / BQ : Bergquist / BK : Bakus / BR : Baer / C : Carter / D : Dendy / DF : Dendy & Frederick / DQ : Desqueyroux / F : Fristedt / H : Hentschel / HO : Hoshino / K : Koltun / L : Lévi / LA : Lambe / LB : Laubenfels / LI : Lindgren / LL : Lévi & Lévi / LU : Lundbeck / NBE : Boury-Esnault / R : Ridley / RD : Ridley & Dendy / RZ : Rezvoi / S : Sarà / SO : van Soest / ST : Stephens / TH : Thomas / T : Topsent / VV : Vacelet & Vasseur / VVL : Vacelet, Vasseur & Lévi.)

2.3. Comparative analysis (correlations, regressions, non-parametric tests).

The scatter diagrams of DW (shaft) and DU (head height) against D1 (Mean length of anisochelae-1) are presented in Figs 8 & 9.

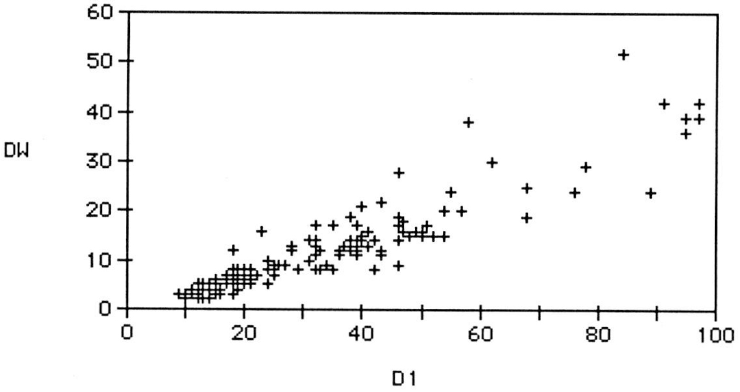

Figure 8 - Scatter diagram of D1 and DW.

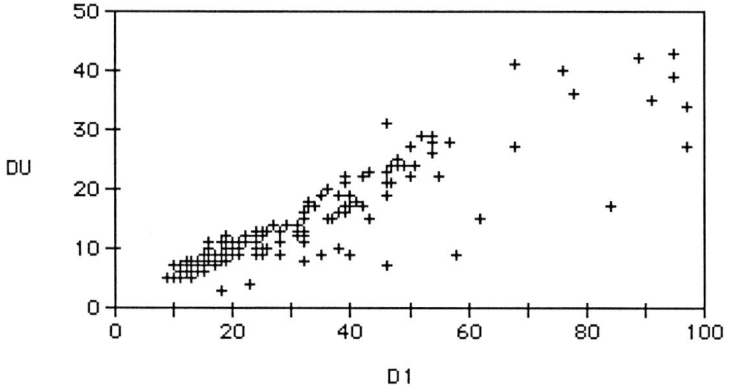

Figure 9 - Scatter diagram of D1 and DU.

All these data (including D2 and D3), are positively correlated. The correlation coefficient for 160 specimens is always higher than 0.91. The best coefficient has been found with DW and D1. The scatter diagram of DU against DV (foot height) is presented in Fig. 10.

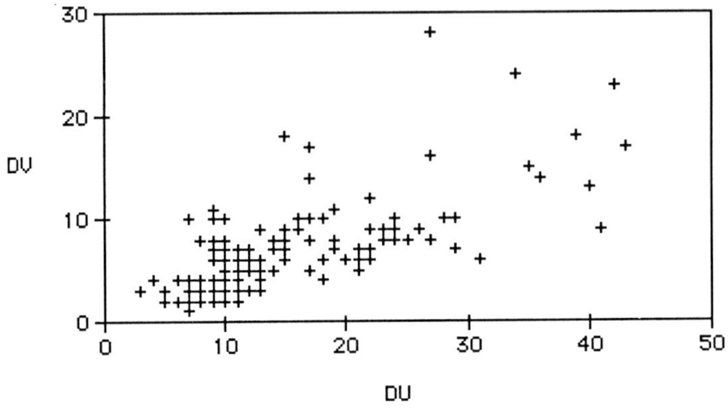

Figure 10 - Scatter diagram of DU and DV.

The correlation coefficient becomes less than 0.70. For the angles A1 and A2, the correlation coefficient is always less than 0.5. The diagram (Fig.11) is, in this case, very scattered.

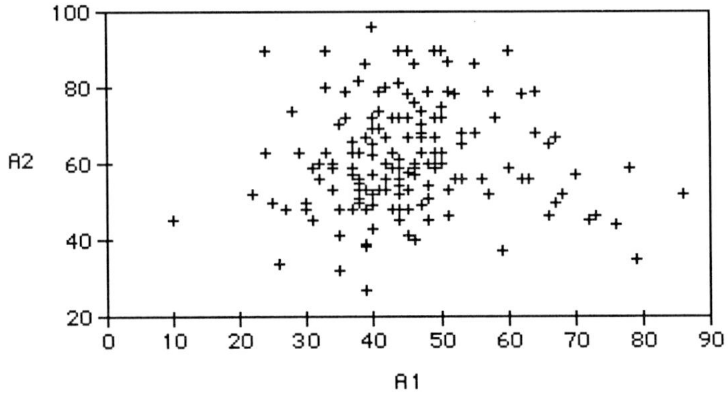

Figure 11 - Scatter diagram of A1 and A2.

From these results two conclusions may be assumed : (1) the longer the anisochela-1, the longer the shaft ; (2) considering that the angles are not correlated but have a modal peak in their distribution (i.e. almost constant), the computerized data useful for anisochelae-1 types discrimination are DU and DV. Other parameters will be used as the breadth of the shaft and the alae outlines for analysis complements.

3. From anisochelae computerized outlines and shape comparisons.

3.1. Tracing the spicular outlines and binary configuration shapes.

Considering that the anisochela-1 elementary geometry defined by the six computerized points is not sufficient to determine anisochelae-1 types, the full face spicules outlines have been studied. All shape analyses are performed using a digitizer and a Goupil microcomputer with the four subprograms written and especially adapted in our laboratory (Doumenc & Toulemont, 1985). A data acquisition subprogram is used to convert data in a coordinate array file. With a second subprogram the shapes are straightened, drawn to scale, and centered. They are then converted into binary coded hatching shapes. The third subprogram computes, by superposition, the similarity index between two binary coded hatching shapes. The fourth subprogram draws the "similarity dendrogram" which is the arranging in groups of anisochelae-1. In this study, five groups of anisochelae-1 have been found. The species belonging to several groups have been described by different authors (i.e. different representations of anisochelae-1 for the same species) :

A) *Mycale adhaerens, angulosa, crassissima, cockburniana, contarenii, curvisigma, erythraeana, euplectellioides, fibrosa, fistulata, fistulifera, graveleyi, imperfecta, longistyla, macrosigma, madraspatana, manaarensis, minima, macginitiei, moluccensis, paradoxa, penicillium, plumosa, profunda, porosa, phyllophila, rotalis, similaris, senegalense, trichela, tylotornata, tunicata, topsenti.*

B) *Mycale acerata, atlantica, bidentata, contarenii, crassissima, curvisigma, lingua, macginitiei, macilenta, massa, meridionalis, nuda, parishi, profunda, sulevoidea, syrinx, toxifera.*

C) *Mycale adhaerens, cofundata, contarenii, digitata, grandis, macilenta, mytilorum, orientalis, plumosa, psila, retifera, strelnicovi, stecarmia, serrulata, stegoderma, sulevoidea, syrinx, tunicata, tenuis, tylota.*

D) *Mycale adhaerens, armata, cleistochela, cucumis, erythraeana, fusca, grandis, graveleyi, isochela, helios, lingua, mammiformis, massa, ovulum, pellucida, quadripartita, similaris, strongylophora, thaumatochela.*

E) *Mycale bellabellensis, cleistochela, dentata, diminuta, fascifibula, fusiformis, grandis, hispida, japonica, intermedia, lindbergi, lobata, myriasclera, novaezelandiae, ochotensis, obscura, papillosa, pectinicola, placoides, richardsoni, sulcata, raphidotoxa.*

The question is to ascertain the taxonomic value of such a classification. The character "anisochela-1" is obviously not sufficient to describe species group. Other characters such as megascleres and other microscleres are necessary (cf "Remarks on the taxonomy of the genus *Mycale*"). As supported by this analysis the five anisochelae groups have a real statistical signification. The next step is the characteristics of the variability within each group.

3.2. The anisochelae-1 types.

The method used for the construction of type is described in Doumenc *et al.*, 1986. The grouping of shapes is based on the hierarchical technique UPGMA (Sneath & Sokal, 1973). The matrix representing the full-face anisochelae-1 drawings are superimposed. For each superimposition a common nucleus, a variation crown and an outside can be defined. It will include as many intersection levels as anisochelae-1 used to constitute the type (for 25 anisochelae-1 there will be 25 intersection levels). The mean value of the similarity coefficient between each section and the original shapes allows its affinity with the group to be measured. The corresponding section to the type will be chosen at the highest value of this mean. The so created types recorded as a matrix will be compared with other shapes. The Fig.12 present the shape of the five types of anisochelae-1.

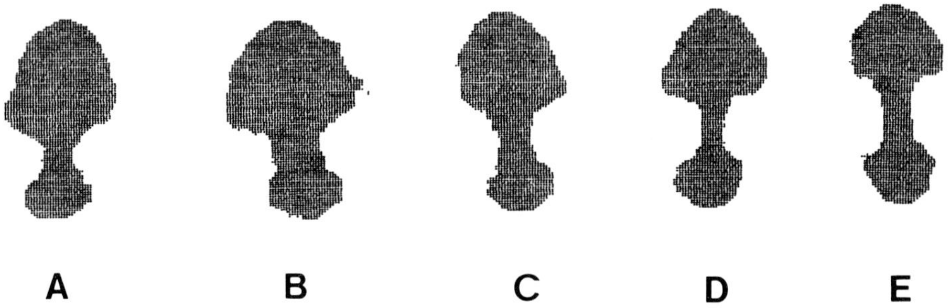

Figure 12 - Anisochelae-1 shape types.

- The **A** type has a mean length of 43.1 μm (min.20, max.88 μm). DW/D1 = 0.29 and DU/DV = 2.96.
- The **B** type has a mean length of 53.5 μm (min.11, max.90 μm). DW/D1 = 0.31 and DU/DV = 3.09.

If these two types have obviously the same biometrical characteristics, they are differentiated by their general shape.

- The **C** type has a mean length of 41 µm (min.18, max.95 µm). DW/D1 = 0.35 and DU/DV = 2.42.
- The **D** type has a mean length of 60 µm (min.18, max.131 µm). DW/D1 = 0.35 and DU/DV = 1.96.
- The **E** type has a mean length of 57 µm (min.20, max.119 µm). DW/D1 = 0.37 and DU/DV = 1.61.

Despite the knowledge that the given data are not sufficient and sometimes rudimentary to study thoroughly all the types of anisochelae-1, enough is known to make some conclusions. If the mean ratio DW/D1 seems to be constant, DU/DV seems to be discontinual. Three ratios have been found : 2 (type D), 2.5 (type C), 3 (types A & B). The last two types are very close and present the most frequent type in the anisochelae-1.

REMARKS ON THE TAXONOMY OF THE GENUS *MYCALE*.

The analysis of morphological diversity of anisochelae-1 allows five groups of species to be separated. However, it remains to be seen whether these clusters are confirmed by other taxonomical characters. All *Mycale* species are classified after their microsclere combination (three size classes of anisochelae (A) : A1, A2, A3 ; three size classes of sigmata (S): S1, S2, S3 ; toxa : T ; raphides : R ; isochelae : I. The results out of 156 data (Fig.13) are :

A : 12 spp.	A1A2 : 9 spp	A1A2A3 : 7 spp.
AS : 32 spp	A1A2S : 11 spp	A1A2A3S : 3 spp.
AS1S2 : 1 sp.	A1A2S1S2 : 6 spp.	A1A2A3S1S2 : 3 spp.
AST : 5 spp.	A1A2ST : 11 spp.	A1A2A3S1S2T : 8 spp.
AR : 1 sp.	A1A2R : 10 spp.	A1A2A3R : 3 spp.
ASR : 4 spp.	A1A2SR : 10 spp.	A1A2A3SR : 8 spp.
A1A2S1S2R : 10 spp.	A1A2A3S1S2R : 8 spp.	A1A2A3S1S2RT : 2 spp.
A1A2S1S2S3R : 1 sp.	A1A2IS1S2T : 3 spp.	

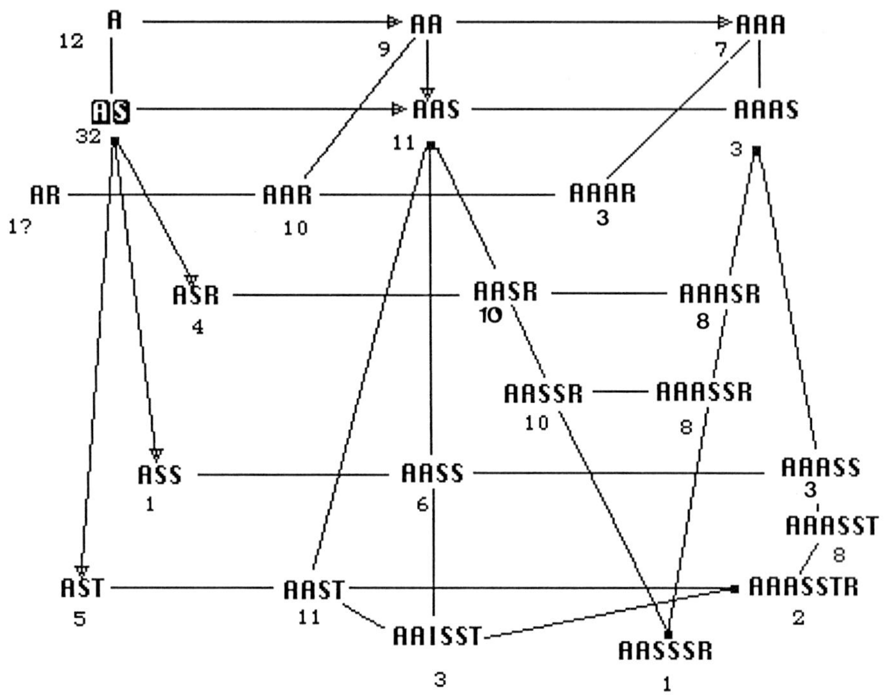

Figure 13 - Number of species for each microscleres combination.

It is interesting that most of microsclere combinations are present, in spite of some errors or omissions. Instead of searching orthogenetic lines, like Hentschel, five main data are compared (Fig.14) : length of megascleres (St) on length of anisochelae-1 ; length of the larger anisochelae (Al) on length of the larger sigmata (Sl) ; presence of toxa (T) ; presence of raphides (R).

Finally the global combination of microscleres and the five shape-types of anisochelae (as defined in the first part of this study) has been considered. At the end of this preliminary work, six correlations are clearly established, one negative and five positive.

1/ The first is negative : toxa and raphides are mutually exclusive with the exception of *M. similaris* and perhaps two other species.

2/ All species with toxa also have sigmata.

3/ The presence of raphides does not involve sigmata.

4/ Species with toxa generally have large sigmata (> 60 μm).

Species	Author code	St/A1	A1/S1	T	R
FUSIFORMIS	L60	3.36	1.61	0	36
STECARMIA	LB54	3.67	1.31	0	90
ARMATA	LB54	3.85	2.76	0	40
ADHAERENS	K59	3.89	1.50	0	250
SULCATA MINOR	H11	4.29	2.70	0	0
INDICA	RAO41	4.49	2.80	0	45
ANISOCHELA	L63	4.55	*5.24*	0	50
OCHOTENSIS	K59	4.57	*49.00*	0	0
GRANDIS	B37	4.68	2.27	0	100
ADHAERENS	K62	4.69	1.67	0	0
LOVENI	K59	4.69	*91.50*	0	0
STRONGYLOPHORA	LB54	4.75	1.00	0	50
PAPILLOSA	K59	4.77	*43.00*	0	0
ADHAERENS	K59	4.78	1.25	0	37
ADHAERENS	BK66	4.83	1.18	0	0
LINGUA	BK66	4.98	1.49	0	0
MACILENTA	LT63	4.99	0.66	221	0
LINGUA	D66	5.05	1.30	0	98
OVULUM	PR33	5.06	*40.00*	0	0
OVULUM	LV93	5.07	*37.50*	0	0
PAPILLOSA	K59	5.13	*40.00*	0	0
ADHAERENS FIB.	K59	5.16	1.22	0	63
HISPIDA	BK66	5.29	1.05	0	0
TYLOTA	K59	5.30	1.38	0	109
ADHAERENS	LB61	5.34	2.11	0	0
LINDBERGI	K59	5.43	2.42	0	67
HISPIDA	K58-59	5.49	1.49	0	0
ADHAERENS	HO81	5.50	1.80	0	0
HELIOS	K59	5.53	1.51	0	0
FIBREXILIS	LBF49	5.56	0.45	100	0
HELIOS	B35	5.63	2.34	0	0
MACGINITIEI	HO81	5.63	0.70	103	0
OCEANICA	T24	5.65	*96.50*	0	0
HISPIDA	LV93	5.65	*51.50*	0	0
SIMONIS	ST15	5.66	0.30	145	0
AEGAGROPILA	RAO41	5.68	0.58	180	0
PECTINICOLA	H11	5.72	0.52	0	0
BELLABELLENSIS	LB32	5.77	*80.00*	0	0
PARISHII	TH73	5.78	1.00	84	20
MERIDIONALIS	L63	5.82	0.75	0	35
OVULUM	R31	5.82	*42.50*	0	0
MADRASPATANA	B37	5.83	0.60	350	0
MADRASPATANA	ALI56	5.84	0.50	356	0
AEGAGROPILA	ALI56	5.84	0.37	149	0
GRANDIS	L58	5.94	2.06	0	35
PAPILLOSA	K59	5:95	*44.00*	0	0
ARNDTI	SO84	6.02	0.52	79	0
MACILENTA	A27	6.03	0.61	87	0
PLUMOSA	D16	6.12	0.52	65	0
SPONGIOSA	B28	6.13	0.71	0	0
HISPIDA	BK66	6.15	1.01	0	0
ANGULOSA	SO84	6.15	0.69	86	48
AEGAGROPILA	W25	6.16	0.54	160	0
JAPONICA	K59	6.20	2.82	395	168
MINIMA	DU82	6.22	*37.00*	400	400
ANGULOSA	SO84	6.24	0.62	72	48
ATLANTICA	L63	6.25	0.28	0	0
SULEVOIDEA	L61	6.25	0.43	40	0
LINGUA	PR33	6.29	4.03	0	0
LAEVIS	H65	6.30	1.96	0	87
PARASITICA	VVL76	6.32	1.23	0	0
SPONGIOSA	D96	6.32	0.38	0	0
LAEVIS	HT65	6.37	1.89	0	102
LAEVIS	HT65	6.40	2.21	0	85

Species	Author code	St/A1	A1/S1	T	R
SULEVOIDEA	SL02	6.55	0.79	140	0
PLUMOSA	HO81	6.56	0.62	180	0
TENUIS	S.78	6.57	0.38	0	0
LAEVIS	H65	6.63	2.28	0	87
MACILENTA	T24	6.64	0.46	195	0
THAUMATOCHELA	K59	6.73	*53.50*	0	0
QUADRIPARTITA	NBE73	6.78	0.44	0	68
PELLITA	T13	6.79	1.56	0	0
MACGINITEIEI	HO81	6.80	0.70	80	0
ORIENTALIS	RDK81	6.88	0.44	300	0
LAEVIS	SO84	6.88	3.30	0	72
LAEVIS	H65	6.89	1.57	0	93
CONTARENII	BB78	6.90	0.74	40	0
MACGINITEIEI	K71	6.92	0.57	115	0
MAGELLANICA	K64	6.94	0.63	0	105
MINIMA	T24	7.03	*29.50*	400	400
MACILENTA	L63	7.04	1.16	45	11
TRIDENS	K64	7.05	0.88	0	0
EUPLECTELLOIDES	R11	7.06	0.47	0	0
PARASITICA	C85	7.10	*46.50*	0	0
SIMILARIS	T24	7.17	0.69	230	30
MACGINITEIEI	HO75	7.19	0.50	70	0
RETIFERA	K59	7.19	0.62	80	0
SYRINX	PF77	7.20	0.48	48	0
LILLIEI	D24	7.23	1.35	0	40
SULCATA	L63	7.24	4.08	0	47
CONTARENII	S60	7.27	*43.00*	50	0
LILLIEI	D24	7.27	2.06	0	52
TUNICATA	T24	7.27	*55.00*	0	45
MACILENTA	ST12	7.30	0.37	230	0
LOBATA	K59	7.32	*32.50*	0	0
LITTORALIS	ST12	7.35	*33.00*	0	400
MASSA	L60	7.36	1.59	0	60
MAGELLANICA	D24	7.37	*76.00*	0	70
CRASSISSIMA	H12	7.40	2.14	0	25
MINIMA	T24	7.42	*33.00*	400	400
CONTARENII	T24	7.45	0.75	70	0
CRASSISSIMA	L61	7.48	1.41	0	25
ROTALIS	S58	7.56	0.56	0	0
MACROCHELA	B32	7.57	0.70	0	35
ADHAERENS PARV.	HO81	7.61	2.41	0	0
TRIDENS	DQ75	7.61	0.74	0	0
SYRINX	BA 22	7.68	0.49	75	0
TUNICATA	T24	7.69	*52.00*	0	45
THAUMATOCHELA	K59	7.76	*53.50*	0	0
ACERATA	H14	7.77	*70.00*	0	40
SYRINX	T24	7.82	*39.00*	50	0
OVULUM	LEV93	7.83	*30.00*	0	0
TOPOROKI	K58-59	7.85	1.04	0	103
MOLLUCCENSIS	H11	7.85	0.62	0	0
SIMILARIS	T24	7.90	0.58	230	30
SULCATA	H11	7.93	4.03	0	0
CONTARENII	T24	7.95	*37.00*	0	0
FIBROSA	NBE-VB82	7.97	0.30	0	0
MACGINITIEI	LB30-32	8.03	0.49	75	0
MASSA	L57	8.04	1.24	0	45
TUNICATA	BA 22	8.09	*51.50*	0	48
DIGITATA	BQ67	8.10	1.93	0	0
SYRINX	L60	8.11	0.42	55	0
CONTARENII	T24	8.12	1.55	0	0
CONTARENII	T24	8.12	1.43	0	0
SUEZZA	ROW11	8.13	0.57	310	0
RICHARDSONI	BK66	8.13	1.43	0	0
SYRINX	T24	8.17	0.50	50	0
RICHARDSONI	BK66	8.22	1.71	0	0

Species	Author code	St/A1	A1/S1	T	R
CONTARENII	BA21	8.25	0.60	74	0
RHAPHIDOTOXA	H12	8.29	0.53	0	300
MODESTA	LA94	8.32	*22.00*	0	0
MACILENTA	A43	8.33	0.39	120	0
GELATINOSA	R84	8.33	1.05	0	75
DOELLO JURAD	B40	8.33	*60.00*	0	0
SP1	VV71	8.35	0.57	130	0
TYLOTA	K58	8.42	1.38	0	109
CONTARENII	SO77	8.42	0.42	160	0
ACERATA	K64	8.46	*85.00*	0	120
THAUMATOCHELA	RZ24	8.53	*51.00*	0	0
LINGUA	HO81	8.57	4.60	0	60
ANTARCTICA	H14	8.58	0.86	0	0
LINGUA	PR33	8.59	7.50	0	60
ROTALIS	PF77	8.59	0.49	0	0
ACERATA	NBE82	8.65	*80.00*	0	46
SULCATA	H12	8.67	*60.00*	0	105
DENDYI	ROW11	8.70	*57.50*	0	80
MASSA	RU65	8.73	1.05	0	45
ROSSI	H14	8.81	*64.00*	0	105
MODESTA	BA22	8.88	0.56	0	0
CAVERNOSA	BQ65	8.88	0.35	0	0
ADHAERENS	HO81	8.95	1.00	0	0
BOLIVARI	FH14	9.00	0.49	180	0
ACERATA MIN.	L61	9.03	*77.50*	0	100
ADHAERENS L.	HO81	9.05	1.47	0	0
DIVERSISIGMATA	SO84	9.10	0.23	180	0
DIVERSISIGMATA	SO84	9.10	0.25	180	0
DIVERSISIGMATA	SO84	9.10	0.32	180	0
MAGELLANICA	HO81	9	*54.00*	0	0
CRASSISSIMA	VV71	9.18	1.21	0	20
NOVAEZELANDIAE	D24	9.20	1.26	0	80
PARASITICA	H11	9.22	*29.50*	0	0
CUCUMIS	K59	9.23	3.64	0	38
SP2	VV71	9.23	*32.50*	65	320
SERRATOHAMATA	LA94	9.31	0.23	39	0
RETIFERA	T24	9.32	0.46	75	0
AMERICANA	SO84	9.33	0.87	0	0
COCKBURNIANA	H11	9.33	*24.00*	0	30
MONANCHORATA	BR32	9.33	*30.00*	0	0
LINGUA	FH14	9.38	3.56	0	60
SULEVOIDEA	H12	9.40	0.62	0	0
REPENS	WH06	9.47	0.63	0	0
FASCIFIBULA	DI45	9.50	0.20	0	0
LAPIDIFORMIS	B32	9.57	0.10	0	0
TRINCOMALIENSIS	RAO41	9.66	0.50	0	0
ROTALIS	AY76	9.66	*29.00*	0	0
TENUISINUOSITYL	HO81	9.70	0.32	100	0
MAGNIRHAPHIDIFERA	SO84	9.73	*26.00*	0	310
MACILENTA	AY76	9.75	0.47	120	0
LINGUA	HO81	9.77	3.83	0	60
POROSA	RD86	9.80	0.21	0	0
NUDA	RD86	9.80	0.21	0	0
PSILA	BK66	9.88	0.25	0	0
ALBANENSIS	H11	9.89	*27.50*	0	26
LINGUA	K59	9.93	2.66	0	85
MYTILORUM	B37	10.00	0.50	0	0
GRAVELEYI	VV65	10.00	0.56	0	0
CRASSA	D96	10.00	*16.00*	0	0
GAUSSIANA	DQ76	10.02	*66.00*	0	44
ACERATA	H14	10.13	*82.50*	0	120
TOPSENTI	B59	10.23	1.26	0	60
TENUISPICULATA	B37	10.25	0.56	0	0
EUPLECTELLOIDES	W25	10.34	0.36	0	0
PARADOXA	LB31	10.34	2.90	0	110
MAGELLANICA	T13	10.35	1.01	0	50
MICRACANTHOXEA	SO77	10.40	1.21	120	0
FUSCA	LE61	10.42	1.47	0	0
STRELNICOVI	RZ24	10.44	*58.50*	0	0
CONTARENII	PF83	10.45	0.64	50	0
ROTALIS	T24	10.51	0.47	0	0
MASSA OCEANICA	T24 B59	10.53	1.53	0	100
MONANCHORATA	BR32	10.61	*24.50*	0	0
FISTULATA	H11	10.66	0.23	0	0
DIMINUTA	S78	10.67	*37.50*	0	60
MACROSIGMA	TA61	10.71	0.08	0	0
MACROSIGMA	LD97	10.71	0.08	0	0
MASSA	T24	10.71	1.44	0	80
PSILA	BK66	10.73	0.24	0	0
DENTATA	S58	10.80	0.27	0	0
TYLOTORNATA	K64	10.83	*66.50*	0	83
ERYTHRAEANA	R11	10.83	0.48	90	150
PHYLLOPHILA	SO80	10.83	0.49	0	0
RAPHIDIOPHORA	H11	10.89	0.34	0	308
LINGUA	K59	11.00	1.10	0	8
DIASTROPHOCHELA	L69	11.05	*31.00*	0	0
ADHAERENS	LA93	11.10	0.46	0	0
ATLANTICA	L63	11.16	0.28	0	0
JAPONICA	K59	11.22	2.82	395	168
CONFUNDATA	LBF54	11.25	0.29	235	120
MICROSIGMATOSA	HT65	11.32	0.53	0	0
SPINOSIGMA	NBE73	11.41	0.34	0	0
MAUNAKEA	LB51	11.43	0.44	0	0
SERRULATA	S60	11.51	0.35	0	0
LAXISSIMA	SO84	11.51	0.26	0	0
SP	B37	11.58	0.71	0	0
ANGULOSA	DK45	11.76	0.39	0	0
PSILA	LB30	11.88	0.15	0	0
PHYLLOPHILA	H11	11.92	*19.50*	0	0
GAUSSIANA	H14	11.94	*64.00*	0	50
SENEGALENSE	L52	12	0.60	0	0
PHYLLOPHILA	L63	12.13	0.61	0	0
CLEISTOCHELA	VV71	12.17	0.58	60	0
CLEISTOCHELA	VV71	12.17	0.66	25	0
MICROSIGMATOSA	SO84	12.22	0.52	0	0
MYTILORUM	ALI56	12.37	0.46	0	0
LINGUA	K59-64	12.40	2.63	0	85
GAUSSIANA	K64	12.46	*67.00*	0	50
TOXIFERA	L63	12.50	0.11	175	0
ANGULOSA	L59	12.50	0.33	0	0
STEGODERMA	LB54	12.50	1.21	40	0
CECILIA	LB50	12.56	0.54	0	0
PROFUNDA	K64	12.75	0.54	0	0
GRAVELEYI	B37	12.75	0.57	0	0
CECILIA	LB36	12.77	0.78	0	0
MICROSIGMATOSA	HT65	12.80	0.55	0	0
RELICTA	AN24	13.02	0.44	0	0
MICROSIGMATOSA	HT65	13.05	0.54	0	0
ISOCHELA	H11	13.11	0.92	0	45
IMPERFECTA	VV71	13.50	0.25	0	0
CONTAX	DI45	13.64	0.29	55	150
TYLOSTRONGYLA	PF82	13.65	0.73	0	0
ROTALIS	ST12	14	0.30	0	0
MASSA	BA21	13.88	1.27	0	80
MURRAYI	RD86	13.89	*72.00*	0	0
MAMMIFORMIS	RD86	13.89	*72.00*	0	0
ATLANTICA	ST17	14.04	0.22	80	0
MICROSIGMATOSA	HT65	14.16	0.48	0	0
LISSOCHELA	BQ65	14.76	0.56	0	0
MICROSIGMATOSA	SO84	15.76	0.44	0	0

Species	Author code	St/A1	A1/S1	T	R
SPINOSIGMA	NBE73	15.79	0.34	0	0
TRICHELA	L63	17.09	*39.50*	0	0
OBSCURA	H12	17.10	*20.50*	0	0
MICROCHAELA	FH21	17.59	0.32	50	0
ANGULOSA	LB36	19.23	0.29	0	0
FISTULATA MI	A27	19.43	0.40	0	0
SULCATA	B59	20.00	0.63	0	0
TOXIFER	D96	20.00	*10.00*	95	0
SANGUINEA	TS59	20.38	0.37	0	0
TOXIFERA	L63	20.71	0.66	110	0
TRIDENS	H14	22.59	0.50	0	0
MICROXEA	VVL76	27.36	0.41	0	0
TOXIFERA	L69	28.75	*8.00*	14	0
INDICA	BR32	37.86	0.28	0	74
LONGISTYLA	K58-59	40.00	2.30	0	0
VOSMAERI	LB42	*640*	0.07	0	0

Figure 14 - St/A1 : length of megascleres (St) on length of anisochelae-1 (A1) ; A1/S1 : length of anisochelae-1 (A1) on length of sigmata (S1) ; T : length of toxa (0 = absence of toxa) ; R : length of raphides (0 = absence of raphides). The numbers in italics correspond to the absence of one spicule type (i.e. A1 for St/A1 and S1 for A1/S1).
(A : Arndt / AlI : Ali / AN : Annandale / AR : Arnesen / AY : Arroyo / B : Burton / BA : Babic / BB : Babio / BI : Bibiloni / BQ : Bergquist / BK : Bakus / BR : Baer / C : Carter / D : Dendy / DC : Descatoire / DF : Dendy & Frederick / DQ : Desqueyroux / DU : Duran / F : Fristedt / FH : Ferrer-Hernandez / H : Hentschel / HO : Hoshino / HT : Hechtel / K : Koltun / L : Lévi / LA : Lambe / LB : Laubenfels / LD : Lendenfeld / LE : Leitao / LI : Lindgren / LL : Lévi & Lévi / LT : Little / LU : Lundbeck / LV : Levinsen / NBE : Boury-Esnault / NBE-VB : Boury-Esnault & Van Beveren / PF : Pulitzer-Finali / PR : Procter / R : Ridley / RAO : Rao / RD : Ridley & Dendy / ROW : Row / RU : Rutzler / RZ : Rezvoi / S : Sarà / SL : Sollas / SO : van Soest / ST : Stephens / T : Topsent / TA : Tanita / TH : Thomas / TS : Tsurnamal / VV : Vacelet & Vasseur / VVL : Vacelet, Vasseur & Lévi / W : Wilson / WH : Whitelegge.)

St/A1=10-17
A1/S1=0,2-0,8
AS
laxissima
bidentata
fistulata
microsigmatosa
imperfecta
lissochela
mytilorum
relicta
sanguinea
senegalense
tenuispiculata
spongiosa
fistulifera
maunakea
cecilia
phyllophila
nuda
fist.macrochela
phillipensis
penicillium
foraminosa
tylosytongyla
atlantica
AAS
phyllophila
modesta
profunda
serrulata
spinosigma
ASR
manaarensis
raphidiophora
arenicola
AAAS-AAASS
graveleyi
rotalis
sulcata
ASS
textilis
AAST
atlantica
toxifera
microchela
curvisigma
AAASST
cleistochela
AAASSTR
erythraeana
contax
AASS
macrosigma

St/A1=5-10
A1/S1=0,2-0,9
AS
euplectellioides
tenuis
trincomaliensis
porosa
moluccensis dichela
AST
suezza
aegagropila
simonis
AAST
aegagropila
bolivari
macginitiei
madraspatana
macilenta australis
sordida orientalis
AAASST
contarenii
macilenta
retifera
sulevoidea
syrinx
arndti
diversisigmata
angulosa
plumosa
parishi
pectinicola
AAASSTR
similaris
AASS
fibrosa
cavernosa
tridens
AASR
toporoki
fascifibula
AASSR
quadripartita
AAASR
meridionalis
macrochela

St/A1=5-10
A1/S1=1-1,7
AS
helios
hispida
richardsoni
AASS
pellita
AAAS
digitata
ASR
fusca
indica
murrayi
AASR
lindbergi
paradoxa
adhaerens
pellucida
crassissima
gelatinosa
cockburniana
laevis
AASSR
indica
japonica
tylota
crassissima
fusiformis
AAASSR
crassissima
grandis-armata
novae-zealandiae
strongylophora
massa
AAASR
anisochela
cucumis
lilliei
lingua

Figure 15 - alpha group (first column), beta group (third column), others (second column).

5/ Species with AS (anisochelae-sigmata) generally have small sigmata (< 60 µm).

6/ It is possible to subdivise the AS (anisochela-sigmata) group of species in :

AS alpha with A < S

AS beta with A > S

Species AS alpha are found more often in warm waters and the shape of their anisochelae-1 belongs to A and B types.

Species AS beta are found more often in cold waters and the shape of their anisochelae-1 belongs to D and E types.

When using the two following ratios (Fig.15) : length of megascleres (St) on length of anisochelae-1 (A1) ; and length of anisochelae-1 (A1) on length of sigmata (S1), alpha and beta groups can be characterized.

alpha	: St/A1 = 10-17	A1/S1 = 0.2-0.8
beta	: St/A1 = 5-10	A1/S1 > 1

Most of species with raphides belong to the beta group (except *M. mannarensis*, *raphidotoxa*, *arenicola*) and most of species without raphides (with toxa or not) belong to alpha group. But all *Mycale* species do not belong to alpha and beta groups nor share ratios of St/A1 and A1/S1 (St/A1 = 5-10 and A1/S1 = 0.2-0.9).

The five species AS (*M. euplectellioides*, *tenuis*, *trincomaliensis*, *porosa*, *moluccensis*, *dichela*) and most of the species with toxa (AST, AAST, AAASST, *Zygomycale* included) appear to belong to the alpha group. Species with raphides (*M. toporoki*, *fascifibula*, *macrochela*, *meridionalis*, *quadripartita*) probably belong to the beta group.

All these conclusions founded on spicular biometry and on anisochelae shape analysis, remain to be completed and detailed, but it is the opinion of the authors that they could already be useful for a new systematic study of Mycalidae.

ACKNOWLEDGMENTS

The assistance of Y. Toulemont and A. Foubert is gratefully acknowledged.

LITERATURE CITED

Doumenc D., Toulemont Y. & Andrade F., 1986 - La banque "CNIDE" du Muséum. In *Compte rendu du séminaire des producteurs de banques françaises de données biologiques factuelles*, Univ. Nice. 200 p.

Doumenc D. & Toulemont Y., 1985 - Apports de la micro-informatique dans la taxonomie des Actiniaires. *Océanis*, 11(40) : 339-347.

Hentschel E., 1913 - Über einen fall von orthogenese bei den Spongien. *Zool. Anz.*, 13 : 255-267.

Ijima I., 1926 - The Hexactinellida of the Siboga Expedition. *Siboga Expedition*, Leiden, VI : 1-382.

Laubenfels M.W. de, 1936 - A discussion of the sponge fauna of the Dry Tortugas in particular and the West Indies in general, with material for a revision of the families and orders of the Porifera. *Pap. Tortugas Lab.*, 30 : 1-225.

Sneath P. & Sokal R., 1973 - Numerical taxonomy. W.H. Freeman & Co, San Francisco XV : 1-573.

Sollas W.J., 1888 - Report on the Tetractinellida collected by HMS Challenger during the years 1873-1876. *Report. Challenger Zool.*, XXV : 1-458.

Topsent E., 1924 - Révision des *Mycale* de l'Europe occidentale. *Ann. Inst. Oceanogr.*, NSI (3) : 77-118.

Topsent E., 1928 - Spongiaires de l'Atlantique et de la Méditerranée provenant des croisières du Prince Albert ler de Monaco. *Résult. Camp. sci. Albert ler Monaco*, 74 : 1-376.

AN ENZYMATIC TECHNIQUE FOR THE SEPARATION OF SPICULES FROM ALCOHOL-PRESERVED SPONGE TISSUE.

P.D. FRY & D. GRAY

Marine Benthos Laboratory
Department of Science, Luton College of Higher Education
Park Square, Luton, Bedfordshire LU1 3JU. U.K.

SYNOPSIS

After removing the alcohol from fragments of preserved specimens of the sponge *Suberites domuncula*, it is possible to free spicules from the material by digesting it with pepsin or trypsin. Sections cut from such alcohol-preserved specimens which have been embedded in wax are also susceptible to digestion by these enzymes after the tissue has been taken down to water. The hydrolysis leaves intact some slender fibrous connections which unite some of the megascleres at the surface of the sponge.

INTRODUCTION

Preparation of spicules necessitates freeing them from the organic part of the sponge, using some agent which destroys the cellular and intercellular components but leaves the spicules intact. Strong oxidizing agents are commonly used, these destroy all of the organic constituents and clear the spicules of debris. However, both nitric acid and sodium hypochlorite which are the most commonly used of these agents, do not allow a progressive destruction of collagenous fibres. The first does not preserve the calcareous concretions which are known in one Demosponge, *Hemimycale columella*, and which may be of a more common occurence (Vacelet *et al.*, this book). Boiling in a sodium hypochlorite solution has been shown in the same species to modify certain delicate siliceous spicules. For these reasons, an enzymatic digestion could be an interesting alternative to these two techniques.

The organic component of sponges is largely proteinaceous, consisting

NATO ASI Series, Vol. G13
Taxonomy of Porifera
Edited by J. Vacelet and N. Boury-Esnault

of cells, mesohyl and collagenous fibres. Cells and mesohyl are susceptible to hydrolysis by proteolytic enzymes, whereas collagen is resistant to a range of such enzymes, as shown by Piez & Gross (1959), Travis (1967), Katzman *et al.* (1970), Junqua *et al.* (1974), Garrone *et al.* (1975), Garrone (1978).

The most commonly used preservative for sponges is ethyl alcohol. This denatures the proteins by removing the water molecules which are associated with the proteins when they are in their natural state. It also changes the orientation of the end-groups on side-chains by breaking hydrogen bonds and salt links, Kauzmann (1959), Okuniki (1961) and Pearse (1968). These changes prevent autolysis and degradation by microorganisms. Once the tissue has been rehydrated and the alcohol removed completely, its susceptibility to proteolytic enzymes is restored.

MATERIAL AND METHODS

Two proteolytic enzymes were tested : mammalian pepsin and mammalian trypsin. Pepsin was used as a 1% solution adjusted to pH 2 with HCl and KCl. Trypsin was used as a 1% solution adjusted to pH 8 with phosphate buffer. (Pepsin hydrolyses proteins by cleaving peptide bonds in which the carbonyl group is derived from the residue of aromatic acids and from those having large hydrocarbon side-chains, trypsin cleaves peptide bonds derived from the basic amino acids lysine and arginine [Traube & Piez, 1971]).

The specimens of *Suberites* used were collected from several locations in the North Sea. In this species there is a marked variation in the density of spiculation within the individual. Different parts of the specimens were sampled so as to give some fragments which were densely spiculated and some in which the spiculation was sparse. Similarly, the sections which were cut from the wax-embedded material were selected so that some were from the surface of the sponge, where the spiculation is dense, and others from areas which were sparsely spiculated. The specimens from which the fragments were taken had been preserved in alcohol for two years. The sections were cut from specimens which had been in alcohol for two years prior to being embedded in wax.

So as to provide a comparison with the efectiveness of the enzymes on alcohol-preserved tissue, a similar series of fragments was taken from frozen tissue. These specimens had been stored in liquid nitrogen for one year prior to being kept at -20°C for one year.

All of the fragments were of the same dimensions : 4 x 4 x 1.5 mm. The sections were all cut at 100 µm.

The sections were brought to water after being de-waxed in the organic solvent CNP 30 (Penetone Chemicals, Cramlington, Northumberland, U.K.).

The de-waxed sections and alcohol-preserved fragments were rehydrated as follows ; using 10 ml distilled water on every occasion, each item was given two 5-minute rinses and then soaked in water for 8 hours, followed by two further 5-minute rinses.

The frozen material was thawed at room temperature for 15 minutes prior to incubation, but not rinsed.

Incubation with either 10 ml pepsin solution or 10 ml trypsin solution was carried out at 37°C for 48 hours.

The solution was removed from the spicule sediment carefully, using a pasteur pipette. 10 ml distilled water was mixed with the spicules and left to settle for 8 hours. After this the water was replaced with a similar volume of 100% ethyl alcohol and left to settle for 4 hours. After a second wash in 10 ml 100% alcohol the spicules were resuspended in 1 ml alcohol and one drop of this suspension transferred to a slide and allowed to dry before mounting and covering.

RESULTS

- **Fragments, alcohol-preserved :**

Those fragments in which the spiculation was sparse had completely disintegrated after only 4 hours of incubation with either pepsin or trypsin, liberating all of the spicules.

Those fragments in which the spiculation was dense had yielded many free spicules after 48 hours of incubation although the fragments had not completely disintegrated ; a square parcel of tightly packed spicules remained unseparated from one another from each of these fragments. An additional 48 hours of incubation with fresh enzyme solution produced no further separation of these parcels, nor were they freed by incubation with trypsin solution for 48 hours following incubation with pepsin.

- **Fragments, frozen :**

These responded to enzyme hydrolysis in exactly the same way as did the

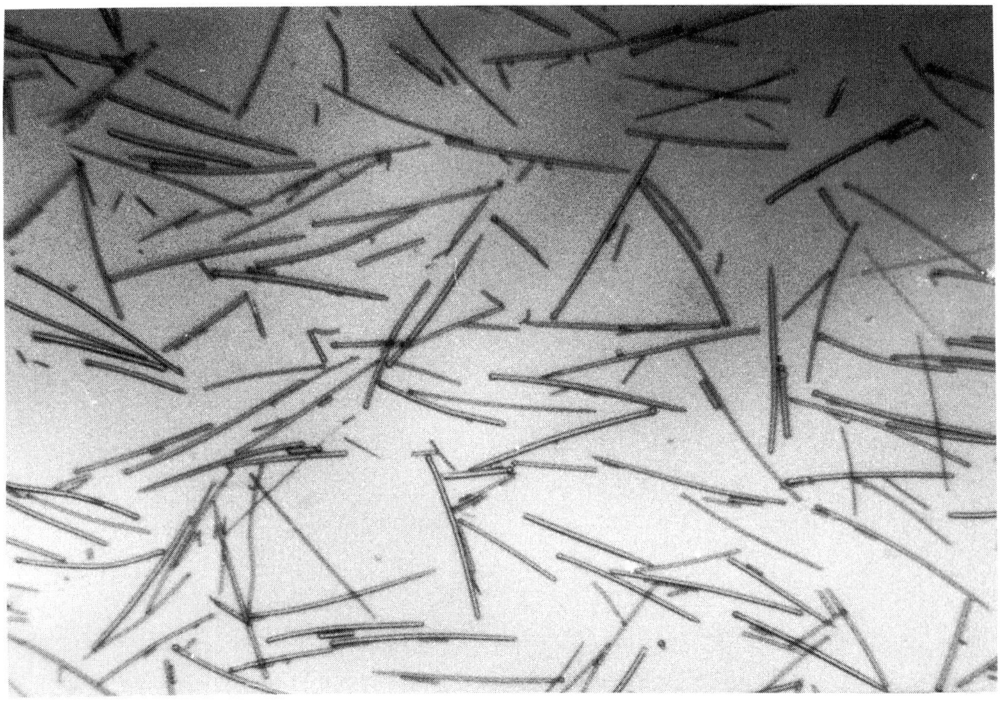

Figure 1 - Spicules released from *Suberites* fragments by digestion with pepsin.

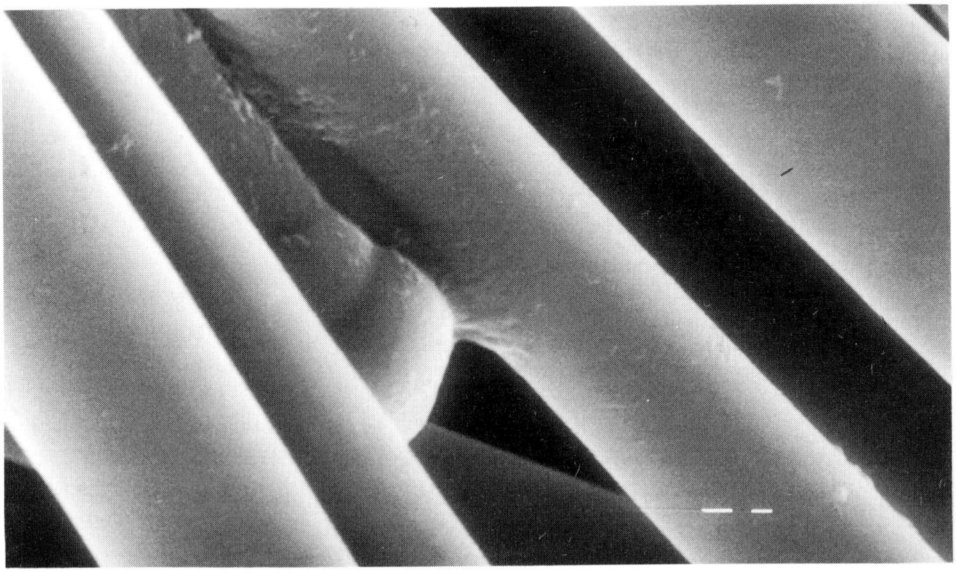

Figure 2 - Fibrous attachments uniting megascleres at the surface of *Suberites* (x 5000).

alcohol-preserved fragments, giving a complete disintegration of the tissue which was sparsely spiculated and leaving a residual square parcel of unseparated spicules from the densely spiculated fragments.

- **Sections cut at 100 µm tissue which had been wax-embedded** :

Numerous whole and broken spicules were liberated but some clusters of spicules remained in the digests of those sections which included any densely spiculated regions.

Digestion with pepsin was found to be preferable to digestion with trypsin because of the lower level of bacterial contamination which developed in the acidic medium used for pepsin.

DISCUSSION

Spicules which were released in this way were free of cellular debris once they had been washed thoroughly (Fig. 1). The square parcels of tightly packed spicules which remained unseparated from one another were from the densely spiculated parts of the fragments and comprised the megascleres from the surface of the sponge, as did the small clusters of spicules which remained in the digests of some of the sections.

Inspection of these square parcels by scanning electron microscopy showed that the spicules were free of organic material save in one respect ; fine strands could be seen on a very few of the megascleres, attaching them to adjacent ones (Fig. 2). These strands, although resistant to the proteolytic enzymes used, were destroyed completely by concentrated nitric acid. Topsent (1900) and Herlant-Meewis (1948) refer to the small amount of spongin which holds the megascleres in bundles at the surface of *Suberites domuncula*. The spongin is in addition to "the muffs of fibrous cells which encircle the bundles", Brien (1973).

By removing the cells and exposing the spicules in this way the association which exists between strands and spicules is revealed. It then becomes possible to distinguish between those spicules which are bound by strands and those which are not, even in those sponges such as *Suberites*, where the fibrous connections are slight. It is also possible to discern the extent of the attachments and it may be that the technique will prove to be helpful in revealing similar associations in other sponge species.

ACKNOWLEDGEMENTS

We thank and most gratefully acknowledge the help given by the following individuals and organisations ; The Ministry of Agriculture and Fisheries Directorate of Fisheries research, Lowestoft, for providing the ship-time which made the collection of material possible ; Dr M.F. Dyer and Dr G.J. Cranmer, who, as N.E.R.C.- and M.A.F.F.-funded research fellows at this laboratory, collected the samples on which these observations were made ; the Leverhulme Trust, for providing funds for the purchase of photographic materials ; Dr G.C. Steele and Mr C. Robinson of this department, who kindly made available and gave essential help with the scanning electron microscope facility ; and Miss S.M. Stone of the Porifera Section, the British Museum (Natural History), who kindly read the draft of this manuscript and made invaluable suggestions for its improvement.

LITERATURE CITED

Brien P., 1973 - Les Démosponges : pp 131-461 in *Traité de Zoologie*, III. P.P.Grassé (ed.), Masson, Paris.

Garrone R.L., Huc A. & Junqua S., 1975 - Fine structure and physico-chemical studies on the collagen of the marine sponge *Chondrosia reniformis* Nardo. *J. Ultrastruct. Res.*, 52 : 261-275.

Garrone R.L., 1978 - Phylogenesis of connective tissue : in *Frontiers of matrix biology* vol.5. Krager, Basel.

Herlant-Meewis H., 1948 - Contribution à l'étude histologique des Spongiaires. *Ann. Soc. Roy. Zool. de Belgique*, L XXIX : 5-36.

Junqua S., Garrone R.L., Pavans de Ceccaty M. & Vacelet J., 1974 - Biochemical and morphological studies on the collagens of horny sponges. *Connect. Tiss. Res.*, 2 : 193-203.

Katzman R.L., Lisowska E. & Jeanloz R.W., 1970 - Invertebrate connective tissue. *Biochem. J.*, 119 : 17-20.

Kauzmann W., 1959 - Some factors in the interpretation of protein denaturation. *Advanc. Protein Chem.*, 14 : 1-63.

Okuniki K., 1961 - Denaturation and inactivation of enzyme proteins. *Advanc. Enzymol.*, 23 : 29-81.

Pearse A.G.E., 1968 - Histochemistry, vol.1. Churchill, London.

Piez K.A. & Gross J., 1959 - The amino acid composition and morphology of some invertebrate and vertebrate collagens. *Biochem. biophys. Acta*, 34 : 24-29.

Topsent E., 1900 - Etude monographique des Spongiaires de France. I. Tetractinellidae. *Archs Zool. exp. gén.*, II : 259-400.

Traub W. & Piez K.A., 1971 - The chemistry and structure of collagen. *Advanc. Protein Chem.*, 25 : 243-342.

Travis D.F., François C.J., Bonar L. & Glimcher M.J., 1967 - Comparative studies of the organic matrices of invertebrate mineralized tissue. *J. Ultrastruct. Res.*, 18 : 519-550.

SKELETAL VARIATION IN EMBRYO-CONTAINING SPECIMENS OF *HALICLONA ROSEA* (BOWERBANK) FROM ANGLESEY, NORTH WALES.

W.C. JONES

School of Animal Biology,
University College of North Wales
Bangor, Gwynedd LL57 2UW, U.K.

SYNOPSIS

Statistical analysis of spicule samples of 36 embryo-containing specimens of *H. rosea* collected in the months of May to September has revealed seasonal trends which are explicable in terms of variations in the rate of production of spicules and in the size of the spicules produced. The trends are correlated with changes in the appearance of the skeleton as seen in vertical and surface sections of the sponge specimens. A similar statistical study of 4 embryo-containing specimens of *H. fistulosa* indicates that there is negligible overlap in the size ranges of the spicules of the two species from Anglesey. Embryo-containing specimens were used to increase confidence in the species indentifications.

A discussion is given on the historical taxonomy of the species *H. rosea*.

INTRODUCTION

Spicule sizes by themselves are insufficient to enable one to sort out the British littoral haplosclerid sponges into species (Jones, 1984). Other diagnostic characters must be used, but the haplosclerids in general are so variable in colour, surface texture, external form and skeletal features that it is not always obvious which characters are truly diagnostic. An exception is *Haliclona elegans* (Bowerbank) sensu Topsent, which can readily be identified in the field by the appearance of interconnecting slime strands when pieces are pulled apart. Using this species the seasonal chan-

NATO ASI Series, Vol. G13
Taxonomy of Porifera
Edited by J. Vacelet and N. Boury-Esnault

ges in appearance of the skeleton and the variation in the lengths and widths of spicule samples throughout the year have been determined (Jones, in press). The seasonal variations are so great that one could well ascribe February and August specimens to different species, were it not for the 'slime strand test' and occurrence of transitional stages between the two extremes. Since the variation is, to some extent at least, phenotypic, spicule widths and possibly spicule production being affected by the temperature and silicate concentration of the ambient water, it is probable that a comparable range of variation would be exhibited by other haplosclerid species from the same collecting site. The problem of whether to split or lump together certain so-called species could thus be lessened or avoided.

Haliclona rosea (Bowerbank) sensu Topsent (1887, as *Reniera rosea* Schmidt (*Isodictia* (sic) *rosea* Bowerbank); 1891, as *Reniera rosea* (Bowerbank)) is another haplosclerid species that is fairly readily identified, thanks to Topsent's description. Information concerning the seasonal changes in spicule lengths and widths (means, standard deviations and coefficients of variation) has already been published (Jones, 1984). However, the range of spicule dimensions approaches and possibly overlaps that of other established species such as *H. fistulosa* (Bowerbank), which is not markedly different from *rosea*. Accurate knowledge concerning the maximum spicule length and width for *rosea* and the smallest maximum length and width in samples of the other species could clearly be helpful when attempting to identify local specimens. To determine these values the species must first be confidently identified. To ensure this, extra diagnostic criteria need to be employed. For *rosea* identified initially using the skeletal criteria given by Topsent, this has now been achieved by restricting the study of spicule samples to those obtained only from specimens containing embryos. One would expect that embryos of different haplosclerid species would be distinguishable by their size and appearance, by whether they are clustered or separated and by the time of year when they occur. Some information on larvae or embryos of Atlantic haplosclerid species is given by Barrois (1876), Topsent (1887, 1911) and Lévi (1956). Those of *rosea* were found in specimens collected from May to September (Pl.1, Fig. 3) ; they appeared to be spherical or ellipsoidal, the latter commonly being at a later stage and containing spicules. In the Anglesey specimens the diameters of apparently round embryos ranged from 159-317 µm, while ellipsoidal forms varied from 349 by 258 to 217 by 156 µm. The embryos are amber-coloured in alcohol-fixed sections mounted in balsam. Usually they are aggregated in clusters, not all in a cluster being necessarily at the same stage of development. Their spicules, when present, are very slender, short, and of apparently uniform size (47-79 µm in different embryos). The spicules at first lie tangentially and oriented at random just under the surface all round the embryo, but at a later stage they migrate to form a stockade-like ring, with the spicules side by side in parallel, near one pole of the developing larva. Four specimens of *Haliclona fistulosa*, identified by the presence of projecting distally-closed off fistulae, long spicules and characteristic surface texture (with irregularly shaped 'pores'), exhibited clustered embryos that were somewhat similar to those of *rosea*, but larger (e.g. 368-478 µm in diameter, and 391 by 433 µm). Their spicules were 108 µm in length when present. These were all collected in August, 1984.

The characters of *H. rosea* described by Topsent (1887) may be listed as follows :

(1) encrusting sponge, or forming an interlacement of rampant or depressed branches of variable width, but of thickness rarely exceeding 1 cm.

(2) Aspiculous dermal membrane, pierced by ostia and covering a continuous subdermal cavity, which is traversed by vertical spicule bundles rendering the surface slightly hispid.

(3) 'Pores', 200-300 µm in diameter and 500-600 µm apart, visible through the dermal membrane and representing the openings of the inhalant canals leading downwards from the floor of the subdermal cavity.

(4) Spicules all oxeote, on average 160 µm by 7-8 µm ; those in the fleshy portions, or membranes traversing the canals, being thinner than those in the skeletal lines.

(5) Skeleton of primary spicule bundles (running to the surface) crossed at diverse angles by feeble secondary lines, with little spongin evident, even at the junctions ; the thickest bundles with rarely more than 3 spicules abreast.

(6) Exhalant canals lying deep down in the sponge.

(7) Colour rose, sometimes pale, often greyish and, at Luc, yellow orange.

(8) Spongin usually colourless, but in places exhibiting a golden yellow to orange colour due to the presence of granules (Topsent, 1891).

(9) Sponge soft when alive, friable when desiccated.

(10) At Luc eggs appear and embryos are released in July ; the 'eggs' are not less than 400 µm in diameter and are grouped under the pinacoderm lining the wide canals. Spicules (70 by 1 µm) appear before the development of the non-flagellated, orange cap at one pole of the embryo.

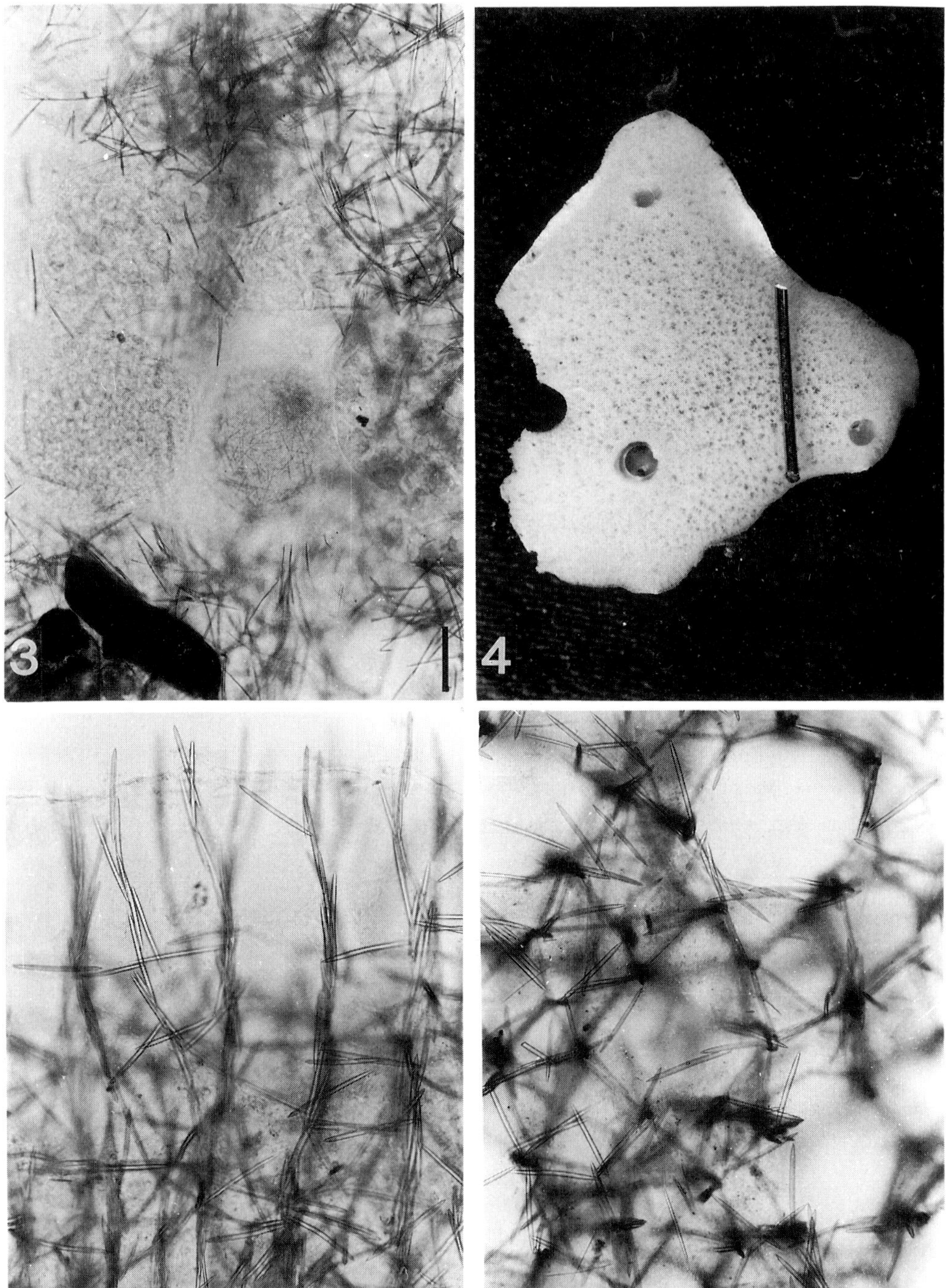

(11) Spherulous cells are large and sparse.
(12) The sponge is common in the English Channel and occurs nearest on the shore compared with all other species of *Haliclona*.

Of these characters I have mostly used the trispicular primary bundles, the aspiculous dermal membrane, the porous appearance of the surface (Pl. 1, Fig. 4) and the absence of slime strands, in the identification of my specimens of *rosea*. The North Wales specimens, particularly those from Rhosneigr, display at times and sometimes in patches the rich amber, almost red, spongin noted by Topsent. Of the specimens exhibiting embryos that are considered in this paper, 3 were collected in May (15th and 16th, 1980), 8 in June (1st, 1981 ; 8th and 11th, 1984), 12 in July (1st and 3rd, 1981 ; 12th, 1983 ; 28th and 30th, 1984), 11 in August (1st, 1984 ; 4th, 1981 ; 27th, 1980 ; and 30th, 1984) and 2 in September (27th, 1984). No embryos have been observed in specimens collected in other months. For convenience of analysis, the specimens were arranged into 6 groups, one for each time of collection, plus or minus a few days, and disregarding the year. The collecting dates for these groups approximated to May 16th (3 specimens), June 9th (8), July 7th (8), August 1st (9), August 30th (6) and September 27th (2).

It could be objected that since embryos only occur in the summer specimens, these groups would not yield the complete annual range of spicule and skeletal variation. However, one commonly finds that parts of summer specimens exhibit the type of skeleton, strongly constructed of thick spicules, that one would have expected from growth in the winter. Some of the spicule samples at least would thus be expected to contain the largest spicules characteristic of the species for the particular locality chosen.

PLATE 1 - Photographs of *H. rosea*.

Figure 3 - Embryos as seen in an hand-cut vertical section. Spicules are visible in the embryo near the centre. The blastomere size is not the same for all embryos in the cluster. The line indicates 100 µm.
Figure 4 - Specimen collected at Rhosneigr on 5th June 1984 and photographed in 90% alcohol. Note the 'pores' and only slightly elevated oscula. The rod indicates 1 cm.
Figure 5 - Vertical section of a specimen collected on 1st June 1981 at Church Island showing the typical orthogonal framework and extensive subdermal cavity. The line indicates 100 µm.
Figure 6 - Surface of a specimen collected at Rhosneigr on 16th May 1980. Spicule tufts protrude upwards through the dermal membrane. A dermal reticulation is not present. The line indicates 100 µm.

METHODS

The specimens were collected from either Rhosneigr or Church Island on the shores of Anglesey. Skeletal characteristics were determined by the methods described by Jones (1984). Vertical sections and surface slices of sponges fixed in 90% alcohol soon after collection were used for the study of skeletal variation after clearing and mounting in balsam. Spicule dimensions were measured using a digitizer connected to a PDP 11 computer. At least 100 spicules were measured at random for each specimen after isolation from the sponge by means of boiling fuming nitric acid.

RESULTS

The coefficient of variation for spicule length and width.

Fig. 1 shows the monthly changes in the average coefficient of variation (100 x standard deviation / mean) for respectively length and width. Significant differences at the 5% level or less occurred between the mean coefficients for width, C.V.(W), for August 1st and August 30th (P>2.5<5%), July 7th and August 30th (P<0.5%) and May 16th and August 30th (P>0.5<1%). For length, C.V.(L), the differences were significant from June 9th to July 7th (P>2.5<5%), July 9th to August 30th (P<1%) and May 16th to August 30th (P>0.1<0.5%). The graph for length follows the same trends as for width, but the swings are much less marked. After the initial rise from May to June, there is a dip in July, followed by a rise to the end of August and a decline in September, but more samples would be required for complete confidence concerning the shape of the graphs between May and the end of August. Essentially similar graphs, with somewhat greater swings, were obtained by aggregating the data per group before determining the overall C.V. The C.V.(W) is always greater than the C.V.(L), partly because more error is involved in width measurement, the digitizer cross-wire being moved from one end to the other of an imaginary line across the centre of the spicule. Also it has been experimentally determined that width is more susceptible to environmental factors (temperature, silicate concentration) than length (Simpson, 1978 ; Simpson et al., 1985).

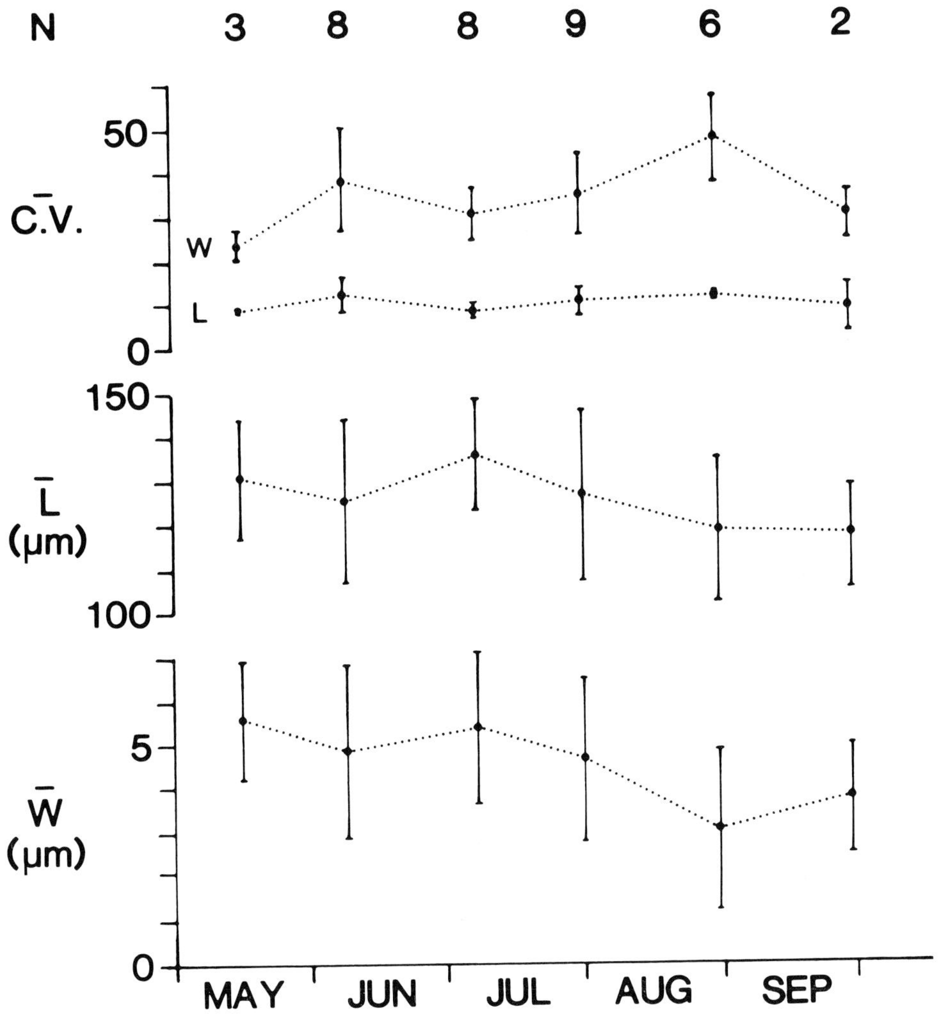

Figure 1 - Graphs of the mean coefficients of variation (C.V.) for width (W) and length (L), and of the mean lengths and mean widths per group of spicule samples plotted against the approximate date of collection of specimens of *H. rosea*. The number of specimens per group (N) is given at the top. Standard deviations are indicated by the vertical lines. In the case of the mean length and width graphs the data for each group were aggregated before determining the standard deviations.

Means and standard deviations.

The means and standard deviations of the lengths and widths of the aggregated data per group are also plotted in Fig.1 . Again the two sets of means show the same trends, with dips in June and late August. These dips contribute to the corresponding high values of coefficients. However the standard deviations tended to be greater for both length and width in June and August, than in May and September, so that the rise in C.V. in June and August is not solely due to the fall in the mean value. To test the statistical significance of the changes in the graphs, one can compare the averages of the sample means per group using Student's 't' test and assuming that the means are normally distributed. The standard deviations of the distributions of the means are less than those for the combined data plotted in Fig. 1, but the mean values per group are the same. Significant differences at the 5% level or less exist between the mean lengths for June and July (P>1<2.5%), and July and September (P>0.1<0.5%). For width, the differences were significant between May and September (P>1<2.5%), July and September (P>1<2.5%) and August 1st and 30th (P>1<2.5%).

Maximum and minimum spicule lengths and widths.

The longest spicule encountered measured 174.5 µm in length and occurred in an early August sample. The maximum lengths for May and September were respectively 159.4 and 142.7 µm. No other month had shorter maximum lengths. The minimum length measured was 59.4 µm (early August), which must be close to the length of the organic axial filament when silicification commences.

The maximum width occurred in the July group and was 10.4 µm. It was 9.4 µm in early August, 8.6 µm in May, 8.5 µm in June, 7.8 µm in late August and 7.7 µm in September. The minimum width recorded was 0.5 µm.

As would be expected high or low mean values of length or width are not strictly correlated with high or low maximum values. From the evidence derived from *H. elegans* (Jones, in press), the largest widths and lengths would be contributed by spicules that developed during the winter months, but whether such spicules occur in a spicule sample would depend on the age of the sponge and possibly the region from which the sample were derived.

% frequency histograms for length and width.

Fig. 2 shows the % frequency distributions for width (1 µm intervals) and length (10 µm intervals) for the aggregated data per group. In May the mode is in the 6-7 µm range for width and the 130-140 range for length and the distributions are negatively skewed, but with little evidence of vigorous new spicule formation. In June spicule production increased abundantly and the mode for width, but not length, shifted somewhat to the left. In July the pattern is much the same, but spicule production had declined and more spicules in the 3-4 µm width category were present. In August, particularly late August, spicule production became abundant once more and the 3-4 µm mode had shifted into the 2-3 µm interval. The mode for length in late August lay in the 110-120 µm interval. In September spicule production appears to have declined and the modes had shifted to the right. The histograms reflect the changes in mean length and mean width displayed in Fig.1,

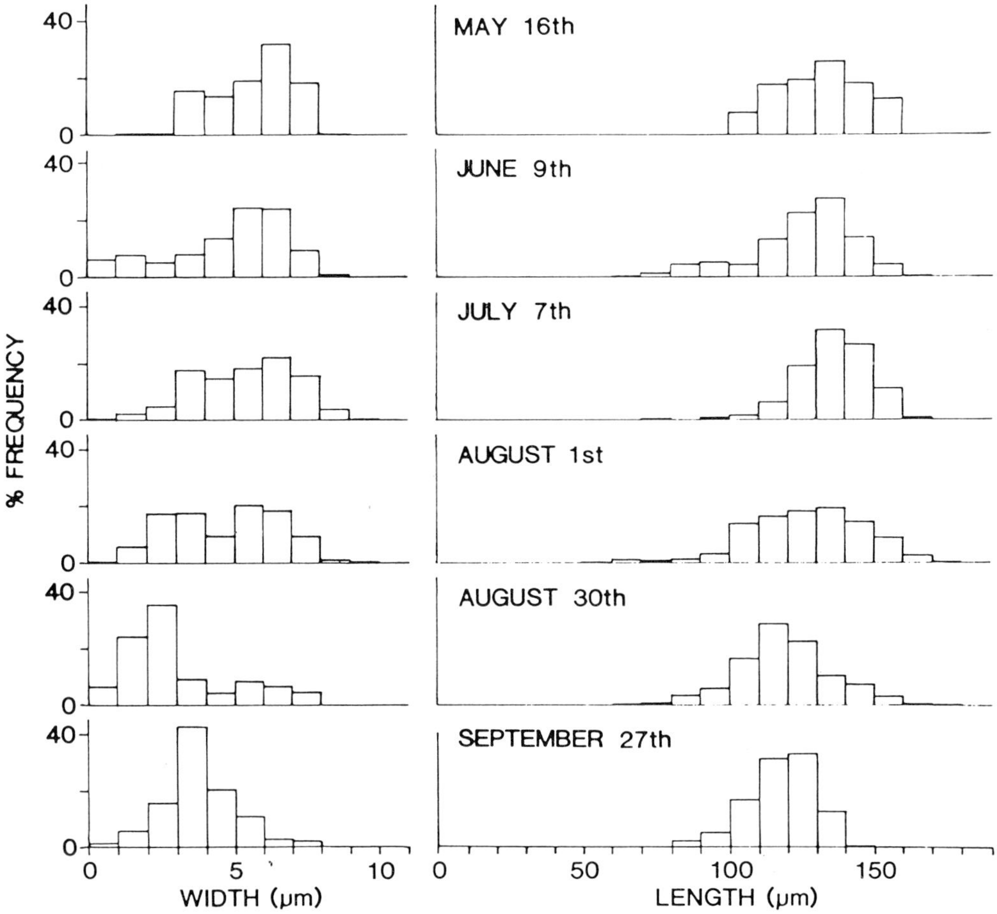

Figure 2 - Relative frequency histograms for the aggregated data for widths and lengths in each group of spicule samples of *H. rosea*.

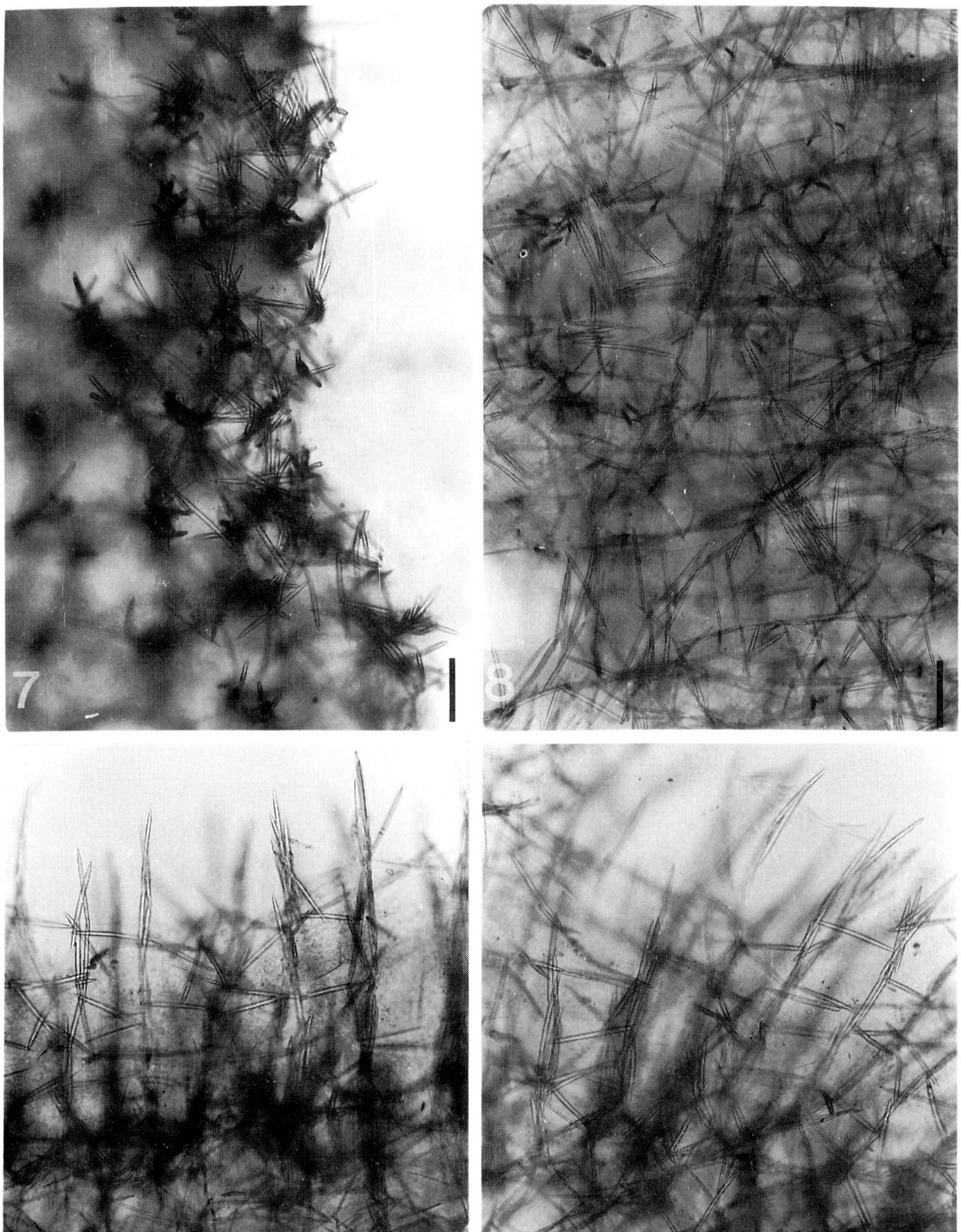
7
8
9
10

and increase confidence in the reality of these changes, even though statistical significance at the 5% level between them was not always achieved.

Sections of specimens of *H. rosea*.

Examination of vertical and surface sections of the sponges with embryos confirms that the skeleton can vary considerably throughout the Summer months. However, not all of the specimens necessarily look alike at the same time. Conditions for growth vary widely in the littoral zone (movements of boulders might result in changes in supply of food, for example), so that the parts of the sponge surface might suggest arrested growth, when other parts exhibit obviously recent growth.

As with *elegans* (Jones, in press), one can usually see an extensive subdermal cavity, traversed by the primary spicule bundles (Pl.1, Fig.5), beneath the dermal membrane. Usually there are 1-3 spicules abreast in these bundles, but sometimes there are more. The projecting ends of the primary bundles appear as tufts when tangential sections through the surface are examined from above using the microscope ; a continuous dermal reticulation, at, or just beneath the surface, is not visible (Pl.1, Fig.6). The surface at the edge of the osculum is much like the rest (Pl.2, Fig.7).

The specimens collected in May exhibit an orthogonal framework of stout spicules at and near the surface, and usually one can distinguish tracts containing loosely-bundled spicules running horizontally, or ascending (or

---

PLATE 2 - Sections of *H. rosea*. The line indicates 100 µm.

Figure 7 - The oscular edge of a specimen collected on 15th May 1980 at Church Island. The spicule tufts are somewhat more concentrated than elsewhere on the surface, because inhalant canals are absent, but otherwise no special features are exhibited.
Figure 8 - Spicule tracts intercrossing in a vertical section of a specimen collected at Rhosneigr on 16th May 1980. The spicules in the tracts vary in thickness and tend to be separated from one another, suggesting that some are developing and all are in transit to regions of growth, or possibly being stored for later incorporation in the skeletal framework.
Figure 9 - Vertical section showing a superficial zone of thick spicules above a zone of thin spicules. The specimen was collected at Rhosneigr on 16th May 1980.
Figure 10 - Vertical section of another specimen collected at the same time and location as the one depicted in Fig.9 . Two zones are evident, the superficial zone exhibiting thick spicules and the deeper zone thinner spicules. The direction of growth of the skeletal framework has altered in between the two zones.

descending) obliquely, particularly in the basal parts of the sponge (Pl.2, Fig.8). As with *elegans* one surmises that these tracts contain developing and fully-grown spicules that are either moving to the growth regions of the skeletal framework, or are being stored for future use. In addition, some of the May sections show conspicuous growth bands, the superficial zone of thick spicules lying above a zone of thinner spicules (Pl.2, Figs 9 & 10). Sometimes an additional zone of thick spicules lies beneath the latter.

In June some of the vertical sponge sections still exhibit the spicule tracts with thick and thin (developing) spicules and they have quite stout spicules in the superficial framework. However, other specimens reveal an abundance of sliver spicules in the meshes of the skeletal framework (particularly in more basal regions) and the presence of thinner spicules in the distal ends of the primary bundles. In these specimens the spicule thickness increases as one follows the bundles downwards from the sponge surface. The thick spicule framework can form a conspicuous band. The superficial zone with thin spicules, on the other hand, can be quite delicate, with relatively few secondary spicules to hold the primary bundles together. The surface is then liable to damage, the projecting ends of the bundles becoming flattened almost into a horizontal direction. In such cases the uniform spacing of projecting tufts in surface sections will not be evident. With the abundant production of sliver spicules in the meshes of the skeletal framework one has difficulty in discerning horizontal and obliquely oriented spicule tracts ; the slivers often appear at random orientation and bundles of spicules in parallel become much less obvious. Also, when there are sliver spicules in abundance one has the impression that the primary bundles can have more than three spicules abreast in places, but counting is difficult.

In July the vertical sections appear with much the same variation as those of June. The presence of thin spicules in the meshes of the skeletal frameworks adds confusion ; spicule tracts can only occasionally be discerned. Some sections exhibit considerable growth, even to the extent of lacking a regular orthogonal spicule framework near the surface, due to inadequate spicule recruitment. In general the superficial spicules are thin, whereas those further down the primary bundles are thicker ; but some specimens can still exhibit thick spicules distally, in parts at least of the surface.

At the end of July and early in August most specimens exhibit a superficial framework of very thin spicules, resting on one of thicker spicules

(Pl. 3 , Figs 11 & 12). Sometimes the surface is patchy, with distinct areas of respectively thick and thin spicules (Pl.3 , Fig.13). There is an abundance of thin spicules in the meshes of the framework and spicule tracts are only vaguely defined. Some surface sections reveal loose bundles of thin spicules running horizontally beneath the subdermal cavity. The organic substance seems densely cellular. With the more delicate construction of the superficial framework the subdermal cavity may be restricted, even lacking in places.

At the end of August there are again thin spicules evident in the meshes (Pl.3 , Fig.14) and fairly thin spicules form the distal parts of the skeletal framework. However, a few specimens still may exhibit thick spicules superficially, at least in parts of the surface, or thick with an admixture of thin.

Finally, the two September specimens both exhibited medium-thin spicules distally and plenty of thin spicules in the meshes of the skeleton. Some evidence of horizontal spicule tracts was afforded by one of them.

For *rosea* then the variations are much the same in general as for *elegans*, neglecting the absence of the dermal reticulation. One can attribute the skeletal variation to two principal factors, the rate of spicule production and the tendency for thin, fully grown spicules to be incorporated in the orthogonal framework in summer months. In winter and spring the newly-forming spicules tend to be thicker and in transit along the spicule tracts leading to the regions of sponge growth, or perhaps being stored in these. In June and August an abundance of thin spicules is produced in the skeletal meshes ; their orientation tends to be at random, so that they confuse the orthogonal arrangement of the skeletal framework. Distinct spicule tracts are less evident.

The striking shift to the left in the modes for both length and width in late August is thus the result of an excess of fully grown, but thin, shorter spicules in the samples. However, in June, when there was little evidence of a shift in the modes, it was the vigorous production of growing spicules that caused the fall in mean values of length and width and the consequent rise in C.V.

Spicule dimensions in the *H. fistulosa* specimens with embryos.

The spicules of the embryo-containing specimens of *H. fistulosa* containing embryos were considerably larger than those in the *rosea* samples.

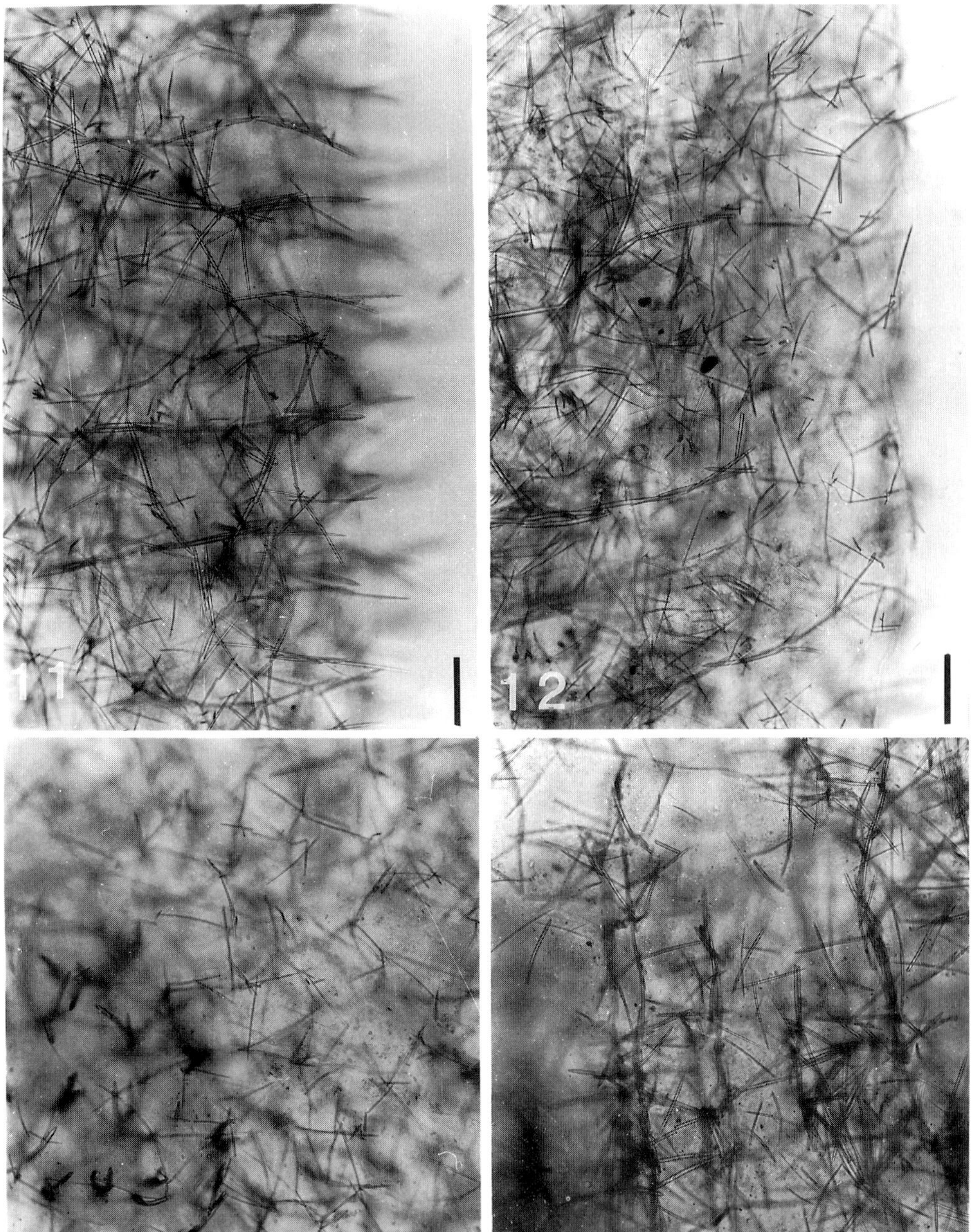
11
12
13
14

Information on the spicule dimensions in the two species is to be found in Table 1. For *fistulosa* the C.V.(L) ranged from 7.4 to 13.1 and the C.V.(W) from 20.7 to 46.3.

Table 1 - Spicule data for embryo-containing specimens of *H. rosea* and *H. fistulosa*. 100 spicules per specimen were measured. There were 36 specimens of *H. rosea* and 4 specimens of *H. fistulosa*. Standard deviations are given in brackets.

Per sample :	*H. rosea* (μm)	*H. fistulosa* (μm)
Largest maximum length	174.5	192.3
Smallest maximum length	118.7	180.1
Largest maximum width	10.4	11.5
Smallest maximum width	5.0	9.4
Largest mean length	145.6 (13.5)	161.3 (12.8)
Smallest mean length	97.7 (15.5)	147.5 (13.8)
Largest mean width	6.3 (1.8)	8.6 (1.9)
Smallest mean width	2.1 (0.9)	5.5 (2.5)
Smallest minimum length	63.9	94.7
Smallest minimum width	0.5	0.7

PLATE 3 - Sections of *H. rosea*. The line indicates 100 μm.

Figure 11 - Vertical section of a specimen collected at Church Island on 28th July 1980, showing a zone of thin spicules above a zone of thick spicules. Note the occurrence of sliver spicules in the interstices of the framework.
Figure 12 - Another example from a Church Island specimen collected on 30th July 1980, showing a few bundles of thick spicules beneath a delicate, incomplete, superficial spicule framework. The thin spicules in the latter appear to have been flattened down near the surface.
Figure 13 - Patchy surface of a specimen collected at Church Island on 30th July 1980, with thick spicule tufts in an area alongside a thin-spicule framework.
Figure 14 - Part of a vertical section of a specimen collected at Rhosneigr on 27th August 1980. Note the abundance of sliver spicules in the interstices. Spicule tracts are not obvious.

DISCUSSION

Spicule dimensions.

Comparison of the spicule data for the embryo-containing specimens of *rosea* and *fistulosa* in Table 1 indicates that the two species are quite distinct from one another in respect of spicule size. It is true that the largest maximum width for *rosea* slightly exceeds the smallest maximum width for *fistulosa*, but reference to the histograms in Fig. 2 confirms that very few spicules of *rosea* occurred in the 9-11 µm width range. Moreover, the length maxima and mean values of the two species do not overlap, so that these could be used to separate them. However, it would be advisable to sample more specimens of *fistulosa*, before categorically stating that overlap rarely occurs. One must, of course, be mindful of the possibility of intrusion of foreign spicules in the samples when comparing the largest sizes. Spicules which appear to be clearly apart from a continuous frequency distribution, such as those depicted in Fig. 2, should be neglected, particularly when few in number.

In a recent review De Weerdt (in press) gives 120-180 by 6-10 µm for the oxea of *rosea* of the British Isles, and correspondingly 150-210 by 7.5-12.5 µm for oxea of *fistulosa*.

Griessinger (1971) states that he has not found *R. rosea* in the Mediterranean, but Garrone (1969) obtained *rosea* from Villefranche-sur-mer. According to Topsent (1925) *R. rosea* (Bowerbank) Schmidt is common at Naples, but later Topsent (1943) claimed synonymy between *R. rosea* (Bowerbank) and *R. aquaeductus* Schmidt, the genotype for *Reniera*. The spicule dimensions of Topsent's Naples specimens were as follows : 82-90 by 4 µm, 92-96 or 95-105 by 4-4.5 µm, 100-110 by 5-6 µm, 130-140 by 4.5 µm (on very large specimens). These measurements are certainly smaller than were found in the Anglesey specimens of *rosea*, but whether a different species is involved (as Griessinger concluded), or whether the differences are the result of phenotypic variation is a problem for further research. According to Griessinger (1971) also, the spicule dimensions of Mediterranean specimens of *fistulosa* (named by him *Pellina fistulosa* (Bowerbank)) are also smaller than in the Anglesey *fistulosa*. In three examples they were 85 - 97.5 ± 1.5 - 107.5 ; 2.5 - 2.75 - 3.75 (fistular spicules, 102 - 120 ± 2 - 132 ; 2.5 - 3.25 - 3.75), 102.5 - 122.5 ± 1.5 - 132.5 ; 2.75 - 5.25 - 6.75 (fistular spicules 110 - 122.5 ± 1.5 - 135 ; 4.5 - 5.0 - 6.0) and 115 - 130

- 152.5 ; 2.85 - 3.5 - 5.25 µm. Again the question arises : are the differences genotypic or phenotypic ?

The variation in the C.V.(W) and C.V.(L) for the Anglesey *rosea* during the summer months was not quite the same as for *H. elegans* (Jones, in press). While the graphs for both species tend broadly to rise and fall as would be expected from the changes in mean temperature and silicate concentration, there was a striking dip in June in the *elegans* graphs and a less obvious dip in July in those for *rosea*. In both species these dips can be explained by a dearth of juvenile spicules in the samples, which results in an increase in the mean spicule dimensions and decrease in the standard deviations. The factors affecting the C.V. are clearly more complex than was envisaged by Griessinger (1971). Besides factors such as silicate concentration, temperature and rate of spicule production, it has recently been claimed that wave force affects the width of spicules of *Halichondria panicea* (Palumbi, 1986). Whether wave force can affect the width of spicules in the Haliclonidae remains to be tested, but if so, since wave action usually varies seasonally, it would also be a factor determining the shape of the annual graphs of coefficients of variation, mean length and mean width. With the exception of *Adocia simulans* and *H. oculata*, however, the Anglesey haliclonid sponges tend to be soft and situated in sheltered places, not exposed to the full force of the waves ; it is thus unlikely that spicule width would be correlated with wave force.

Taxonomy.

Bowerbank (1866) described *Isodictya rosea* as a rose-pink, encrusting species with a minutely hispid, uneven surface ; the oscula were large, dispersed, simple or slightly elevated and sub-fistulous. The 'pores' (i.e. lumina of the vertically descending inhalant canals ; Topsent, 1887) were inconspicuous ; the dermal membrane was aspiculous and the skeleton slender and delicate, composed of oxea of 127 µm in length full size ; tension spicules of the interstitial membranes (i.e. not incorporated in the skeletal framework) were slender and not abundant ; reproductive bodies abounded and measured 1-5 times the length of the larger spicules (i.e. about 130 to 635 µm). In his illustrations of the spicules (Bowerbank, 1874, Plate XLIX, Figs 13 & 14) the skeletal and tension oxea are 118 by 9 µm and 102 by 4 µm respectively. However, I have a photograph of a convenient area of Bowerbank's spicule preparation of the type (Tenby, Bk682) containing about 175 spicules and the largest of these measured 140.8 by 6.7 µm. These spicules are similar in shape to those of the species described in this paper, whereas Bowerbank's Fig. 13 suggests that the oxea should be short and stout, somewhat like the spicules of *Haliclona elegans*. In Volume IV (Bowerbank, 1882, p. 11) it is stated that the spicule bundles are bi- or trispiculous. All in all there can be little doubt that Bowerbank's *Isodictya rosea* is the same as the species described in this paper, bearing in mind the range of skeletal variation found.

The use of the generic name requires justification. Topsent (1887) named the species *Reniera rosea* Schmidt (*Isodictia* (sic) *rosea* Bowerbank). However, according to Burton (1934, p. 535) the generic name *Reniera* Nardo must be abandoned because it was inadequately defined by Nardo and there is confusion as to whether *Rayneria* Nardo 1833 (type *R. typus*) is synonymous with *Reniera* Nardo 1841 (type *R. typica*). Burton (1934), Topsent (1938) and Griessinger (1971) thought that there had been a mere change of spelling. However, Wiedenmayer (1977a, p. 86) is of the opinion that Nardo originally had two different genera in mind. An additional source of confusion concerns the authorship of the generic name *Reniera*. Contrary to Burton, Wiedenmayer asserts that Schmidt should be regarded as the author, *Reniera* Nardo being a 'nomen nudum'.

After its rejection by Burton the name *Reniera* was revived by Griessinger (1971) and Lévi (1973), to denote haplosclerid sponges which have relatively little spongin, no tangential ectosomal skeletal and a somewhat untidily ordered skeletal framework resulting from variability in spicule length. The genus contrasts with *Adocia* Gray 1867, in which a tangential ectosomal skeleton is exhibited (one must neglect the obvious error in the generic description : Gray stated that there is a 'skin without spicules', whereas the stated genotype, *I. simulans* Bowerbank, has a dermal reticulation). It also contrasts with *Haliclona* Grant 1835-1841, originally used for the species *oculata*, in which there is an isodictyal unispicular framework, of short, uniform spicules, which tend to be enveloped by a greater amount of spongin. However, Wiedenmayer (1977b) and Bergquist & Warne (1980) have recently re-emphasized the difficulty in distinguishing between *Reniera* and *Haliclona*, the characters used to separate them being relative rather than absolute, but while Wiedenmayer prefers to retain *Reniera*, Bergquist & Warne unite *Reniera* and *Haliclona* under the same genus *Haliclona*, thus following de Laubenfels (1936). They retain the family name Haliclonidae de Laubenfels, whereas, according to Wiedenmayer, Renieridae Ridley could be used and has priority. A recent review of the taxonomic history of the Haplosclerida has been given by de Weerdt (1985). Further studies (see de Weerdt & van Soest, 1986) have led to the conclusion that *Haliclona*, *Reniera*, *Adocia*, and *Gellius* are synonymous. Consequently, I have used the name *Haliclona rosea* for the sponge discussed in this paper, despite the species being renieroid.

Barrois (1876, footnote on p. 60) considered *Isodictia* (sic) *rosea* Bowerbank to be a variety of *I. cinerea* (Grant) Bowerbank (1866). Topsent (1887, p. 94 and p. 156) was inclined to agree, although later (1932) he listed the two as separate species. The relationship between them must be considered, because *cinerea* is an older name than *rosea*. The original *Spongia cinerea* Grant (1827) was found in the Firth of Forth and was described as a blackish-grey specimen resembling a dark putrid sponge even though healthy. It had an irregular outline and a smooth, convex, fleshy, transparent surface. The pores could not be seen without a lens and the oscula were few, very large, regularly circular and lying deeper than the general surface. The spicules were siliceous, remarkably uniform in size, rather short, curved, equally thick throughout, and pointed suddenly at both ends. Judging from the illustration (Plate II, Fig. 3 in comparison with Fig. 9, which was magnified 50 times) the spicules were about 150 µm long. This description is reminiscent of the plaque form of *Adocia simulans* Johnston, both in regard to the external surface and the uniform skeleton of short, stout, sharply pointed oxea. Moreover, I have seen healthy, blackish, putrescent-looking specimens of this species at Carriganorana, Southern Ireland and hence cannot agree with Topsent's statement (1938) that exteriorly Grant's specimen scarcely resembled a plaque of *simulans*. However, Johnston (1842) regarded the two as distinct species, describing the one under *Halichondria cinerea* and the other under *H. simulans*. In his description of the former, however, he altered somewhat the characters given by Grant, *cinerea* now being of uniform hair-brown or ash-grey colour, soft and friable when dry, of very fine sponge-like texture and with indistinctly marked oscula. There is no indication that Johnston actually saw the unique specimen of *cinerea* collected by Grant, which, as stated above, I believe could have been what Johnston called *simulans*.

Four of Johnston's specimens labelled *Halichondria cinerea* were later studied by Bowerbank, who interpreted them as four distinct species, mainly on the basis of spicule size and the presence or absence of a tangential ectosomal reticulation. The four specimens were named *Isodictya permollis* Bowerbank (smaller specimen 17c), *I. peachii* Bowerbank (larger specimen

17c), *I. varians* Bowerbank (17e) and *I. cinerea* (17d). The last specimen was identified by comparison with a portion of the type specimen which Bowerbank (1866, p. 275) obtained from Grant. Later, however, Bowerbank stated (1874, p. 121) that the type specimen could not be found by Grant when it was required for the purpose of making an illustration. It is thus impossible to check that Bowerbank, or Johnston before him, had been correct when identifying specimens as the original *cinerea*.

Bowerbank's description of *S. cinerea* Grant, based presumably on the specimen 17d, his own specimens and possibly one of the specimens 17c, because this is figured under *I. cinerea* (1874, Plate XLVIII, Fig. 2), indicates that it is an encrusting form, lilac in colour when alive, with a smooth, even surface, and an aspiculous, pellucid dermal membrane (1866, p. 274). The spicules are 'rather long and stout', being 152 by 9 µm and the spicule framework is unispiculous. The absence of a dermal reticulation would appear to preclude the possibility of this species being *simulans*. However, Burton (1934) reckoned that the so-called *cinerea* depicted by Bowerbank (1874) in Plate XLVIII Fig.1, was identical to Bowerbank's specimen of *I. simulans* and in the absence of a type specimen for *cinerea*, he nominated this specimen as the neotype of the species *I. cinerea*. Then, because of the similarity to *I. simulans*, he concluded that the two were synonymous and should be named *Adocia cinerea*. Topsent (1938) disagreed, but his discussion contains errors. He stated that Burton had taken the sponge depicted in Bowerbank's Plate XLVIII Fig. 2, as the neotype, instead of Fig. 1. He also stated that the oxea of Grant's *cinerea* were of a smaller size than those of *Spongia oculata*, whereas I calculate that they measured 150 µm in length. At all events he concluded that nothing permitted one to suppose that *Spongia cinerea* Grant had been an *Adocia*. I would dissent from this, because, as stated above, I believe that Grant's *cinerea* could have been a specimen of what we now call *Adocia (Haliclona) simulans*. Nevertheless, one would agree with Topsent that Burton had been somewhat arbitrary in selecting one of Bowerbank's specimens as the neotype for *cinerea* and then arguing that, because this had a dermal reticulation, the generic name should be *Adocia*. One would also agree that Burton behaved arbitrarily in rejecting the generic name *Reniera*, while maintaining the specific name *cinerea* ; the former could reasonably be retained even though its meaning were to change somewhat as more or less species became incorporated, whereas the latter had an original description so incomplete as to render it meaningless. De Laubenfels (1934) confirms this by stating that *cinerea* should be *permollis*. I might add that I have examined the type slide of *cinerea* prepared by Burton (labelled *Isodictya cinerea* Bowerbank, Fowey, Bowerbank collection type, 32.1.5.7a) and found the sections quite similar to those of the type for *elegans* (Jones, in press).

Initially Bowerbank believed that *cinerea* was rare, but later he obtained specimens from various localities exhibiting considerable variations in form (1874, p. 121). The variability led Burton (1926) to restore Johnston's four specimens to *cinerea* and in addition to incorporate a range of other so-called species within the same category. Within this range was included *elegans*, which, sensu Topsent (1887, 1894), is quite distinct and easily recognized owing to the presence of slime strands, and *simulans*, which, though a variable species, is easily distinguished by its rigidity, its uniform skeleton, its dermal reticulation and the form of its spicules. Histologically it is different from *elegans* (Tuzet, 1932). It is thus certain that Burton lumped too many species together within his concept of *cinerea*, so that one cannot affirm that the *rosea* described in this paper is synonymous with *cinerea* sensu Burton, 1926.

From the above it is clear that the name *cinerea* is of little taxonomic value. It is a 'nomen nudum'. The characteristics of *cinerea* described by Johnston and Bowerbank are incomplete and not necessarily applicable to the *cinerea* of Grant. Also, while others have recorded *cinerea* from various localities (e.g. Lundbeck, 1902 ; Stephens, 1915 ; de Laubenfels, 1932) to the extent that the species is regarded as cosmopolitan, it has been pointed out that the identifications have not always been reliable (Bergquist & Warne, 1980), and that the *cinerea* specimens recorded from different parts of the world are not necessarily all of the same species (Lundbeck, 1902 ; Griessinger, 1971). Clearly, it is preferable to retain the specific name *rosea* for the species described in this paper.

The relationship between *H. rosea* and *H. fistulosa* remains to be discussed. The two might be confused when specimens of the latter do not exhibit hollow fistulae, sealed off at their distal ends. However, the spicule sizes do not overlap, as pointed out above, and Topsent (1932) could not recall having observed on any one of these sponges oscula exhibiting the characters of both *fistulosa* and *rosea*. A clear description of *fistulosa* from the Mediterranean, under the genus *Pellina*, has recently been given by Griessinger (1971).

ACKNOWLEDGMENTS

I am grateful to Miss Carol F. Grundy for measuring spicule samples from some of the specimens of *rosea*.

LITERATURE CITED

Barrois C., 1876 - Mémoire sur l'embryologie de quelques éponges de la Manche. *Ann. Sci. nat. (Zool. Paleontol.)*, sér. 6, Art. 11 : 1-84.

Bergquist P.R. & Warne K.P., 1980 - The marine fauna of New Zealand : Porifera. Demospongiae. (3) (Haplosclerida and Nepheliospongida). *New Zealand Oceanographic Institute Memoire*, 87 : 1-42.

Bowerbank J.S., 1866, 1874, 1882 - A monograph of the British Spongiadae. 2: 1-388 ; 3 : 1-367 ; 4 : 1-250, Ray Society, London.

Burton M., 1926 - Observations on some British species of sponges belonging to the genus *Reniera*. *Ann. Mag. nat. hist.* ser. 9, 17 : 415-424.

Garrone R., 1969 - Collagène, spongine et squelette minéral chez l'éponge *Haliclona rosea* (O.S.) (Démosponge, Haplosceride) *J. Microsc.* 8 : 581-598.

Grant R., 1827 - Notice of two new species of British sponges. *Edinburgh new philosoph. J.*, 2 : 203-204.

Grant R., 1835-41 - Outlines of comparative Anatomy. H. Bailliere, London. 656 p.

Gray J.E., 1867 - Note on the arrangement of sponges, with the description of some new genera. *Proc. zool. Soc. Lond.*, 492-558.

Griessinger J.-M., 1971 - Etude des Réniérides de Méditerranée (Démosponges Haplosclérides). *Bull. Mus. natl Hist. nat.* sér. 3, *Zool.* : 97-175.

Johnston G., 1842 - A history of British sponges and lithophytes. W.H. Lizars, Edinburgh. 202 p.

Jones W.C., 1984 - Spicule dimensions as taxonomic criteria in the identification of Haplosclerid sponges from the shores of Anglesey. *Zool. J. Linn. Soc.* 80 : 239-259.

Jones W.C., in press - Seasonal variations in the skeleton and spicule dimensions of *Haliclona elegans* (Bowerbank) sensu Topsent (1887) from two sites in the North Wales. In *European contributions to the Taxonomy of Sponges*. W.C. Jones (ed.), publications of the Sherkin Island Marine Station, n°1.

Laubenfels M.W. de, 1932 - The marine and fresh-water sponges of California. *Proc. U.S. natl Mus.*, n°2927 81(4) : 1-140.

Laubenfels M.W. de, 1936 - A discussion of the sponge fauna of the Dry Tortugas in particular and the West Indies in general, with material for a revision of the families and orders of the Porifera. *Paps Tortugas Lab.*, 30,n°467 : 1-225. Carnegie Institution, Washington.

Lévi C., 1956 - Etude des *Halisarca* de Roscoff. Embryologie et systématique des Démosponges. *Archs Zool. exp. gén.*, 93 : 1-184.

Lévi C., 1973 - Systématique de la classe des Demospongiaria (Démosponges) : pp 577-631 in *Traité de Zoologie. Anatomie, Systématique, Biologie.* (3). P.P. Grassé (ed.), Masson, Paris.

Lundberk W., 1902 - The Danish Ingolf-expedition. VI. Porifera. (Pt.1) Homorrhaphidae & Heterorraphidae : 1-108. H. Hagerup, Copenhagen.

Nardo G.D., 1833 - Auszug aus einem neuen System der Spongiarien, wonach bereits die Aufstellung in der Universitäts-Sammlung zu Padua gemacht ist. *Isis Oken* : 519-523.

Nardo G.D., 1847 - Prospetto della fauna marina volgare del veneto estuario: pp 113-156 in *(Estratto dell'opera) Venezia e le sue lagune*, 2, Pt.1 (1-45 in reprint). Antonelli, Venice.

Palumbi S.R., (1986) - How body plans limit acclimation : responses of a demosponge to wave force. *Ecology* 67(1) : 208-14.

Schmidt O., 1862 - Die Spongien des Adriatischen Meeres. Engelmann, Leipzig. 88 p.

Simpson T.L., 1978 - The biology of the marine sponge *Microciona prolifera* (Ellis & Solander). III. Spicule secretion and the effect of temperature on spicule size. *J. exp. mar. Biol. Ecol.*, 35 : 31-42.

Simpson T.L., Gil M., Connes R., Diaz J.-P. & Paris J., 1985 - Effects of germanium (Ge) on the silica spicules of the marine sponge *Suberites domuncula* : transformation of spicule type. *J. Morphol.*, 183 : 117-128.

Stephens J., 1915 - Atlantic sponges collected by the Scottish National Antarctic Expedition. *Trans. roy. Soc. Edinburgh*, 50 : Pt II, n°15 : 423-467.

Topsent E., 1887 - Contributions à l'étude des Clionides. *Archs Zool. exp. gén.* sér. 2, 5 bis Mem. IV (supplement), 1-165.

Topsent E., 1891 - Essai sur la faune des spongiaires de Roscoff. *Archs Zool. exp. gén.*, sér. 2, 9 : 523-554.

Topsent E., 1894 - Etude sur la faune des spongiaires du Pas-de-Calais suivie d'une application de la nomenclature actuelle à la monographie de Bowerbank. *Rev. biol. Nord France* 7 : 16-28.

Topsent E., 1911 - Sur les affinités des *Halichondria* et la classification des Halichondrines d'après leurs formes larvaires. *Archs Zool. exp. gén.*, Notes et Revues, n°1, sér. 5, 7 : 1-15.

Topsent E., 1925 - Etude des spongiaires du Golfe de Naples. *Archs Zool. exp. gén.* 63 : 623-725.

Topsent E., 1932 - Remarques sur des éponges de l'estuaire de la Rance. *Bull. Lab. St-Servan*, fasc. VIII : 1-7.

Topsent E., 1938 - Commentaires sur quelques genres d'éponges marines. *Bull Inst. Océanogr. Monaco.*, N. 744 : 1-23.

Topsent E. & Olivier L., 1943 - Eponges observées dans les parages de Monaco. *Bull. Inst. océanogr. Monaco*, N.854 : 1-12.

Tuzet O., 1932 - Recherches sur l'histologie des éponges *Reniera elegans* (Bowerbank) et *Reniera simulans* (Johnston). *Archs Zool. exp. gén.*, 74 : 169-192.

Weerdt W.H. de, 1985 - A systematic revision of the North Eastern Atlantic shallow-water Haplosclerida (Porifera, Demospongiae), Part 1 : Introduction, Oceanapiidae and Petrosiidae. *Beaufortia*, 35, n°5 : 61-91.

Weerdt W.H. de, in press - The marine shallow-water Haliclonidae (Haplosclerida, Porifera) of the British Isles. In *European contributions to the Taxonomy of Sponges*. W.C. Jones (ed.), publications of the Sherkin Island Marine Station, n°1.

Weerdt W.H. de & Soest R.W.M. van, 1986 - Marine shallow-water Haplosclerida (Porifera) from the South-Eastern part of the North Atlantic Ocean. *Zool. Verh.*, n° 225 : 3-49.

Wiedenmayer, F., 1977a - Shallow-water sponges of the Western Bahamas. *Experientia*. Supplementum 28 : 1-287. Birkhauser, Basel & Stuttgart.

Wiedenmayer F., 1977b - The Nepheliospongiidae Clarke 1900 (Demosponges, Upper Devonian to Recent), an ultraconservative, chiefly shallow-marine sponge family. *Eclogae Geol. Helv.*, 70(3) : 885-918.

THE HAPLOSCLERID SPONGE FAUNA OF BANYULS-SUR-MER (MEDITERRANEAN), WITH THE DESCRIPTION OF A NEW SPECIES.

F. van LENT & W.H. de WEERDT

Institute of Taxonomic Zoology, University of Amsterdam
P.O. Box 20125, 1000 HC Amsterdam, The Netherlands.

SYNOPSIS

The Haplosclerid sponge fauna of Banyuls-sur-mer (Mediterranean) is the subject of this study. Fifteen species were collected among which one is described as new to science, viz. *Haliclona griessingeri* n.sp. A review of the Haplosclerid fauna of the Banyuls region is given and comparisons with neighbouring areas are made.

INTRODUCTION

During a previous workshop on taxonomy of N-E Atlantic sponges held in S-W Ireland the question was raised whether sponges, reported both from the N-E Atlantic and the Mediterranean Sea, were in fact the same species. A small project was set up to compare several Chalinid species mentioned from both areas.

The outcome of this research was expected to reveal more information on the geographical distribution of sponges and marine benthic organisms in general.

To get an impression in a relatively short period of time and to benefit from extensive work done by others in the past (Topsent, 1892b, 1893 ; Boury-Esnault, 1971), Roscoff (N-E Atlantic) and Banyuls-sur-mer (Mediterranean) were chosen as sampling sites. However, as the work advanced, it became clear that a comparison of specimens from the N-E Atlantic and Mediterranean was not possible for most of the species selected below, because the northern Atlantic fauna at Roscoff and the Mediterranean fauna at Banyuls appeared to be quite dissimilar. Only in the case of one species,

NATO ASI Series, Vol. G13
Taxonomy of Porifera
Edited by J. Vacelet and N. Boury-Esnault

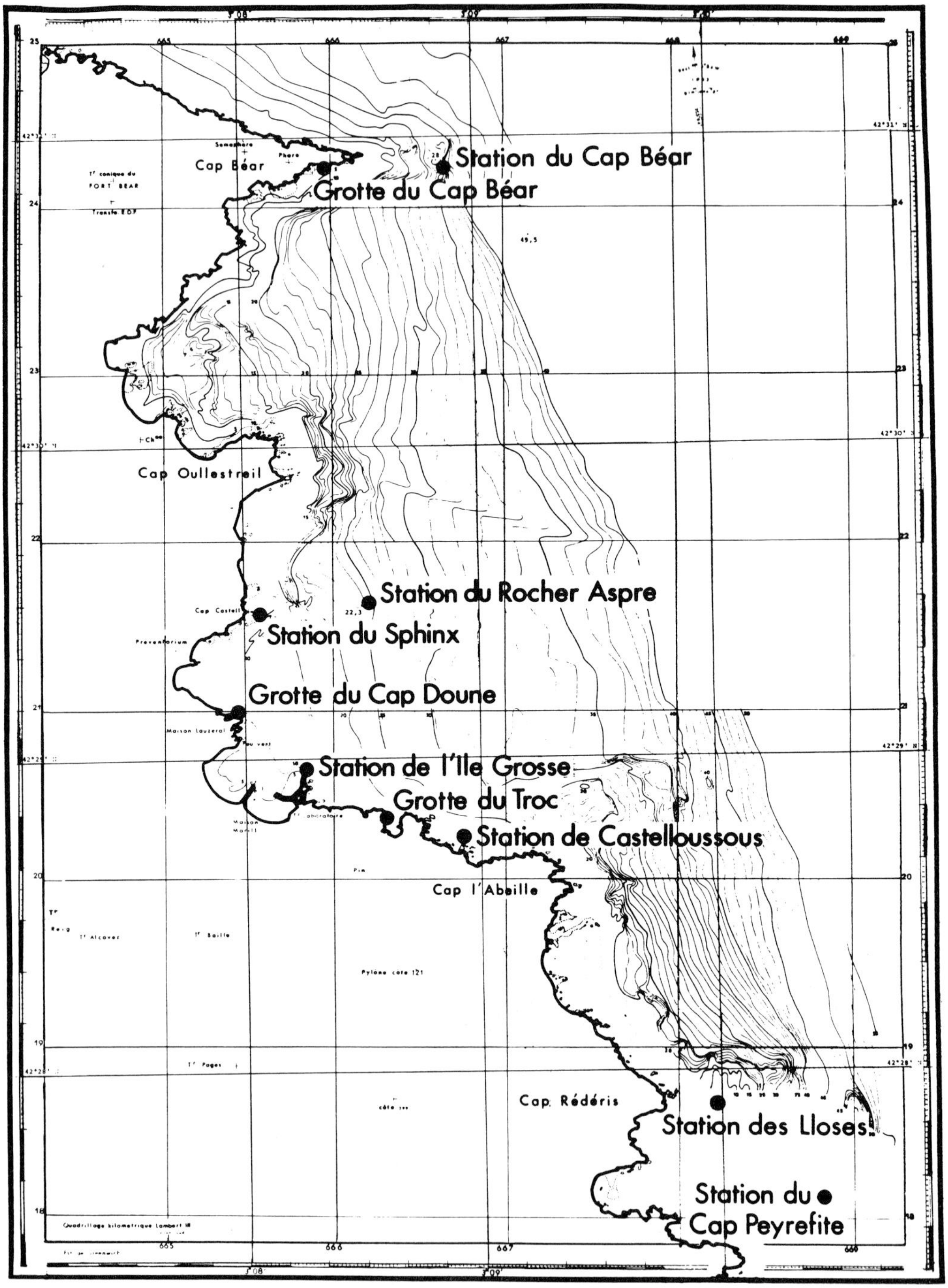

Figure 1 - Situation of sampling stations (From : Boury-Esnault, 1971).

Haliclona fibulata, could a comparison be made.

Hence, this report gives a list and detailed description of all the chalinid sponges (with their habitats) found during this research in the region of Banyuls. This information is then compared with earlier records of the same species from the Mediterranean and N-E Atlantic. Also presented are theoretical considerations on the Mediterranean-Atlantic distribution of the Chalinid species.

Following de Weerdt (1986), *Reniera, Gellius,* and "*Pellina*" are considered as synonyms of *Haliclona*.

MATERIAL AND METHODS

In order to become acquainted with the species from the N-E Atlantic, a short visit was made to the "Station Marine de Roscoff" in the N-W of France, at the beginning of May 1984. The following Chalinid sponges were

Table 1 - Description of collecting-stations (see Fig. 1).

- **Banyuls-sur-mer** :
  - Ile Grosse. Vertical and overhanging walls, with many cavities and large boulders. In between these boulders there is sand. (Diving depth : 14-15 m).
  - Cap du Troc. Vertical and overhanging walls, with many cavities and very large boulders. The bottom consists of sand. (Diving depth : ~ 25 m).
  - Cap l'Abeille. Coralligene and boulders, with sand in between. (Diving depth : ~ 40 m).
  - Cap Rédéris. Vertical and overhanging walls, also very large boulders. A lot of cavities are present, while sand is at the bottom. (Diving depth : ~ 30 m).
  - Canadell. Boulders and coralligene, with sand in between. Some vertical and overhanging walls with cavities are present. (Diving depth : ~ 25 m).
  - Cap Oullestrell. Vertical and overhanging walls, also very large boulders. Many cavities and a sandy bottom. Sheltered from the "tramontane" (N.W. wind). (Diving depth : ~ 25 m).
  - Cap Béar. Coralligene and large boulders, also vertical and overhanging walls with cavities. A sandy bottom. Facing Grotte du Cap Béar : very sheltered from the "tramontane". (Diving depth : ~ 20 m). Cap Béar : very exposed. (diving depth : ~ 40 m).
- **Marseille** :
  - Ile PLane. Very large cave with many cavities. Sandy bottom. (Diving depth : ~ 30 m).

Table 2 - Haplosclerid species which have been found at different stations (see Fig.2) in the region of Banyuls-sur-mer in different studies. * : stations visited or species found at a certain station during this study (1984). + : stations visited or species found at a certain station by Boury-Esnault (1968, 1971). # : stations visited or species found at a certain station by Topsent (1892, 1893, 1925). $: reported by others authors.

	*+# cap/gr. Béar	* cap Oullestrell	+# rocher Aspre	+ Sphinx	+ cap Doune	*+# île Grosse	*+ cap/gr. du Troc	+ Castelloussous	* # cap Abeille	*+ Cap Lloses Rédéris	+ Cap Peyref.	* Canadell	#$ Site Unknown
H.cratera		*				*		+	*	*+		*	#
H.fulva		*		+		*+	*		* #	*			
H.mamillata									*				
H.mucosa	*	*				*	*		*	*		*	
H.sarai	*	*		+		*	*+		*	+			
H.valliculata	*						*						
H.citrina		*		+		*+	*+	+					#
H.mediterranea							*		*	*		*	
H.implexa						*			*			*	
H.flavescens													#
H.arenata													#
H.plana													#
H.cinerea						#							
H.aquaeductus													#
H.griessingeri	*												
H.fibulata	+		+	+	+	*+	+	+		+	+		#
H.angulata			+							+			#
"*H.luridus*"	+			+	+	+	+	+		+	+		
H.apertus	+									+			
H.microsigma	+					+							
H.lacazei	+							+	* #				
H.parietalis									#				#
H.fistulosa									#				
H.magna						Facing Cap Creus							
H.varia			#										
H.simulans						#							#
H.ficiformis	*+	*		+	+	*+	*	+	*	*+		*	$
D.lenis	#						*		#				#
C.nicaeensis	+		+	+						+			

collected by means of wading through the intertidal area, diving and dredging : *Haliclona simulans*, *H. cinerea*, *H. indistincta*, *H. rosea*, and *H. fistulosa*. Habitat, depth, colour etc., of every specimen, were noted and sections were made for microscopic examination. One cross-section and if possible, one tangential section of the ectosome were taken. These sections were immersed in alcohol, dried and embedded in Canada Balsam. Afterwards, the sponges were preserved in alcohol.

The research was continued from mid-May to mid-September 1984, at the "laboratoire Arago", Banyuls-sur-mer, in the South of France. A list of collecting stations is given in table 1 and in Fig. 1. In September 1984 the "Station Marine d'Endoume" in Marseille, was visited for some extra collecting and to study the preserved collections present at the station.

- Most species were identified by using Griessinger's (1971) revision on Mediterranean Chalinids. Part of his type-material, which is present in the Museum National d'Histoire Naturelle (MNHN), was studied for comparison.
- All the described sponges were collected from solid substrata.
- The colour of the species was described for the living sponge. Unless stated otherwise, they all lost their colour in alcohol and became white, cream, yellowish or grey.
- The additions "very common", "common" and "not common" were only valid for the region of Banyuls in 1984.
- The average length of the spicules (based on 25 measurements) is given as well as the maximum and minimum length and the average width.
- The descriptions are only valid for the sponges found in the region of Banyuls and mentioned under the given numbers.
- The maximum depth of diving was 40 m.

SYSTEMATIC DESCRIPTIONS

Family CHALINIDAE sensu de Weerdt, 1986.

Genus *Haliclona* Grant, 1835.

Haliclona cratera **(Schmidt, 1862) (n. comb.)**

Reniera cratera Schmidt, 1862 : 73 ; Griessinger, 1971 : 128, Fig. 4f.

Material. Banyuls-sur-mer : ZMA POR 5429, île Grosse, 14 m, on overhanging cliff ; 5720, île Grosse, 14 m, on steep wall ; 5719, cap Oullestrell, 16 m, on steep wall and edge of crevice ; 5425, cap l'Abeille, 23 m, on boulder ; 5428, Canadell, 25 m, on boulder.

Description.

Shape and size : the sponges have an irregular and cushion-shaped encrusting base of approximately 0.5 cm thickness, with rounded edges. The apical oscules, 0.3 to 0.5 cm in diameter, are situated on irregularly shaped, thick-walled (0.3 cm) tubes, which rise from the joint base. These tubes have a height of up to 5 cm and are often found growing together forming ridges.

Consistency : spongy, compressible and rather soft. The surface is slightly rough. When broken, a large amount of mucus is secreted.

Colour : rosy, (soft)orange. The majority are yellowish at the base.

Ectosome : a unispicular, regular, isotropic reticulation with 3-sided meshes ; spicules are bound by spongin at the nodes.

Choanosome : a unispicular, ladder-like, 4-angled reticulum which is cemented by spongin at the nodes. The reticulation is often confused.

Spicules : slightly curved strongyles. 239-274 ± 16-300/8 µm.

Remarks. They are found at depths to 25 m, but never in very obscure places. Common. Topsent (1925b) reported this species from the region of Banyuls, while Boury-Esnault (1971) has found this species at Lloses and Castelloussous, between 17 and 20 m, always in obscure places (Tab. 2 & Fig. 1). Pulitzer-Finali (1978) reports *H. cratera* from a depth of 2-3 m to 70 m (Bay of Naples). It is a common species in the Mediterranean.

Distribution. Mediterranean, Cape Verde Islands and West Africa.

Haliclona citrina (Topsent, 1892)

Reniera citrina Topsent, 1892 : 19

Haliclona citrina Griessinger, 1971 : 154, Fig. 11e.

Material. Banyuls : ZMA POR 5473, cap Oullestrell, 11 m, cavity in overhanging cliff ; 5474, cap du Troc, 13 m, overhanging cliff ; 5475, île Grosse, 14 m, overhanging cliff ; 5476, île Grosse, 14 m, in cavity ; 5477, cap Oullestrell, 16m, on wall ; 5478, île Grosse (right), 15m, in cavity.

Description.

Shape and size : this species has an encrusting base with an average thickness of 0.4 cm. Thick-walled tubes vary in height from 0.5 to 5 cm. The apical oscules can reach a diameter of 0.9 cm.

Consistency : limp, soft and compressible. Very "spongy" and fragile. The surface is smooth and even.

Colour : lemon-yellow.

Ectosome : unispicular, isodictyal reticulation.

Choanosome : unispicular, isodictyal reticulation, which develops at many places into unispicular primary lines. A small amount of spongin is present at the nodes.

Spicules : Oxea : regular in size, slightly curved with tapering points. 114-135 ± 9-160/5 µm.

Remarks. Found only above approximately 20 m. Common. Topsent (1892) reported this species from the region of Banyuls, while Boury-Esnault (1971) found it in the same region at Grotte du Troc, Castelloussous, île Grosse and Sphinx at a depth of 15 m, (Tab. 2 & Fig. 1). This species has never been found at another locality in the Mediterranean, but Boury-Esnault & Lopes (1985) have recently reported it from the Azores (at a depth of 9 m).

Distribution. Mediterranean, Azores.

Haliclona mediterranea Griessinger, 1971

Haliclona mediterranea Griessinger, 1971 : 153, Fig. 11a, Pl. III, Fig. 1.

Material. Banyuls : ZMA POR 5479, cap du Troc, 23 m, on ceiling in cavity ; 5480, cap l'Abeille, 25 m, in cavity ; 5481, Canadell, 25 m, in cavity ; 5482, cap Réderis, 27 m, in cavity.

Description.

Shape and size : the specimens have a thinly encrusting base of 0.2 cm, with tubes rising up to a height of 3 cm. The oscules are apical and have a diameter up to 0.5 cm.

Consistency : soft, limp and compressible. Very fragile. The surface is smooth and even, the exhalant orifices are very obvious.

Colour : rosy, lilac, dark red.

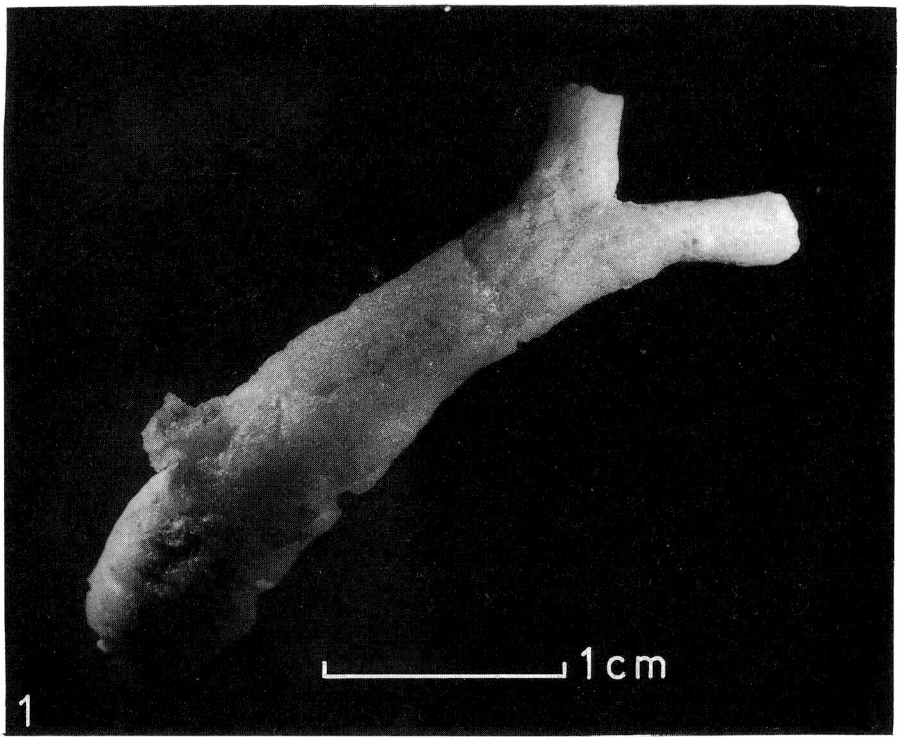

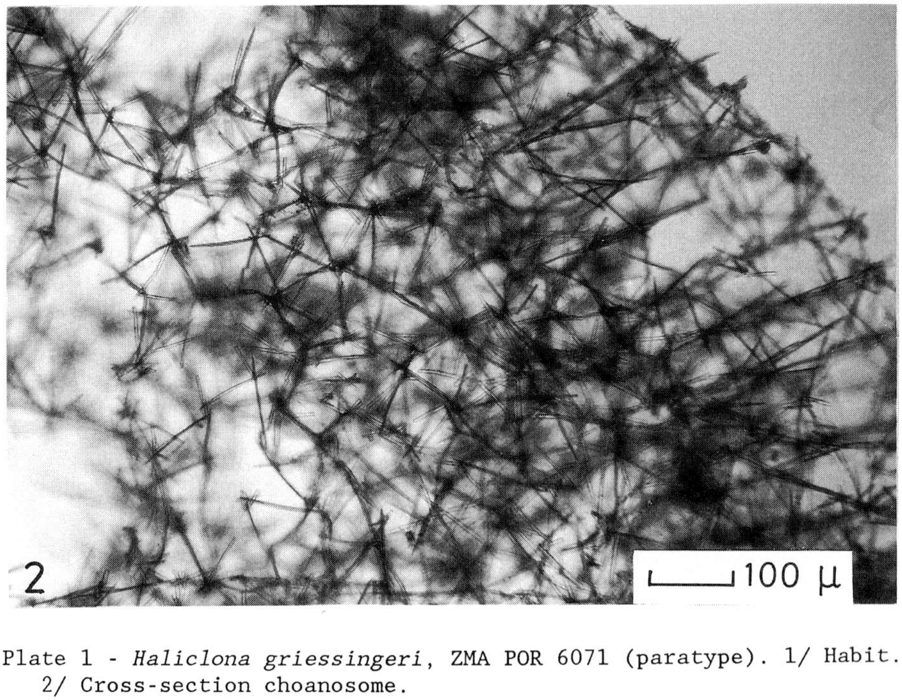

Plate 1 - *Haliclona griessingeri*, ZMA POR 6071 (paratype). 1/ Habit. 2/ Cross-section choanosome.

Ectosome : unispicular, isodictyal reticulation.
Choanosome : unispicular, isodictyal reticulation, with nodal spongin.
Spicules : Oxea : Small, slim and slightly curved, with tapering points. They are very regular. 87-96 ± 4-103/4 μm.

Remarks. Found below approximately 20 m. Always in obscure places. Common. This is the first record for the region of Banyuls. Boury-Esnault & Lopes (1985) have found *H. mediterranea* at a depth of 10 and 12 m at the Azores and Pulitzer-Finali (1978) recorded this species at 45 m (Bay of Naples).

Distribution. Mediterranean, Azores.

Haliclona griessingeri **spec. nov.**

Haliclona sp. Griessinger, 1971 : 157

Material. Banyuls : ZMA POR 6070 (holotype), cap Béar, 17 m, near opening of cavity ; 6071, 6073 (paratypes), cap Béar, 17 m, in opening of cavity ; 6072 (paratype), cap Béar, 15 m, in cavity.

Description. (Pl. 1 Figs 1, 2 ; Fig. 2a, b).
Shape and size : ZMA POR 6070 (holotype) is a thin (to 0.25 cm) incrustation on the bryozoan *Pentapora*, approximately 4 cm in length and up to 1 cm in width. It has few, small (1 to 1.5 cm in diameter), flush oscules, hardly visible and irregulary scattered. The exhalant system is very conspicuous. POR 6072 (one of the paratypes) is thin (up to 0.2 cm), encrusting a small rock (debris), of 2.5 to 3.5 cm in extension.
Consistency : rather firm, but fragile. The surface is smooth, in some places slightly punctate.
Colour : cream-white, greenish.
Ectosome : absent.
Choanosome : a regular, sub-isotropic reticulation, with unispicular primary and secondary lines. Nodal spongin (Fig. 2a).
Spicules : Oxea : slightly curved (Fig. 2b). 79-121 ± 9-144/3.5 μm.

Remarks. Always found at cap Béar. This is the first record for the region of Banyuls.

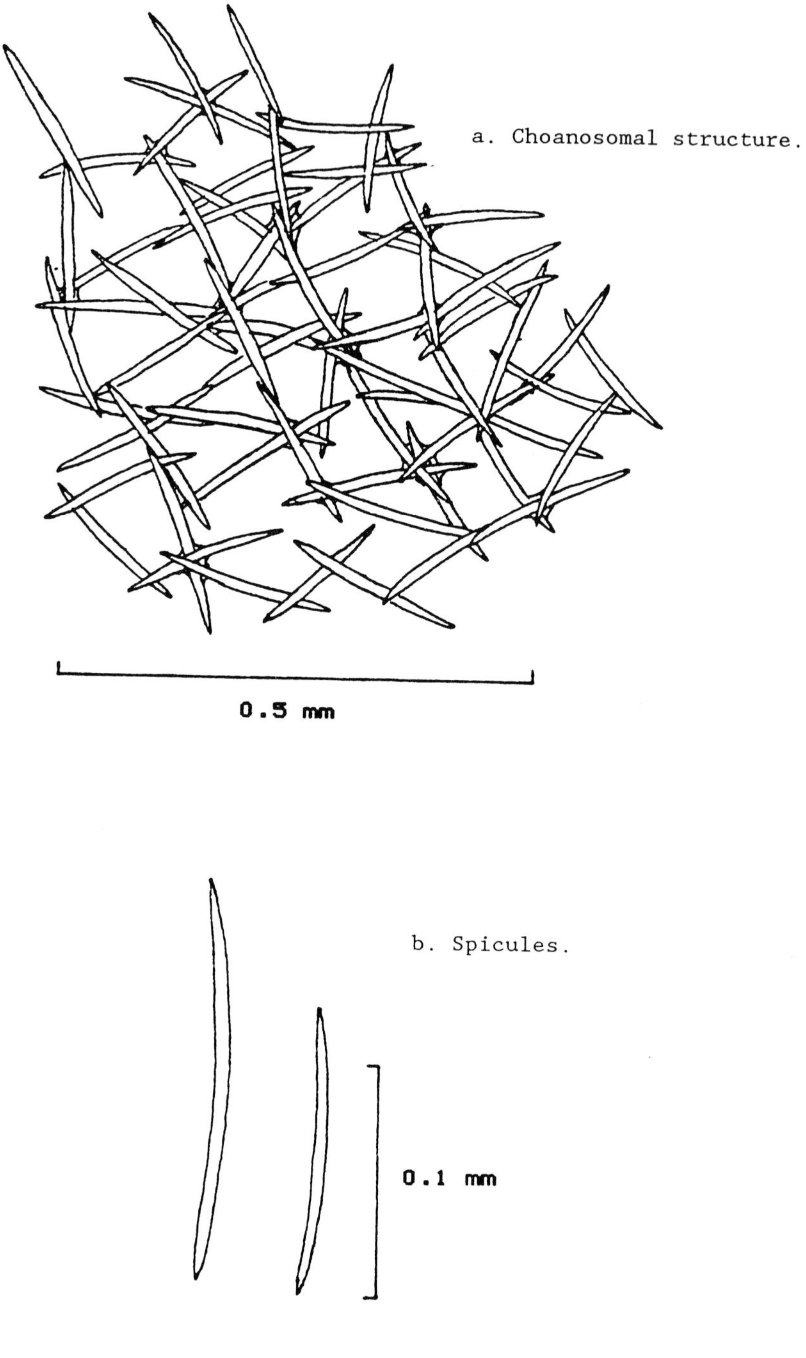

Figure 2 - *Haliclona griessingeri*, choanosomal structure and spicules.

Distribution. Mediterranean.

Discussion. This species is very similar to the one described by Griessinger (1971). His description of habit and skeleton coincides closely with the above described specimens. One of the specimens reported by Griessinger (1971) covered a sponge (*Ircinia*) and was found at a depth of 40 m, while the other was found at 106 m. In habit, the new species may be quite similar to *H. valliculata*, but it differs by possessing a unispicular skeleton and smaller oxea. In skeletal architecture it is hardly distinguishable from *H. mediterranea* and *H. citrina*. The three species are quite certainly closely related but smaller spicules, colour, habit and absence of the conspicuous exhalant canals at the surface in the latter two species are sufficiently discriminating characters.

Haliclona mamillata **(Griessinger, 1971) (n. comb.)**

Reniera mamillata Griessinger, 1971 : 132, Fig. 5b, 6c, j, Pl. 1 Fig.3.

Material. Banyuls : ZMA POR 5460, cap l'Abeille, 32 m. Marseille : 5461, île Plane, 25 m, in cave.

Description.

Shape and size : both specimens consist of a thickly encrusting base, 7.5 cm across, with several mammiform and vulcano-shaped, thick-walled oscules. These have an average height of 2.5 cm. The inner walls of the oscules (0.2 to 0.5 cm in diameter) are pierced with exhalant holes, often arranged in circles.

Consistency : moderately soft and brittle, compressible. A little mucus is secreted, when broken or squeezed. The surface is smooth and even.

Colour : rose-orange, with yellowish tinges at the base.

Ectosome : irregular reticulation.

Choanosome : rather irregular reticulation of paucispicular primary lines and single, interconnecting spicules, with many choanosomal spaces. Spongin at the nodes.

Spicules : Oxea : with tapering, short points. 129-153 ± 18-182/7 µm.

Remarks. Presumably only present below approximately 20 m. Not common. This is the first record for the region of Banyuls. Griessinger (1971) reports *H.*

mamillata from a depth of 52 m (Aegean Sea).

Distribution. Mediterranean.

Discussion. Specimen ZMA POR 5460 has spicules with a mean length of 169 ± 8 µm and specimen POR 5461 has spicules with a mean length of 138 ± 7 µm. Both lengths are within the range given by Griessinger (1971) for the same species. Under water *H. mammillata* is easily mistaken for *H. cratera*, because form, colour and consistency can be very similar.

Haliclona mucosa (Griessinger, 1971) (n. comb.)

Reniera mucosa Griessinger, 1971 : 140, Fig.8b, 9a, Pl.1 Fig. 5, Pl.3 Fig.4.

Material. Banyuls : ZMA POR 5441, île Grosse, 14 m, in cavity ; 5442, île Grosse, 14 m, in cavity ; 5443, cap l'Abeille, 25 m, in cavity ; 5444, cap du Troc, 20 m, in cavity ; 5445, cap l'Abeille, 23 m, in cavity ; 5446, cap l'Abeille, 26 m, in cavity ; 5447, cap l'Abeille, 32 m, in cavity. Marseille : 5448, île Plane, 25 m, in cave.

Description.

Shape and size : the shape of the specimens varies from thin incrustations to thick lumps up to 3 cm. The oscules (0.1 to 0.3 cm in diameter) are frequently elevated with a maximum height of 0.5 cm. Long, thin processes are often present.

Consistency : rather firm, but very brittle. Extremely mucous. The surface is very irregular.

Colour : varying from cream, yellowish to brownish. After preservation the colour is a shade of brown.

Ectosome : irregular reticulation.

Choanosome : paucispicular primary lines, with mostly oblique, interconnecting spicules. Spongin scarce.

Spicules : Oxea : often curved with tapering points. 220-247 ± 14-270/7µm.

Remarks. Always found in obscure places, to a depth of 35 m. Very common. First record for the region of Banyuls.

Distribution. Mediterranean.

Haliclona sarai **(Pulitzer-Finali, 1969) (n. comb.)**

Reniera viscosa Sarà, 1961 : 50 ; not : Topsent, 1888 : 149.

Reniera sarai Pulitzer-Finali, 1969 : 97 ; Griessinger, 1971 : 136, Fig. 6i, l, m, Pl. 1 Fig. 6, Pl. 3 Fig. 3.

Material. Banyuls : ZMA POR 5449, île Grosse, 14m, in cavity ; 5450, cap Oullestrell, 25 m, in cavity ; 5451, cap du Troc, 20 m, in cavity ; 5452, île Grosse, 15 m, in cavity ; 5453, cap Béar, 17 m, in cavity ; 5454, cap l'Abeille, 32 m, in cavity ; 5455, cap l'Abeille, 23 m, in cavity ; 5456, cap l'Abeille, 32 m, in cavity ; 5457, cap Oullestrell, 16 m, in cavity.

Description.

Shape and size : the specimens manifest a massive shape, often large irregular lumps (up to 10 cm across) with a maximum thickness of 3.5 cm. Some sponges (POR 5450 and POR 5455) have thick-walled tubes (up to 0.6 cm), which reach a height of 4 cm. The oscules are flush, elevated or if tubes are present, apical. They have a diameter of 0.2 to 0.7 cm.

Consitency : compact, firm, inelastic and brittle. The surface varies from smooth to very irregular. It sticks to the fingers if touched. When broken, some specimens secreted mucus.

Colour : cream, (dark)grey. The interior was often full of orange spots.

Ectosome : a rather dense, irregular reticulation, with many spicules interdispersed.

Choanosome : pauci- to multispicular primary lines. Many interconnecting spicules in confusion. No spongin.

Spicules : Oxea : slightly curved. They are sometimes asymmetric and have long tapering points. 152-170 ± 10-194/5 µm.

Remarks. Commonly found to a depth of 35 m, always in obscure places. In the region of Banyuls, Boury-Esnault (1971) has reported this species from Lloses, Sphinx and the Grotte du Troc, at a depth between 13 and 30 m (tab.2 & Fig. 1). Griessinger (1971) reports a finding depth of 150 m for *H. sarai* (Costa Brava, Spain). It is a common species in the Mediterranean.

Distribution. Mediterranean.

Haliclona valliculata **(Griessinger, 1971) (n. comb.)**

Reniera valliculata Griessinger, 1971 : 134, Fig. 5d, 6e, Pl. 1 Fig. 4, Pl. 3 Fig. 2.

Material. Banyuls : ZMA POR 5458, cap du Troc, 20 m, in cavity ; 5459, cap Béar, 17 m, in cavity.

Description.

Shape and size : ZMA POR 5458 is a small, thin crust (0.3 cm) with an oscule of 0.3 cm in diameter. POR 5459 is a rather massive lump (1.5 cm across) with the aquiferous canals clearly visible. These canals rise above the surface. It has oscules of 0.1 to 0.3 cm in diameter.

Consistency : rather firm, but fragile, brittle. A little mucus was secreted when broken. The surface is very irregular.

Colour : ZMA POR 5458 is pale cream, while POR 5459 is yellowish verging to light-brown.

Ectosome : an irregular reticulation with many loose spicules.

Choanosome : paucispicular primary lines, which are characteristically winding. Transverse interconnecting spicules. There is a small amount of spongin at the nodes.

Spicules : Oxea : slim and rather curved, sometimes even flexuous, with long, tapering points. Centrotylotes are quite common. 190-226 ± 28-281/ 6 µm.

Remarks. Not common, only found twice. This is the first recorded observation for the region of Banyuls. Pulitzer-Finali (1978) records this species at 3-5 m depth.

Distribution. Mediterranean.

Discussion. The two specimens demonstrate a difference in spicule length. The oxea of ZMA POR 5458 have a mean length of 203 ± 16 µm, while the oxea of POR 5459 are 249 ± 15 µm. Griessinger (1971) gives a maximum mean length of 192.5 µm. Furthermore, nearly all the spicules of POR 5458 are centrotylotes, while the majority of the spicules for POR 5459 are simple oxea. Nevertheless, all habit and skeletal structures taken into account, it is quite certain that both POR 5458 and POR 5459 conform to the same species. Thus, a considerable variation in oxea between specimens is possible.

Haliclona implexa **(Schmidt, 1868) (n. comb.)**

Reniera implexa Schmidt, 1868 : 27 ; Griessinger, 1971 : 133.

Material. Banyuls : ZMA POR 5484, cap l'Abeille, 25 m, in cavity ; 5487, cap l'Abeille, 23 m, in cavity ; 5490, cap l'Abeille, 30 m, in cavity ; 5491, cap l'Abeille, 32 m.

Description.

Shape and size : ZMA POR 5487 is a tube, 1.5 cm in height, with apical oscules of 0.2 to 0.3 cm in diameter. POR 5484, 5490 and 5491 are erect, irregular forms to a height of 2 cm, with apical oscules which have a diameter of 0.2 to 0.5 cm. Small, thin processes are often present.

Consistency : Soft and compressible. Very fragile. The surface is smooth and even and the exhalant orifices can be clearly seen. When squeezed, a very small amount of mucus was secreted.

Colour : Rose, purple-red (violet), white-cream or brownish. Sometimes white-cream with violet and/or rose tinges or a mixture of violet and rose.

Ectosome : semi-dense, irregular, unispicular reticulation.

Choanosome : semi-dense, irregular reticulation, with ill-defined pauci-spicular primary lines and many single, confused interconnecting spicules. The skeleton is reinforced by irregular multispicular fibres. A small amount of spongin is present at the nodes.

Spicules : Oxea : regular and slightly curved. They have tapering points. 122-139 ± 8-171/5.5 µm.

Remarks. They are found below 20 m, always in obscure places. Common. First record for the region of Banyuls. It is a common species in the Mediterranean (Griessinger, 1971).

Distribution. Mediterranean, Canary Islands, Madeira Archipelago, Azores.

Discussion. The specimens were not stalked, but have irregular growth forms with many proliferations. Comparisons made with a specimen of *H. implexa* (stalked) collected from the Azores, revealed no essential differences.

Haliclona fulva (Topsent, 1893) (n. comb.)

Reniera fulva Topsent, 1893 : XXXIX ; Griessinger, 1971 : 138, Fig. 7, 8al, Pl. 2 Fig. 3.

Material. Banyuls : ZMA POR 5430, cap du Troc, 23 m, in cavity ; 5431, île Grosse, 14 m, overhanging cliff ; 5432, cap Oullestrell, 16 m, in cavity ; 5433, cap l'Abeille, 25 m, in cavity (ceiling) ; 5434, île Grosse (right), 15m, in cavity ; 5435, cap Oullestrell, 17 m, in cavity ; 5437, cap Oullestrell, 15 m, in cavity. Marseille : 5436, île Plane, 25 m, on wall of cave .

Description.

Shape and size : all the specimens found are incrustations (0.2 to 1.5 cm thick) with oscules of 0.2 to 0.4 cm in diameter. These oscules can be elevated up to a height of 1 cm.

Consistency : they are very firm ; only the specimen found in the region of Marseille, ZMA POR 5436, was spongy and much softer. All of them are brittle. The surface is smooth and slightly hispid.

Colour : bright orange.

Ectosome : detachable, possessing a rather semi-dense, sub-isotropic reticulation of spicules, with little nodal spongin.

Choanosome : paucispicular primary lines, often confused because of dense spicules. Spongin (at the nodes) may be rather developed.

Spicules : Oxea : sometimes strongly curved or sinuous. Shape and size are rather variable. They have long, tapering points. 232-276 ± 23-312/7 μm.

Remarks. The sponges are always found in dark places, often covering a large area. Very common, but not found deeper than 25 m. In the region of Banyuls, Topsent (1893) has found *H. fulva* at cap l'Abeille, while Boury-Esnault (1971) reported this species from île Grosse and Sphinx, between 10 and 15 m (Tab. 2 & Fig. 1). Pulitzer-Finali (1978) reported it from a depth of 70 m (Bay of Naples). It is a common species in the Mediterranean.

Distribution. Mediterranean, Canary Islands, Madeira Archipelago.

Haliclona magna (Vacelet, 1969) (n. comb.)

Reniera magna Vacelet, 1969 : 211
Pellina magna ; Griessinger, 1971 : 148, Fig. 10b.

Material. Banyuls : ZMA POR 5467, facing cap Creus, 280 m ; 5468, facing cap Creus, 285 m.

Description.
Shape and size : both specimens are large, broken-off pieces (the largest piece is 12x6x5 cm). The inner wall of the oscule (1.5 to 3.5 cm in diameter) is pierced by large holes, which are 0.5 to 1 cm in diameter. The aquiferous system is clearly visible and consists of many canals, up to 0.5 cm in diameter.
Consistency : firm, rather brittle. The surface is smooth, but rough to the touch.
Colour : white, cream.
Ectosome : detachable. A sub-isotropic, somewhat confused reticulation.
Choanosome : a sub-isotropic reticulation, in places consisting of paucispicular primary lines. Many spicules in confusion. Very little nodal spongin.
Spicules : Oxea : very regular and abruptly pointed. 171-198 ± 8-209/7.5 μm.

Remarks. First record for the region of Banyuls. Together with the sponges, coral fragments were pulled up. Vacelet (1969) notes that, at approximately 150m, *H. magna* has been observed often from a small submarine. It may be considered as a species that occurs only in deeper water (below 130 m).

Distribution. Mediterranean.

Haliclona fibulata (Schmidt, 1862)

Reniera fibulata Schmidt, 1862 : 73
Gellius fibulatus ; Topsent, 1890 : 201
Haliclona fibulata ; de Weerdt, 1986 : 137

Material. Banyuls : ZMA POR 5469, île Grosse, 14 m, in cavity.

Description.
Shape and size : the specimen is a cushion-shaped incrustation, 0.5 cm

thick. One flush oscule, 0.2 cm in diameter, is present.
Consistency : rather soft, brittle and very fragile. The surface has the characteristic reticulated appearance.
Colour : rose.
Ectosome : a confused mass of oxea and sigmata.
Choanosome : paucispicular lines with interconnecting spicules, all very dense and confused. Spongin at the nodes.
Spicules : the megascleres are slightly curved oxea, with long points. 186 ± 5/7.5 µm. The microscleres are thin sigmata (16 µm), which are abundant.

Remarks. Only found once ; however Boury-Esnault (1971) noted that, in the region of Banyuls, *H. fibulata* had often been found at all stations visited to a depth of 30 m (Tab.2 & Fig.1). Earlier, Topsent (1892b) had already noted the species in the same region (Tab. 2).

Distribution. Mediterranean, Canary Islands, North Atlantic.

Discussion. There is no difference between this specimen and a specimen from Lough Ine, S.W. Ireland, which is present in the ZMA collection (ZMA POR 5520). The oxea are of the same shape and in the same size category. The habit is identical. The sigmata of the Mediterranean specimen are slightly smaller (16 µm) than those of the S.W. Ireland specimen (19 µm), the shape is the same.

Haliclona lacazei Topsent, 1893 (n. comb.)

Gellius lacazei Topsent, 1893 : XXXV.

Material. Banyuls : ZMA POR 5470, cap l'Abeille, 23 m, ceiling in very dark cavity ; 5471, cap l'Abeille, 32 m, in cavity ; 5472, cap l'Abeille, 26 m, in cavity.

Description.
Shape and size : the specimens are irregular incrustations, up to 1 cm thick, with rounded edges. They have thick-walled oscules, up to 0.7 cm in diameter, which are elevated to a height of 1.5 cm, but flush oscules are also present.
Consistency : variable, ZMA POR 5471 was rather soft, while POR 5470 and POR

5472 were rather firm. The surface is slightly rough and fine-grained.
Colour : cream, yellowish.
Ectosome : an irregular, sub-isotropic reticulation.
Choanosome : irregular reticulation of uni- to paucispicular primary lines, with confused interconnecting spicules.
Spicules : the megascleres are large, robust, curved oxea. 369-394 ± 13-418/12 µm. The microscleres are thin raphides with an average length of 70 µm, not organized in dragmata and very abundant.

Remarks. All the specimens were found in one area (coralligene), below 20 m. Not common. Boury-Esnault (1971) has found this species in the region of Banyuls, at Grotte de Béar and Castelloussous, between 3 and 5 m (Tab. 2 & Fig. 1). Topsent (1893) found one specimen at cap l'Abeille, at a depth of 40 m (Tab. 2 & Fig. 1).

Distribution. Mediterranean, Cap Verde Islands, West Africa.

Genus *Dendroxea* Griessinger, 1971

Dendroxea lenis **(Topsent, 1892)**

Reniera lenis Topsent, 1892 : XIX
Dendroxea lenis, Griessinger, 1971 : 152

Material. Banyuls : ZMA POR 5462, cap du Troc, 19 m, in cavity.

Description.
Shape and size : the specimens are thin incrustations, up to a thickness of 0.25 cm. The oscules are indistinct (very small).
Consistency : soft. When broken a small amount of mucus is secreted. The surface is irregular and sticks to the fingers when touched.
Colour : grey-white.
Ectosome : no special dermal skeleton, only a membrane without spicules is present.
Choanosome : branching, multi-spicular tracts, which rise up from a common, basal, densely reticulated mass of spicules. The tracts are connected by transverse single spicules, but also groups of loose, disoriented spicules are observed. Very little nodal spongin.

Spicules : Oxea : thin and often slightly curved. Tapering points. 118-135 ± 12-163/3.5 µm.

Remarks. Found in obscure places. Not common. Topsent (1892) has found this species in the region of Banyuls at cap Béar and cap l'Abeille (Tab.2 & Fig.1).

Distribution. Mediterranean, Canary Islands, Azores.

Family PETROSIIDAE van Soest, 1980

Genus *Petrosia* Vosmaer, 1885

Petrosia ficiformis **(Poiret, 1789)**

Spongia ficiformis Poiret, 1789 : 61

Material. Banyuls : ZMA POR 5463, île Grosse, 14 m, in very dark cavity ; 5464, cap l'Abeille, 32 m, in cavity ; 5465, cap l'Abeille, in cavity ; 5466, cap l'Abeille, 29 m, in cavity.

Description.

Shape and size : the shape of the specimens varies considerably. ZMA POR 5463 is a knoll of 2.5 cm in diameter and 1.5 cm in height, whereas POR 5465 and POR 5466 are ramose, rampant forms up to 2 cm in diameter. POR 5464 resembles a straight pole rising up from the surface (approximately 4 cm high and 2 cm in diamater), thickened at the summit, with an apical oscule. The oscules, 0.1 to 0.4 cm in diameter, are flush or whith a slightly raised, thin rim.

Consistency : very firm. The surface appears smooth, but is rough to the touch.

Colour : cream, white. Sometimes (partly) violet.

Ectosome : detachable. A dense reticulation of different sized spicules, with rounded meshes. Many loose spicules in confusion.

Choanosome : dense, multispicular tracts. They form an irregular reticulation within the coarser meshes. Loosely scattered spicules are abundant.

Spicules : Oxea and strongyles : in ZMA POR 5463 and POR 5465 only oxea were present. They vary in shape, category and size. The smallest spicules were generally found in the ectosome.

Remarks. They are found to a depth of approximately 35 m, and are very common. Boury-Esnault (1971) records this species from nearly all her stations in the region of Banyuls (grotte de Béar, grotte du cap Doume, Sphinx, île Grosse, Castelloussous, Lloses) to a depth of 25 m (Tab.2 & Fig.1).

Distribution. Mediterranean, Canary Islands, Cap Verde Islands, West Africa, Azores.

Discussion. Topsent (1928) suggested that *P. ficiformis* is conspecific with *P. crassa* Carter, 1876. Although the species are very similar, it is rather certain that they represent two species with a different geographical distribution (de Weerdt, 1985).

GENERAL DISCUSSION

Of the approximately 51 Haplosclerid species (genera : *Haliclona, Dendroxea, Rhizoniera, Petrosia, Acervochalina, Siphonochalina, Callyspongia, Calyx*) known from the Mediterranean (Griessinger, 1971 ; Pulitzer-Finali, 1983) approximately 29 species are reported from the region of Banyuls (Topsent, 1892b, 1893, 1925a & b ; Boury-Esnault, 1971 ; present study) (Tab. 2).

During this research 15 species were found, one of which is a new species and six species are recorded for the first time from this region (Tab. 2). Of the 13 species, which have not been found during the present study, nine (*H. simulans, H. varia, H. fistulosa, H. parietalis, H. cinerea, H. plana, H. flavescens, H. arenata, H. aquaeductus*) have been recorded before 1925 (Topsent, 1892b, 1893, 1925a & b). Of the species allegedly shared with the N.E. Atlantic, only *H. fibulata* has been found in Banyuls. As comparative studies revealed (de Weerdt & Stone, personal communication), *H. rosea* is not found in the Mediterranean. The same very likely applies to *H. indistincta* and *H. viscosa*. Both are described from the Mediterranean by Topsent & Olivier(1943), but they were never again reported from that region, which makes their occurence doubtful.

H. simulans and *H. cinerea* have been found earlier in the region of Banyuls by Topsent (1892b, 1925a & b). Neither Boury-Esnault (1971) nor present researches (in this study) have found these species afterwards (Tab.2). The descriptions of Topsent (who knew both species from the N.W. coasts of

France), Vacelet (1969), Griessinger (1971) and Pulitzer-Finali (1978) for specimens from the Mediterranean give no indication that they are wrongly named and so they presumably have a distribution as far as the southern border of the boreal province (Briggs, 1974). With respect to *H. simulans* this is also established by *Siphonochalina crassa* Topsent (1925), described from Naples, which is conspecific with *H. simulans* (de Weerdt, 1986).

The same presumed distribution is found in *Acervochalina limbata* (Topsent, 1892b, 1925a ; Topsent & Olivier, 1943 ; Griessinger, 1971 ; Pulitzer-Finali, 1978), *H. fistulosa* (Topsent, 1892b ; Griessinger, 1971 ; Pulitzer-Finali, 1983), *H. fibulata* (Topsent, 1892 ; Topsent & Olivier, 1943 ; Pulitzer-Finali, 1978, 1983) and *H. angulata* (Topsent, 1925 b ; Boury-Esnault, 1971 ; Pulitzer-Finali, 1978, 1983). Only *H. fibulata* and *H. angulata* have been previously found in the region of Banyuls (Tab. 2). *Gellius luridus* sensu Topsent, 1928 is very likely synonymous with *H. angulata*, while *G. luridus* sensu Boury-Esnault,1971 is *Gellius marismedi* Pulitzer-Finali, 1978 (de Weerdt & van Soest, 1986). *Gellius luridus* Lundbeck, 1902 is very probably an *Oceanapia*, occuring only in Arctic waters.

It is not quite clear whether or not *H. mediterranea* and *H. implexa* were previously observed in the region of Banyuls. In 1943, Topsent & Olivier describes *H. aquaeductus* from the region of Monaco and suggests it is synonymous with *H. rosea*. He mentioned an entirely unispicular skeleton. Griessinger (1971), confirmed by de Weerdt & Stone (personal communication), stated that the synonymy of *H. aquaeductus* and *H. rosea*, suggested by Topsent & Olivier (1943), is not valid and that the *H. rosea* described from the Bay of Naples (1925b) is in fact *H. mediterranea*. Topsent & Olivier remarked that *H. aquaeductus* has often been found in the region of Banyuls. Because of the resemblance of *H. aquaeductus* to *H. mediterranea*, it is likely that the *H. aquaeductus* reported from Banyuls, was also a *H. mediterranea*. Another possibility is that *H. aquaeductus* found in Banyuls was mixed, with *H. implexa*. Both species are quite similar in surface structure. *H. implexa* usually occurs rather deep. Griessinger (1971) mentions a depth of more than 40 m, while Pulitzer-Finali (1978, 1983), Topsent (1904, Azores) and de Weerdt & van Soest (1986, Azores, Madeira Archipelago, Canary Islands) are noting depths between 65 and 320 m. Topsent reports *H. aquaeductus* from Banyuls as occuring in rather deep open sea.

Most of the Chalinid species occuring in the Mediterranean have a distribution in the southern neighbouring areas (Madeira Archipelago, Canary Islands, Cape Verde Islands, Azores, West African coast). The Mediterranean has a distinct sponge fauna, with little relation to the fauna of the boreal province of the Atlantic. The same accounts for the fauna as a whole (Ekman, 1953).

ACKNOWLEDGEMENTS

Our sincere thanks are due to the sailors of the "Laboratoire Arago" and the divers of the C.N.R.S. for their helpful assistance in the field, as well as the management, who provided accomodation. Also we want to thank Dr J. Vacelet who gave access to his collection and arranged accomodation and diving in Marseille.

Dr R.W.M. van Soest is thanked for his suggestions and critical reading of this report and Drs Wapstra and Drs E. Verheij for their assistance in the field at Roscoff and Marseille.

The commissie "Buitenlandse Marien-Biologische Stations", of the Koninklijke Nederlandse Academie van Wetenschappen, supported the first author with a grant.

LITERATURE CITED

Boury-Esnault N., 1968 - Les spongiaires des fonds rocheux de Banyuls-sur-mer. Ecologie et systématique. *Th. 3e cycle : Océanogr. biol., Paris.*

Boury-Esnault N., 1971 - Les spongiaires de la zone rocheuse littorale de Banyuls-sur-mer. 2. Systématique. *Vie Milieu* 22 : 287-350.

Boury-Esnault N. & Lopes M.T., 1985 - Les démosponges littorales de l'archipel des Açores. *Ann. Inst. océanogr.*, 61(2) : 149-225.

Bowerbank J.S., 1866 - A monograph of the British Spongiadae. Ray Society, London, 2 : 1-388.

Briggs J.C., 1974 - Marine Zoogeography. McGraw-Hill, New York, 1-475.

Cruz T. & Bacallado J.J., 1983 - Contribución al conocimiento de los espongiarios de las Islas Canarias. I. Demosponjas, Homosclerophorida y Astrophorida del litoral de Tenerife. *Bol. Inst. Oceanogr.*, 6 : 76-87.

Ekman, S., 1953 - Zoogeography of the sea. Sidgwick and Jackson, London.

Grant, R.E., 1835 - Animal Kingdom : pp 107-118 in , *The encyclopedia of the anatomy and physiology, 1.* Todd, Robert Bentley (ed.), Sherwood, Gilbert & Piper, London.

Griessinger J.M., 1971 - Etude des Réniérides de Méditerranée (Démosponges, Haplosclerida). *Bull. Mus. natl Hist. nat.*, 3(3) (Zoologie 3) : 97-182.

Pulitzer-Finali G., 1978 - Report on a collection of sponges from the Bay of Naples. III. Hadromerida, Axinellida, Poecilosclerida, Halichondrida, Haplosclerida. *Boll. Ist. Biol. Univ. Genova*, 45 : 7-89.

Pulitzer-Finali G., 1983 - A collection of Mediterranean Demospongiae (Porifera) with, in appendix, a list of Demospongiae hitherto recorded from the Mediterranean Sea. *Ann. Mus. Civ. Storia nat. Genova*, 84 : 445-661.

Sarà M., 1971-1972 - Guida ai Poriferi della Fauna Italiana. *Quad. Civ. Staz. Idrobiol. Milano*, 3-4 : 53-97.

Schmidt O., 1892 - Die Spongien des Adriatischen Meeres. Engelmann, Leipzig, 1-88.

Topsent E., 1888 - Contribution à l'étude des Clionides. *Archs Zool. exp. gén.*, 2(5) : 1-166.

Topsent E., 1890 - Eponges de la Manche. *Mém. Soc. zool. France*, 3(1) : 195-205.

Topsent E., 1892a - Contribution à l'étude des Spongiaires de l'Atlantique Nord. *Rés. Camp. sci. Prince Monaco*, 2 : 1-165.

Topsent E., 1892b - Diagnoses d'éponges nouvelles de la Méditerranée et plus particulièrement de Banyuls. *Archs Zool. exp. gén.*, 2(10), Notes & Revue, XVIII-XXVIII.

Topsent E., 1893 - Nouvelles séries de diagnoses d'éponges de Roscoff et de Banyuls. *Archs Zool. exp. gén. sér. 3(1)*, Notes & Revue, XXXIII-XLIII.

Topsent E., 1904 - Spongiaires des Açores. *Rés. Camp. sci. Prince Monaco*, 25 : 1-280.

Topsent E., 1925a - Eponges de l'étang de Thau. *Bull. Inst. océanogr. Monaco*, 452 : 1-19.

Topsent E., 1925b - Etude des Spongiaires du golfe de Naples. *Archs zool. exp. gén.*, 63(5) : 623-725.

Topsent E., 1928 - Spongiaires de l'Atlantique et de la Méditerranée, provenant des croisières du Prince Albert 1er de Monaco. *Rés. Camp. Sci. Prince Monaco*, 74 : 1-373.

Topsent E. & Olivier L., 1943 - Eponges observées dans les parages de Monaco. *Bull. Inst. océanogr. Monaco*, 854 : 1-12.

Vacelet J., 1969 - Eponges de la Roche du Large et de l'Etage bathyal de Méditerranée (récoltes de la soucoupe plongeante Cousteau et dragages). *Mém. Mus. natl Hist. nat., A Zoologie*, 59(2) : 1-75.

Weerdt W.H., de, 1986 - A systematic revision of the north-eastern Atlantic shallow-water Haplosclerida (porifera, Demospongiae) part II : Chalinidae. *Beaufortia*, 36 : 81-165.

Weerdt W.H., de, & van Soest R.W.M., 1986 - Marine shallow-water Haplosclerida (Porifera) from the south-eastern part of the North Atlantic Ocean. *Zool. Verh. Leiden*, 225 : 1-49.

LITTORAL DEMOSPONGES FROM THE BANKS OF THE STRAIT OF SICILY AND THE ALBORAN SEA.

M. PANSINI

Istituto di Zoologia dell'Università
Via Balbi 5, I-16126 Genova, Italy.

SYNOPSIS

The material of this study has been collected from the South-Western Mediterranean in the course of two expeditions by the Italian Research Vessel "Bannock". The primary collection method employed was SCUBA diving. The sampled area is affected by an important inflow of water from the Atlantic. Fifty-nine species have been studied. One of them, *Stylostichon equiosculatus*, is new to science. Some ecological considerations are also discussed.

INTRODUCTION

The littoral sponge fauna of the Strait of Sicily and of the Alboran Sea has never been the object of a systematic survey, even though part of the material collected by the "Calypso" expedition of 1958 has been partially studied (Griessinger, 1971). Therefore sponge samples have been taken between 0 and 80 m deep. The sampling has been made primarily by SCUBA diving during the course of two expeditions by the Italian Research Vessel "Bannock". The first, in 1981, surveyed parts of the Graham, Terribile and Pantelleria Vecchia banks off the South-Western Sicilian coast. The second expedition, in 1985, reached the Alboran Sea and the Spanish-Moroccan coast near Ceuta (Fig. 1).

Due to the difficult conditions that are normally encountered when working offshore, most of the dives were made in the proximity of the ship in the Strait of Sicily. More ample sampling was possible in the insular and coastal stations. Around the Island of Alboran the sponge population has been sampled from the surface to a depth of about 80 m by one dive along the coast, one dive off the Island and one haul on a neighbouring detritic

NATO ASI Series, Vol. G13
Taxonomy of Porifera
Edited by J. Vacelet and N. Boury-Esnault

bottom by a trawling net. The same technique was unsuccessful on the Spanish-Moroccan coast (Punta de la Almina) due to the presence of a strong current (4 knots) flowing eastward. Therefore a comparison between the collection stations could not be conducted.

METHODS

Most of the studied material was collected by means of SCUBA by the author. Whenever possible, specimens were photographed underwater using a Nikonos III camera with macrotube attachment and strobe. Therefore the observations "in vivo" refer to the sponges in the field and to their colour pictures. Some of the most common species where not preserved but directly identified in the field, so they are included in the species list only. Type material was incorporated in the collections of the Museo Civico di Storia Naturale of Genova.

LIST OF STATIONS

1. Island of Pantelleria, Punta Limarsi, 0-30 m, rock and boulders, SCUBA.
2. Graham Bank, 9-27 m, (position of the summit of the submerged volcano : Lat 37°10.2' N, Long 12°43' E), oval terrace of sand and volcanic scoriae, with ridges, grooves and small overhangs, SCUBA.
3. Pantelleria Vecchia Bank, one of the more prominent parts of Adventure Bank which is part of the continental shelf, 22 m, rock, SCUBA.
4. Terribile Bank, samples in the quadrat bordered by Lat 37°04'-37°18' N and Long 12°31'- 12°58' E, 42-46 m, irregular calcareous bottom, SCUBA.
5. Island of Alboran. One half hour haul by a rigid mouth trawling net. Initial position 35°52.4' N - 02°57.8' W, final position 35°59.4' N - 02°58.2' W, 70-80 m, detritic bottom with calcareous organic conglomerates.
6. Island of Alboran, south-east coast, 7-10 m, flat rocky bottom with boulders, scattered rocks, roofs, SCUBA.
7. 500 m east of Alboran, 32 m, irregular rocky bottom with *Laminaria ochroleuca*, SCUBA.
8. Punta de la Almina, 150 m off the point, 3-25 m, rocky shoal with irregular steps, SCUBA.

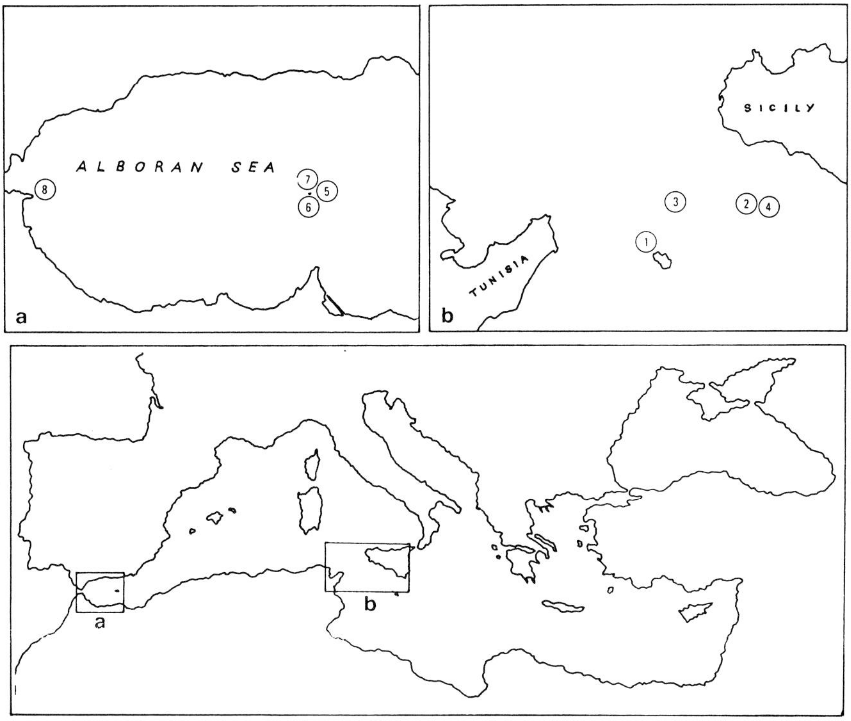

Figure 1 - Approximate position of the collecting localities.

LIST OF SPECIES RECORDED

Stelletta hispida (Buccich)
Stelletta mediterranea (Topsent)
Stryphnus mucronatus (Schmidt)
Erylus euastrum (Schmidt)
Dercitus plicatus (Schmidt)
Poecillastra compressa (Bowerbank)
Pseudosuberites hyalinus (Ridley & Dendy)
Pseudosuberites sulphureus (Bowerbank)
Myxilla rosacea (Lieberkühn)
Tedania anhelans (Lieberkühn)
Anchinoe coriaceus (Fristedt)
Anchinoe fictitius (Bowerbank)
Anchinoe mercator (Schmidt)
Anchinoe paupertas (Bowerbank)
Anchinoe tenacior Topsent
Stylostichon equiosculatus sp.n.
Acarnus tortilis Topsent

Suberites carnosus (Johnston)
Suberites domuncula (Olivi)
Polymastia mammillaris (Müller)
Spirastrella cunctatrix Schmidt
Cliona viridis (Schmidt)
Aaptos aaptos (Schmidt)
Chondrosia reniformis Nardo
Diplastrella bistellata (Schmidt)
Axinella damicornis (Esper)
Axinella polypoides Schmidt
Agelas oroides Schmidt
Batzella inops (Topsent)
Halichondria aurantiaca (Schmidt)
Dictyonella incisa (Schmidt)
Hemimycale columella (Bowerbank)
Hymeniacidon sanguinea (Grant)
Mycale syrinx (Schmidt)
Biemna variantia (Bowerbank)
Crambe crambe (Schmidt)
Crella elegans (Schmidt)
Crella pulvinar (Schmidt)
Pytheas rosea (Topsent)
Antho involvens (Schmidt)
Microciona assimilis (Topsent)
Haliclona subtilis Griessinger
Reniera cratera Schmidt
Reniera perlucida Griessinger
Reniera mamillata Griessinger
Reniera valliculata Griessinger
Reniera sp.
Siphonochalina coriacea Schmidt
Siphonochalina balearica Ferrer-Hernandez
Petrosia ficiformis (Poiret)
Pellina cf. semitubulosa (Lieberkühn)
Pellina sp.
Spongia officinalis Linné
Ircinia variabilis (Schmidt)
Ircinia foetida (Schmidt)
Dysidea avara (Schmidt)
Spongionella pulchella (Sowerby)
Aplysilla sulfurea Schulze
Hexadella racovitzai Topsent

ECOLOGICAL AND BIOGEOGRAPHICAL NOTES

The distribution of the sponge fauna is undoubtedly bound to the peculiar characteristics of the environment investigated, even if a precise relation cannot be found at the moment.

On the banks of the Strait of Sicily the most outstanding characteristics are the clarity of the water which allows the development of very rich photophilic flora at unusual depths (40 m and more), and the high intensity water-movement which affects the growth of many massive and erect sponges. Such effects are displayed in the growth of massive sponge species such as *Agelas oroides* which is found with an encrusting habit, while erect species such as *Axinella* sp., *Acanthella* sp., *Siphonochalina* sp., are absent or may be found deeper than usual in furrows, pits or sheltered positions on the bottom.

Another characteristic of the Strait of Sicily is the presence of water of Atlantic origin. The surface water flows eastward while the bottom water flows westward through the channel which cuts the ridge between the continental shelf of Sicily and Tunisia at a depth of about 400 m (Sverdrup *et al.*, 1942). This exchange is very similar to that taking place through the Strait of Gibraltar. Such hydrological conditions allow the development of the characteristic *Laminaria* beds whose presence is restricted to very peculiar areas in the Mediterranean. These beds are very dense in the Alboran Sea where both the insular and coastal stations are characterized by an increase in turbidity, by a drop in water temperature and salinity and by the presence of steady and rather strong, constant currents. In effect this Atlantic water flow is the most important shared character between two areas, the Alboran Sea and the Strait of Sicily. Though they appear rather different from other ecological conditions.

From a biogeographical point of view, the composition of the sponge fauna seems to be remarkably affected by this water exchange. This is supported by the distribution data. The most numerous group, counting 27 species, that is 47.3 % out of the total of 57 (not including the specimens identified to a generic level only), shows a Mediterranean-Atlantic distribution. Of these species, eight (*Stelleta hispida*, *Dercitus plicatus*, *Hemimycale columella*, *Crella pulvinar*, *Myxilla rosacea*, *Anchinoe mercator*, *Antho involvens*, *Spongia officinalis*) occur both to the North and to the South of Gibraltar, eight to the South and eleven to the North only. The distribution of these species in the Eastern-Atlantic is limited to the Atlanto-Mediterranean and to the Boreal Provinces. They do not occur on the Western-Atlantic coast or in the West Indian region. A few species of recent description, namely *Reniera perlucida* and *Reniera valliculata* have been reported only from the Azores and Canary Islands and Canary Islands respectively. This group of Atlanto-Mediterranean species appears unevenly distributed in the Eastern-Atlantic, probably due to the scarce data available for several areas of this region. A better knowledge of the Mauretanian and Senegalese regions, as suggested by de Weerdt & van Soest (1986), might result in a more uniform distribution pattern of the Atlanto-Mediterranean species. This group formed mostly by littoral species of temperate waters (except perhaps *Poecillastra compressa* and *Biemna variantia*) may have had its distributional center in the Western Mediterranean and have subsequently colonized more or less uniformly the neighbouring Atlantic regions. The outflow of water of Mediterranean origin is so important and so widespread in the Atlantic (Boury-Esnault & Lopes, 1985) that this possibility does not seen inconceiveable.

A second group is formed by 15 species (*Stryphnus mucronatus*, *Erylus euastrum*, *Pseudosuberites hyalinus*, *Suberites domuncula*, *Polymastia mammillaris*, *Cliona viridis*, *Chondrosia reniformis*, *Hymeniacidon sanguinea*, *Acarnus tortilis*, *Ircinia variabilis*, *Ircinia foetida*, *Dysidea avara*, *Spongionella pulchella*, *Aplysilla sulfurea*, *Hexadella racovitzai*) which are cosmopolitan or show a distribution which goes beyond the boundaries of the Atlanto-Mediterranean or Boreal Provinces. All the data reported for these widespread species result from the author's personal evaluation of records and synonymies, some of which are not generally accepted.

As usual, a rather small number of species (*Pseudosuberites sulphureus*, *Suberites carnosus*, *Aaptos aaptos*, *Tedania anhelans*) were found to display an amphi-Atlantic distribution, while only *Spirastrella cunctatrix* is recorded from both the Mediterranean and the Red Sea.

The last group consisting of only ten species (*Stelletta mediterranea*, *Diplastrella bistellata*, *Halichondria aurantiaca*, *Agelas oroides*, *Anchinoe tenacior*, *Stylostichon equiosculatus*, *Microciona assimilis*, *Haliclona subtilis*, *Reniera mamillata*, *Siphonochalina balearica*) is formed by Mediterranean endemic sponges which are a mere 17.2 % of the total. This proportion of endemisms is far lower than the 44.6 % reported by Vacelet (1980) for the Mediterranean Demosponges. The actual number of endemisms for the present study, however, could be even lower considering that two Haplosclerid species (*Haliclona subtilis* and *Reniera mamillata*) are of relatively recent description and that the new *Stylostichon* species recorded from Alboran would most probably also occur outside the Strait of Gibraltar. The very high proportion of shared sponge fauna between Mediterranean and Atlantic in the present study is due also, in my opinion, to the peculiar characteristics of the zones investigated which have typically open sea characters, as explained above. A more detailed study, covering the same region along with the littoral zones and their numerous microenvironments, where the influence of the Atlantic water is certainly weaker, should have given quite different results.

Notwithstanding the latter consideration, it may be concluded that Vacelet's (1980) statement, that the strongest affinities of the Mediterranean sponge fauna are with the neighbouring Eastern-Atlantic, is supported once more by the present data.

DESCRIPTION OF THE SPECIES

D E M O S P O N G I A E

TETRACTINOMORPHA

ASTROPHORIDA Lévi, 1973

STELLETTIDAE Carter, 1875

Stelletta hispida (Buccich)

Ancorina hispida Buccich, 1886

Material. BN 37 and BN 39, st.5, off Alboran, laminarians, 70-80 m, 10.9.85.

Description. Both specimens are globose, rather hard, hispid. They measure 4 and 8 cm across respectively. The basic colour is grey with different shades : light brown for BN 37 and violet for BN 39. They live among the rhizoids of the laminarians and BN 37 appears extremely entangled with these stems which are embedded in the sponge tissue. Their surface is covered by other epizoans which thrive at the base of the alga.

Spicules. Oxea up to 2250 by 55 µm. Plagotriaenes : mean size about 1400 by 60 µm. Chiasters : 9-11.5 µm in diameter. Oxyasters : about 14 µm in diameter.

Remarks. The reduction of the clads is more apparent in BN 37 which also shows more slender spicules than BN 39.

Stelletta mediterranea (Topsent)

Pilochrota mediterranea Topsent, 1893

Material. BN 34, st. 5, off Alboran, laminarians, 70-80 m, 10.9.85.

Description. The specimen, 8 cm long and 2.5 cm wide, consists of two massive, more or less cylindrical bodies which are separated by a constriction (Pl. Ia). Even this specimen is tightly entangled and sometimes crossed by the laminarian stems. The surface appears irregularly hispid, without evident oscula. The violet colour of the upper parts of the sponge fades along the sides and turns to white inside. The consistency is very hard.

__Spicules__ (fig. 2). Stout oxea up to 2750 by 60 µm ; slender oxea, slightly flexuous : 700-800 by 4-6 µm. Orthotriaenes : rhabdome about 950 by 30 µm, clads reach 115 µm with a thickness at the base of 35 µm. Anatriaenes : 1600 by 14 µm. Oxyasters, sometimes with a small centrum, 8-17 µm. Rhaphides have not been observed.

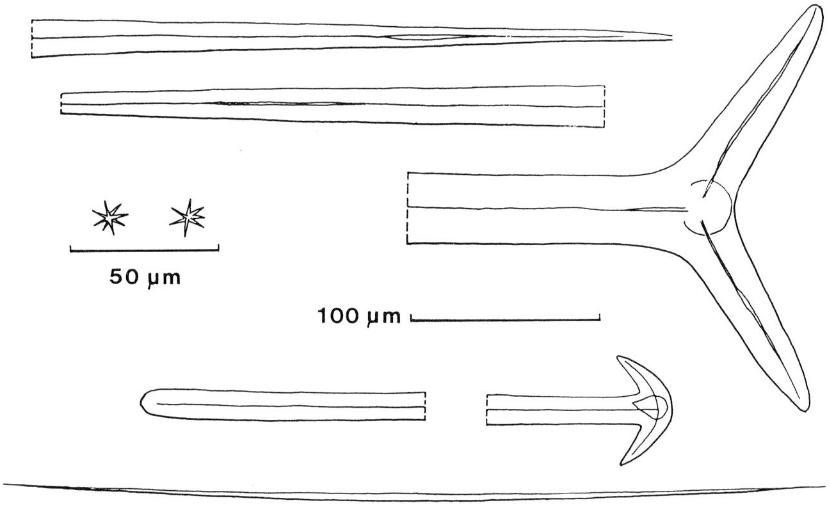

Figure 2 - Spicules of *Stelletta mediterranea* (Topsent).

__Remarks__. As far as I know this is the second record of this species which for the first time has been found in its massive form. Topsent's (1894) careful description fits the sponge rather well except for the slightly larger size of the bigger oxea and for the absence of trichodragmata in the present specimen.

Stryphnus mucronatus (Schmidt)

Stelletta mucronata Schmidt, 1868

Material. BN 31, st. 5, off Alboran, detritic bottom, 70-80 m, 10.9.85.

Description. The specimen is conical, 6 cm high, hispid, hard and black.

Spicules. Oxea up to 2450 by 42 µm. Dichotriaenes : rhabdome 300-650 by 18-30 µm ; cladome up to 250 µm in diameter. Oxyasters up to 55 µm with the actine number generally reduced to four. Amphiasters to sanidasters up to 18 µm.

PACHASTRELLIDAE Carter, 1875

Dercitus plicatus (Schmidt)

Corticium plicatum Schmidt, 1868

Material. BN 24, st. 7, off Alboran, rock and laminarians, 32 m, 9.9.85.

Description. An amorphous, whitish mass insinuating among the laminarians stems. Dichotriaenes are missing in this specimen.

Poecillastra compressa (Bowerbank)

Ecionemia compressa Bowerbank, 1866

Material. BN 35, st. 5, off Alboran, detritic bottom, 70-80 m, 10.9.85.

Description. Several laminar fragments, 1-1.5 cm thick, of bigger specimens broken by the trawling net. The colour in vivo is beige with pink shades. The sponge is slimy even before preservation. Spicules as usual.

HADROMERIDA Topsent, 1898

SUBERITIDAE Schmidt, 1870

Pseudosuberites hyalinus (Ridley & Dendy)

Hymeniacidon hyalina Ridley & Dendy, 1887

Material. BN 40, st. 5, off Alboran, detritic bottom, 70-80 m, 10.9.85.

Description. Amorphous specimen embedded with Serpulid tubes with a very irregular, grooved surface (Pl. Ib). The consistency is soft and the texture very loose. The colour in vivo is pale yellow. A translucid ectosome with tangentially arranged single spicules is supported by an halichondrioid choanosomal skeleton where tylostyles may be occasionally condensed in spicular tracts.

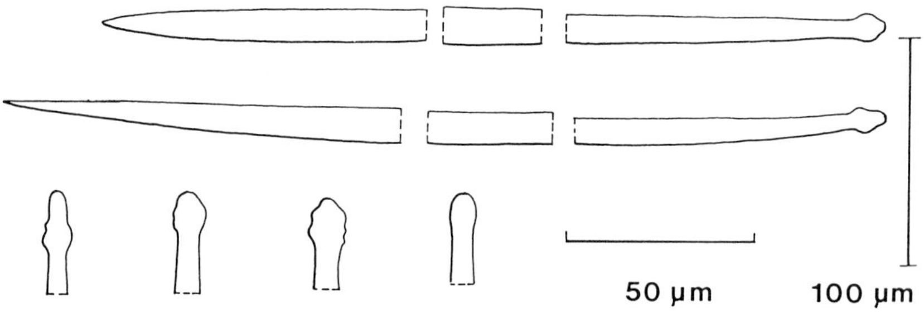

Figure 3 - Spicules of *Pseudosuberites hyalinus* (Ridley & Dendy).

Spicules (Fig. 3). Tylostyles to subtylostyles, fusiform, straight or curved and occasionally slightly flexuous. All intermediate sizes occur between minimum and maximum values : 198-725 by 5-15 μm.

Pseudosuberites sulphureus (Bowerbank)

Hymeniacidon sulphurea Bowerbank, 1866

Material. BG 12, BG 13, st. 2, Graham Bank, volcanic scoriae, 25 m, 17.7.81.

Description. Two specimens rather irregular in shape, about 10 cm across in their maximum diameter. Lobate or mamillate processes sprawl out from an encrusting base and develop repently. Oscula are flush, round and scattered. The consistency is soft and the colour bright yellow. The choanosomal skeleton is typically halichondrioid.

Spicules (fig. 4). Tylostyles to styles, measuring 140-170 by 2-7.5 μm.

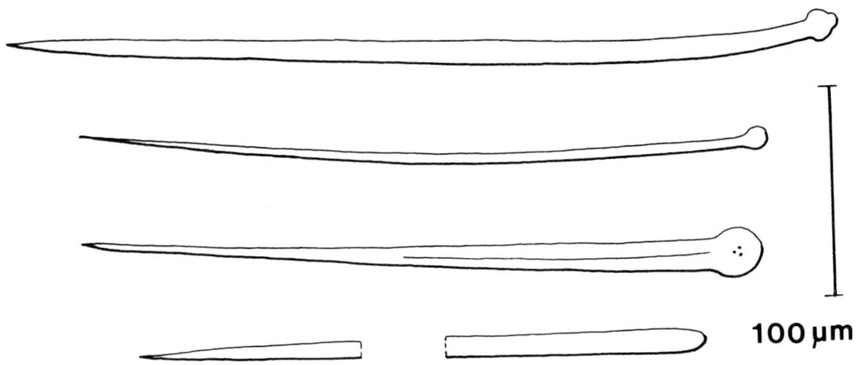

Figure 4 - Spicules of *Pseudosuberites sulphureus* (Bowerbank)

Remarks. The presence of relatively short, straight tylostyles with rounded and long conical shaft is worthy of note.

Suberites carnosus (Johnston)

Halichondria carnosa Johnston, 1842

Material. BN 31 bis, st. 5, off Alboran, detritic bottom, 70-80 m, 10.9.85.

Description. A small encrusting specimen on *Stryphnus mucronatus*.

Suberites domuncula (Olivi)

Alcyonium domuncula Olivi, 1792

Material. BG 15, st. 2, Graham Bank, rock, 20 m, 17.7.81.

Description. Globose specimen with hermit crab, orange in colour. Spicules are only styles and oxea ; tylostyles and microscleres are lacking completely.

POLYMASTIIDAE Gray, 1867

Polymastia mammillaris (Müller)

Spongia mammillaris Müller, 1806

Material. BN 28, st. 5, off Alboran, detritic bottom, 70-80 m, 10.9.85.

Description. Two flattened specimens, 2-3 mm thick, with roundish borders, bearing papillae up to 1 cm high. The colour in vivo is dark yellow.

Spicules. Tylostyles and subtylostyles, fusiform, mostly straight : 540-1230 by 10-19 µm. Short tylostyles with well formed tyle, gently curved : 90-160 by 2-4 µm.

TIMEIDAE Topsent, 1928

Diplastrella bistellata (Schmidt)

Tethya bistellata Schmidt, 1862

Material. BN 38, st. 5, off Alboran, detritic bottom, 70-80 m, 10.9.85.

AXINELLIDA Lévi, 1953

AXINELLIDAE Ridley & Dendy, 1887

Axinella damicornis (Esper)

Spongia damicornis Esper, 1794

Material. BN 30, st. 5, off Alboran, detritic bottom, 70-80 m, 10.9.85.

CERACTINOMORPHA

HALICHONDRIIDA Vosmaer, 1887

HALICHONDRIIDAE Vosmaer, 1887

Halichondria aurantiaca (Schmidt)

Reniera aurantiaca Schmidt, 1864

Material. BN 5, BN 8, st.7, off Alboran, rock and laminarians, 32m, 9.9.85 ; BN 32, BN 33, st. 5, off Alboran, detritic bottom, 70-80 m, 10.9.85 ; BN 43, BN 44, st. 8, Punta Almina, rock, 15 m, 13.9.85.

Description. Thickly encrusting or cushion shaped according to different substrate and water-movement conditions. Oscula are slightly elevated, somewhat regular in shape. The surface is uneven, often undulated. Subdermal canals are clearly detectable in the living specimens. The colour is deep yellow to light orange.

Spicules. Oxea straight, slightly curved or bent, measuring 315-990 by 3-14 µm. Differences in spicule length have been observed among the specimens, while thickness appears to be more constant.

Batzella inops (Topsent)

Halichondria inops Topsent, 1891

Material. BN 10, st. 7, off Alboran, 32 m, 9.9.85.

Description. Small, amorphous, specimen of light brown with green shades, growing entangled with a *Sphaerococcus* alga. It was soft and not mucoidal in vivo.

Spicules. Strongyles regular, 160-200 by 2-4 µm.

HYMENIACIDONIDAE de Laubenfels, 1934

Dictyonella incisa (Schmidt)

Phakellia incisa Schmidt, 1880

Material. BN 2, BN 9, st. 7, off Alboran, rock, 32 m, 9.9.85 ; BN 42, st. 8, Punta Almina, rock, 12 m, 13.9.85.

Description. Cushion shaped or globular. Rather consistent but not elastic. Bright orange in colour.

Spicules. Straight, gently curved, occasionally flexuous styles, measuring 600-1620 by 3-9 µm.

Hymeniacidon sanguinea (Grant)

Spongia sanguinea Grant, 1826

Material. BN 47, st. 9, Punta Almina, rocky boulders, 6 m, 13.9.85.

Description. Encrusting to insinuating, emitting slender papillae 0.3-0.5 mm in diameter and up to 3 cm high. The colour is light orange.

Spicules. Gently curved styles, rather regular, 180-350 by 2-6.5 µm.

Ecology. The species is very common in the upper parts of the rocky substratum, which is covered by algae and abundant sandy sediment. The sponge creeps into the microcavities of the rock.

POECILOSCLERIDA Topsent, 1928

MYCALIDAE Lundbeck, 1905

Mycale syrinx (Schmidt)

Esperia syrinx Schmidt, 1862

Material. BT 1, st. 4, Terribile Bank, rock, 12 m, 18.7.81 ; BN 36, st. 5, off Alboran, detritic bottom, 70-80 m, 10.9.85.

Description. Short, stout tubes about 5 cm high and 0.8 cm thick.The maximum diameter is 3 cm. They show apical, cloacal vents surrounded by a thin rim. Surface hispid, covered by an easily detachable ectosome. Colour in life is rose. The skeleton is formed by a tangential net and a choanosomal reticulation of polyspicular fibres as described by Topsent (1924, p.96) and Pulitzer-Finali (1978, p.48).

Spicules. Subtylostyles straight or slightly flexuous : 270-330 by 4-6 µm. Anisochelae : as figured by Topsent (1924) from 12 to 42 µm long, all intermediate sizes being found. Toxa : 30-51 by 0.5-1 µm. Sigmas : up to 117 by 5 µm, found only in BN 36.

Ecology. Specimen BT 1 was collected on a subvertical cliff in a shaded position.

BIEMNIDAE Hentschel, 1923

Biemna variantia (Bowerbank)

Desmacidon peachii Bowerbank, 1866

Material. BN 29 bis, st. 5, off Alboran, detritic bottom, 70-80 m, 10.9.85.

Description. Small, whitish, hispid incrustation on a fragment of organic conglomerate.

Spicules. Styles straight or gently curved, occasionally transformed to oxea, 305-440 by 6-9 µm. Microxea straight or slightly curved, 38-77 by 1-2.5 µm. Sigmas of only one category, 42-103 by 4-6 µm.

ESPERIOPSIDAE Hentschel, 1923

Crambe crambe (Schmidt)

Suberites crambe Schmidt, 1862

Material. BN 19, st. 6, Alboran, S-E coast, rock, 7-10 m, 10.9.85.

CRELLIDAE Hentschel, 1923

Crella elegans (Schmidt)

Cribrella elegans Schmidt, 1862

Material. BN 41, st. 7, off Alboran, detritic bottom, 70-80 m, 10.9.85.

Description. Thin incrustation on rhizoids of dead laminarians. Deep pink in life.

Spicules. Tornotes, straight with well pointed ends, 225-249 by 4-6 µm. Acanthoxea, straight or slightly curved, 84-126 by 5-7.5 µm. Acanthostyles, almost straight, uniformly spined, reaching up to 160 by 4-5 µm. Since there are extraneous spicules in the preparation the acanthostyles may not be proper.

Pytheas rosea (Topsent)

Yvesia rosea Topsent, 1892

Material. BG. 4, st. 2, Graham Bank, volcanic rock, 7 m, 17.7.81 ; BN 7, st. 7, Alboran, rock, 32 m, 9.9.85.

Description. The specimens are erect, about 5 cm high, not branching but with conical (BN 7) and even flattened processes (BG 4). The consistency is soft but the sponge is not fragile. The colour in vivo is carmine red. Small, round oscula, about 1 mm in diameter, are detectable in slightly raised positions only on BG 4.

Spicules. Basal acanthostyles : 117-131 by 3-4 µm. Dermal acanthostyles, rather rare, 85-96 by 2.5-3 µm. Tornotes : 230-290 by 2.9-6.5 µm. Isochelae : 19-21 µm.

Remarks. The erect habit in this species seems to be unusual, since other records (Mediterranean and Azores) and the original description refer to encrusting specimens.

MYXILLIDAE Topsent, 1928

Myxilla rosacea (Lieberkühn)

Halichondria rosacea Lieberkühn, 1859

Material. BN 11, BN 25, st. 7, off Alboran, rock, 32 m, 9.9.85 ; BN 48, st. 8, Punta Almina, rock, 20 m, 13.9.85.

Description. BN 11 and BN 25 are supported and entangled by the stout stems of *Laminaria ochroleuca*. They show a massive and amorphous form, conglomerating extraneous materials such as sand, tubes of polychetes, bryozoans etc. The colour in vivo is orange. The round oscula are occasionally detectable. The surface is covered by a translucent, very fragile ectosome showing the underlying crumbly texture. The sponge produces abundant mucus even before preservation. BN 48 is a small fragment, supported by some algae too.

Spicules. Tornotes, 136-178 by 2-4 µm. Acanthostyles, 141-169 by 5-7 µm. Isanchorae, 18-28 µm. Sigmas, 14-37 µm.

ANCHINOIDAE Topsent, 1928

Anchinoe coriaceus (Fristedt)

Stylopus coriaceus Fristedt, 1885

Material. BN 18, st. 6, Alboran, S-E coast, rock, 6 m, 10.9.85.

Description. Rather thick incrustation on a subvertical rock exposed to light and strong water-movement. The ectosome appears particularly tough and tenacious as observed by Topsent (1936). The colour in vivo is pale blue with violet shades, while the choanosome is mustard yellow.

Spicules. Acanthostyles, straight with curved spines : 90-165 by 3-6 µm. Only fragments of the bigger acanthostyles are present. Ectosomal subtylotes are slender, straight, with slightly different extremities, and measure 180-225 by 2-3 µm.

Remarks. According to this record, which is the first for the southern Mediterranean, and to the recent one for the Azores (Boury-Esnault & Lopes, 1985) the species appears more uniformly distributed in the Atlanto-Mediterranean Province.

Anchinoe mercator (Schmidt)

Suberotelites mercator Schmidt, 1868

Material. BG 1, BG 2, Graham Bank, volcanic scoriae, 25 m, 16.7.81.

Description. The two specimens are massive, with large lobes and more slender, sometimes contort processes, both bearing apical oscula. The surface is even with subdermal canals mostly evident in the living sponge. The colour in vivo is pink with orange shading, while the choanosome is yellow. The consistency is soft but not fragile. Ectosomal skeleton consists of tangentially arranged subtylotes, while in the choanosome subtylotes and acantostyles form ill defined multispicular tracts.

Spicules (Fig. 5). Subtylotes, 210-260 by 2-4 µm. Acanthostyles, 130-190 by 4-6 µm.

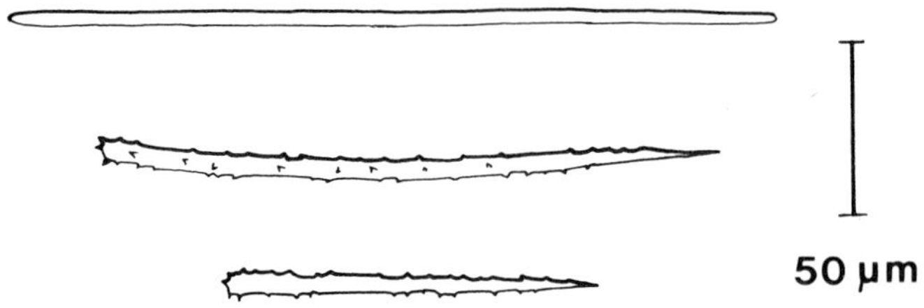

Figure 5 - Spicules of *Anchinoe mercator* (Schmidt).

Ecology. The species, which is fairly abundant on the bank, lives in rather shaded positions : in this particular case under the longitudinal overhangs generated by the outcropping fault.

Anchinoe paupertas (Bowerbank)

Hymeniacidon paupertas Bowerbank, 1866

Material. PT 2, st.1, Pantelleria Island, Punta Limarsi, rock, 2 m, 15.7.81.

Description. Thin incrustation on a vertical rock, purple red in vivo, with ostioliferous areas and exhalant canals well evident. The specimen was in a reproductive period in July since a section revealed the presence of many embryos.

Spicules. Anisostrongyles to subtylotes, straight, 235-280 by 3-5 µm. Acanthostyles slightly curved, completely even or with a few scattered spines in the basal third, 270-320 by 4-6 µm. The second category of acanthostyles is lacking in this specimen. Arcuate isochelae, numerous, 19-24 µm.

Ecology. The species is mostly abundant on rocky cliffs and slopes exposed to light and water movement, in a band between the surface and about ten meters deep. The same distribution was observed by the author around the Island of Elba.

Anchinoe fictitius (Bowerbank)

Microciona fictitia Bowerbank, 1866

Material. BN 13, BN 22, off Alboran, rock and laminarians, 32 m, 9.9.85 ; BN 17, st. 6, Alboran S-E coast, rock, 7-10 m, 10.9.85 ; BN 45, st. 8, Punta Almina, rock, 20 m, 13.9.85.

Description. The first three specimens (BN 13, 17, 22) are thin, brick red incrustations showing the typical ostioliferous areas. BN 45, however, shows a completely unusual aspect being massive, amorphous, with a very soft texture, light brown to yellow in colour, mucous after preservation. It exhibits a plumose reticulation with branching polyspicular tracts bound by spongin. A few isochelae are embedded in these spicular tracts, while they are more numerous in the ectosome. Rare isochelae are embedded in these spicular tracts, while they are more numerous in the ectosome. Abundant spicules of all types are loosely interposed into the reticulations. Shape and size of the spicules are identical in all the specimens.

Spicules. Tornotes : 280-310 by 3-5 µm. Acanthostyles I : 250-360 by 5-7 µm. Acanthostyles II : 90-180 by 3-5 µm. Arcuate isochelae : 17-24 µm.

Remarks. Specimen BN 45 is tentatively attributed to this species.

Stylostichon equiosculatus sp.n.

Material. BN 1, st. 7, off Alboran, rock, 32 m, 9.9.85 (holotype, MSNG 47904) ; BN 15, st. 6, Alboran, S-E coast, rock, 7-10 m, 10.9.85 (paratype, MSNG 47905).

Description. Wide incrustations stretching for several square decimeters on the upper and well lighted parts of the rocks. Their thickness may reach 1cm (Pl. Ic). Numerous round oscula open at the top of small elevations. They are regularly disposed, round, slightly elevated, with a visible net of converging canals. The surface appears cribrous, hispid, uneven, covered by a tough but not detachable ectosome. The colour in vivo is generally dark orange, sometimes with brown shades.

The plumose skeleton is formed by ascending columns of densely packed acanthostyles bound by spongin, which are strongly echinated by the smaller acanthostyles (Fig. 6). These columns, giving a fibrous consistency to the choanosome, start from a basal layer of spongin with erect acanthostyles and end in bunches of tornotes hispidating the sponge surface. Isochelae are dispersed among the columns.

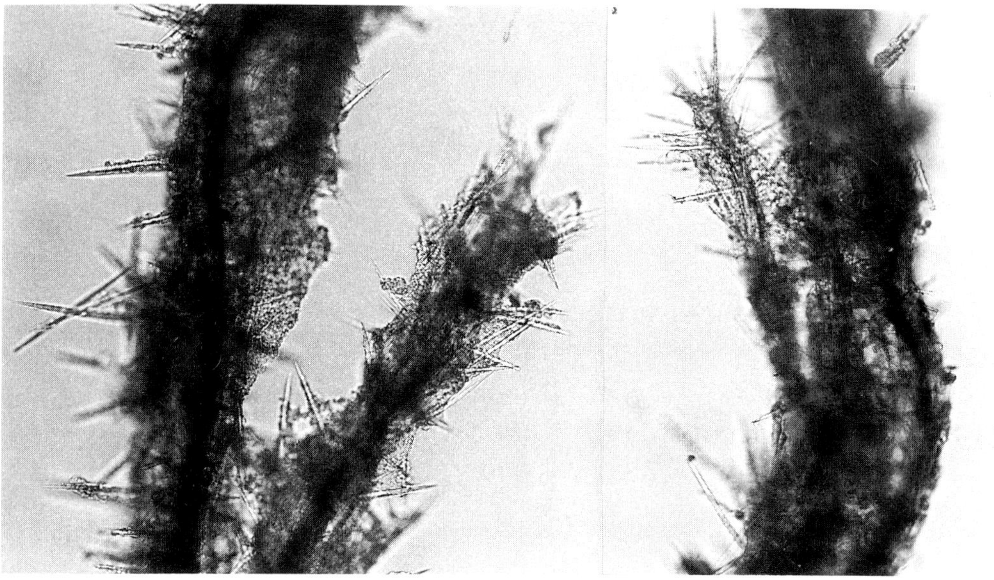

Figure 6 - Ascending columns in the choanosomal skeleton of *Stylostichon equiosculatus* sp. n.

Spicules (Fig. 7). Tornotes, fusiform, generally straight or slightly flexuous, with well pointed but dissimilar extremities : 130-190 by 4.5-6 µm. Arcuate isochelae, rather slender, not abundant, 13-15 µm. Acanthostyles may be divided in two categories but all terms of passage are easily found. The smaller ones (65-103 by 4-5 µm) are straight and well spined whereas the larger ones (140-188 by 6-7.5 µm) are slightly bent and show shorter and more scattered spines. Both have inconspicuous heads.

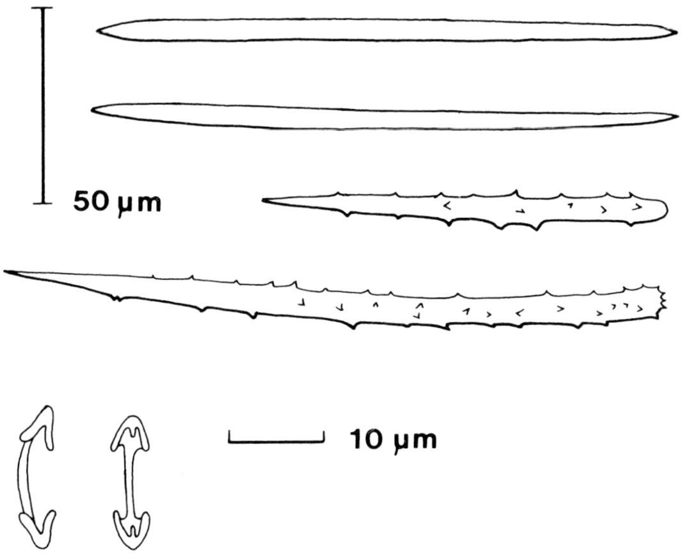

Figure 7 - Spicules of *Stylostichon equiosculatus* sp. n.

<u>Remarks</u>. The skeletal characteristics of this species are typical of the genus *Stylostichon*, even though at first glance, a certain similarity in the form and size of spicules was found with *Hymedesmia castanea* Sarà of the Ligurian Sea. The closest *Stylostichon* species, which is devoid of sigmas, seems to be *lieberküehni* (Burton). It differs, however, from the present one in colour (being red), the bigger sizes of spicules and by the presence of two categories of isochelae.

<u>Ecology</u>. The species is very abundant on the shallow rocky platform on the S-E side of the Island of Alboran. The thickness of the incrustations appears to be related to water movement intensity : it is reduced in the exposed positions. The species is present also off the island among laminarians at a depth of 32 m.

CLATHRIIDAE Hentschel, 1923

Microciona assimilis (Topsent)

Clathria assimilis Topsent, 1925

Material BG 8, st. 2, Graham Bank, volcanic scoriae, 25 m, 16.7.81.

Description. Thin incrustation, orange in vivo.

Spicules. Principal acanthostyles, 216-282 by 3.5-6 µm, smooth or occasionally with a few spines in their basal end. Auxiliary subtylostyles, 188-265 by 2.5-3.5 µm. Accessory acanthostyles, straight or slightly curved, uniformly spined, 65-122 by 3-5 µm. Toxas, slender or stout; with deep curved flexion and spined extremities, 89-343 by 1-2.5 µm. Isochelae palmate, not rare, 5.2-8.7 µm.

Remarks. The spicule dimensions correspond to those of *Microciona assimilis* (Topsent) as reported by Lévi (1960). In spite of the great similarity of this species with *Microciona spinarcus* Carter & Hope, I did not find a complete identity between the two as did Descatoire (1967) and Boury-Esnault & Lopès (1985). Considering however the smaller size of the principal acanthostyles and the relative abundance of chelae in *Microciona assimilis* I prefer to attribute the present specimen to this species.

Antho involvens (Schmidt)

Myxilla involvens Schmidt, 1864

Material. BG 3, BG 7 bis, BG 19, st. 2, Graham Bank, volcanic scoriae, 20 m, 16.7.81 ; BN 14, st. 7, off Alboran, rock, 32 m, 16.7.81.

Description. Three specimens are encrusting on rock or calcareous tubes of serpulids (BG 7) while the fourth one (BG 19) is bushy, not more than 1.5 cm high. The colour in vivo is bright orange except for BN 14 which is red.

Remarks. The specimen coming from Alboran shows a notable reduction of the spines which are apparent in the principal acanthostyles and very few in the accessory acanthostyles. Toxas of all sizes are remarkably abundant and reach 245 µm in length.

Acarnus tortilis Topsent

Acarnus tortilis Topsent, 1892

Material. BG 11, st. 2, Graham Bank, volcanic scoriae, 25 m, 17.7.81.

Description. The specimen is massive (Pl. 1d), cushion shaped, 6 cm wide and 4 cm high. Three main oscula, 1 cm wide, more or less elliptical, bordered by rims, are slightly elevated on the upper part of the sponge. The colour in vivo is bright red changing to a dull pink in alcohol. The consistency is soft, fragile. A distinct, not easily detachable ectosomal membrane, formed by a layer of disorderly arranged tylotes covers wide vestibular spaces. In the choanosome ascending tracts of styles appear echinated by the cladotylotes. Abundant arcuate toxas and isochelae are dispersed among the main skeletal structures. The choanosome aggregates a large quantity of coarse volcanic sand.

Spicules. Styles, 352-451 by 7-9 µm. Tylotes, 253-370 by 4-6 µm. Cladotylotes, 98-197 by 2-4.5 µm. Arcuate toxas, 103-173 by 2-2.5 µm. Open toxas, rare, 220 by 1.5 µm. Isochelae palmate, 12-16 µm.

Remarks. I dwelt upon the morphology of this species because it is the first time, as far as I know, that it is recorded in its massive form. The other known specimens (Mediterranean, Azores, Amboine) were all encrusting and enveloping. Their colour was described as pink, brown or blackish. The habit of agglomerating sand is common to *Acarnus polytylus* Pulitzer-Finali (1983).

HAPLOSCLERIDA Topsent, 1928

HALICLONIDAE de Laubenfels, 1932

Reniera cratera Schmidt

Reniera cratera Schmidt, 1862

Material. PV 1, PV 3, st. 3, Pantelleria Vecchia Bank, rock, 22 m, 21.7.81 ; PT 7, st. 1, Pantelleria Island, Punta Limarsi, rocky cliff, 27 m, 20.7.81.

Remarks. This species quite common and seemingly well known, has been almost always recorded with a spiculation of mere strongyles. Only Sarà (1958, p. 238) found oxea and intermediate spicules between strongyles and oxea in specimens from Liguria. Since slender, straight or gently curved oxea are to be found in the material examined, other specimens coming from several localities were checked in order to ascertain the presence of these spicules. The results are reported below.

SPECIMENS	OXEA (µm)	STRONGYLES (µm)
227 Bogliasco (Liguria)	210-272 x 2-5	192-305 x 5-11
J 9 Velj losinj (Kres)	235-330 x 1.5-5	190-280 x 6-12
5C 7 Lacco Ameno (Ischia)	319-338 x 2-3.5	250-330 x 7-13
ET 3/5 Island of Elba	250-286 x 1-3	170-296 x 5-11
ML 12 Zaffiro cave (Naples)	257-316 x 1-3	234-304 x 6-11
PT 7 Island of Pantelleria	291-330 x 1.5-3	210-327 x 8-23
PV 1 Pantelleria Vecchia	263-282 x 1.5-4.5	197-272 x 5-10
PV 3 Pantelleria Vecchia	249-263 x 1-3.5	202-282 x 7-10

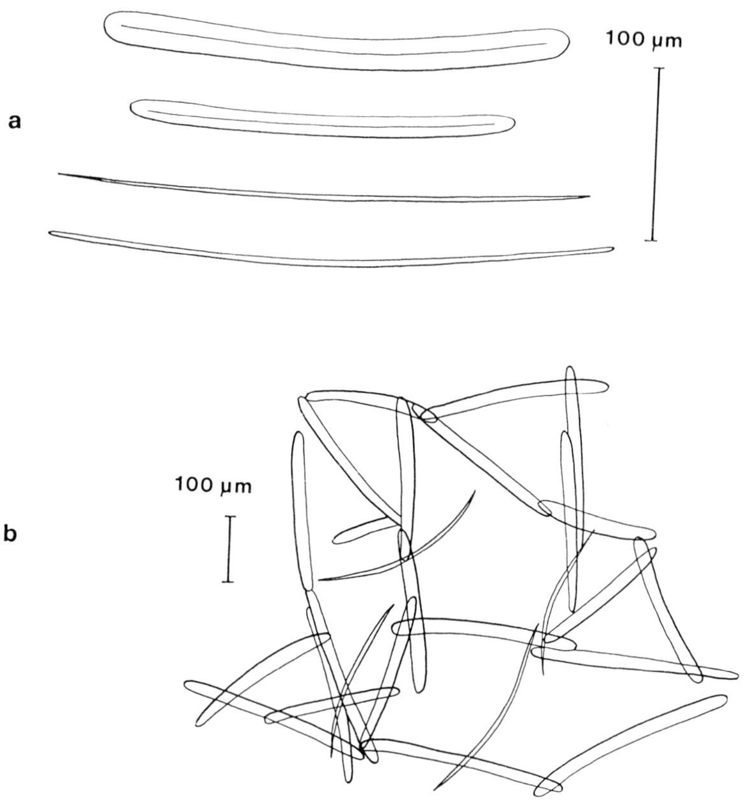

Figure 8 - *Reniera cratera* Schmidt : a/ spicules, b/ skeleton in a cross section.

These oxea (Fig. 8a) are rather constant in form. They usually have sharpened extremities but stepped or rounded ends are frequent. They are always clearly distinguishable from the small and juvenile strongyles and cannot be considered, in my opinion, as a spicule category evolving to stongyles. Regarding the skeleton (Fig. 8b) the oxeas seem to be inserted in the reticulation of strongyles but they may often be found in transverse position, unbound by the nodal spongin. A few intermediate spicules, resembling malformed strongyles (with ill rounded ends) may always be found.

Specimen PT 7 (Pl.Ie) is worthy of note because it displays an unusual, erect form with coalescent and single tubes (about 7 cm high with a diameter of 1-1.5 cm). The colour in vivo is dark orange, and the consistency is firmer than in other specimens. Fairly slimy when living, it produces abundant mucus after preservation in formalin. PT 7 has stronger strongyles than the other specimens, but no other differences in the skeletal pattern have been noted. I think the described form fits in with the range of variability of the species and agree with Pulitzer-Finali (1978, p.76) who considered the distinction of a similar specimen unjustified.

Haliclona subtilis Griessinger

Haliclona subtilis Griessinger, 1971

__Material__. BG 18, st. 2, Graham Bank, volcanic scoriae, 20 m, 16.7.81.

__Description__. More or less cylindrical eminences, 2 to 3 cm high, bearing apical oscula and starting from a common base. The colour in vivo is pink lilac and the consistency very soft. The skeletal reticulation exhibits small amounts of spongin at the nodes and is generally unispicular.

__Spicules__. Straight or slightly curved oxea usually with sharpened ends, but stylote modifications are common. They measure 94-173 by 1.5-5 µm. The variation of thickness is most evident.

__Remarks__. The specimen differs from those described by Griessinger (1971) in colour and spicule size, but the skeletal arrangement that was checked against Griessinger's preparations is identical. Since I have found *H. subtilis* off Monaco with oxea up to 150 µm and at Montecristo, (in the Tuscan Archipelago), a bright orange specimen, it may be assumed that the range of variation in this species may be greater than previously thought.

Reniera perlucida Griessinger

Reniera perlucida Griessinger, 1971

Material. BG 9, BGQ 1, st. 2, Graham Bank, volcanic scoriae, 25 m, 16.7.81 ; BN 4, st. 7, off Alboran, laminarians, 32 m, 9.9.85.

Description. Many specimens have been observed on the Graham Bank in well lit positions : the largest of them (BGQ 1) is massive, cushion shaped, with slightly raised, irregularly rounded oscula (5-15 mm wide) circled by a thin rim. Specimen BN 4 shows some holes, piercing the whole thickness of the sponge, as noted by Griessinger (1971). The largest specimen is 25 cm across and 6 cm high. The surface is even, with gentle swellings and depressions. The colour underwater is pale wine-rose but it rapidly fades changing to white when the sponge dries. The consistency is firm, the sponge is very fragile even in the water ; dried specimens however, become hard and lose their brittleness. A hyaline pseudocortex with a unspicular reticulation forming more or less quadrangular meshes, allows the vestibules underneath to be clearly visible. The crumbly choanosome shows paucispicular ascending tracts with transversal spicules slightly disordered.

Spicules. Most oxeas are slightly curved but straight ones are common. Ends are generally sharpened. They measure 141-178 by 1.5-6 µm in BG 9 and BGQ 1 while BN 4 has slightly stouter oxea reaching 190 by 9 µm.

Remarks. According to very recent data (Pulitzer-Finali, 1983 ; de Weerdt & van Soest, 1986) the species seems to thrive both on deep rocky bottoms (more than 100 m deep) and in the intertidal zone.

Reniera valliculata Griessinger

Reniera valliculata Griessinger, 1971

Material. BG 10, st. 2, Graham Bank, volcanic scoriae, 25 m, 16.7.81.

Description. The specimen is massive, 6 cm wide and up to 1.5 cm thick. The surface is very irregular, unhispid, with small lobes (1-1.5 cm high) bearing round oscula at their summit. The colour in vivo is whitish. The skeleton is compact, fairly confused, with paucispicular tracts and spicules interposed at different angles.

Spicules. Thin, curved and sometimes slightly flexuous oxea, often displaying irregular swellings, which may be central or subterminal. The ends are sharpened. They measure 141-169 by 1.5-3.5 µm.

Reniera mamillata Griessinger

Reniera mamillata Griessinger,1971

Material. BT 4, BT X, st. 4, Terribile Bank, rock, 42 m, 18.7.81.

Description. The sponges are massive to tubular, made of a series of lobes and tubes, up to 5 cm high, with round apical oscules (3-12 mm wide). The tubes are often coalescing and show irregular swellings. Their thickness averages 5 mm. The colour in vivo is yellow-rose. The surface is even, finely cribrous, with a fine hispidation detectable at the microscope. The consistency is soft but elastic and the specimens are slightly mucous when alive. An aspicular exopinacoderm, not detachable, covers the sponge. In a tangential section the ostiol areas, aspicular but divided in smaller parts by spicular tracts, are clearly detectable. The choanosomal skeleton is constituted by primary paucispicular tracts connected by transverse or oblique spicules, forming quadrangular or triangular meshes. Spongin is fairly scarce but the high spicular density makes the skeleton rather compact.

Spicules (Fig. 9). Slender (1.5-2 µm thick) and stout oxea (4-4.5 µm thick), both 94-132 µm long. The stouter ones often display blunt extremities while the smaller are more sharpened.

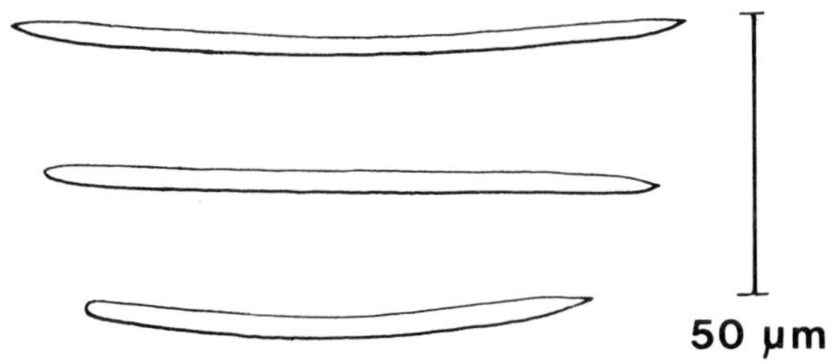

Figure 9 - Spicules of *Reniera mamillata* Griessinger.

Remarks. Awaiting availability of ampler material, the species is here tentatively attributed to *Reniera mamillata*, even if the crosschecking with Griessinger material raised some doubts because of the shape and size of the oxea.

Reniera sp.

Material. BN 48, st. 8, Punta Almina, rock, 20 m, 13.9.85.

Description. Small, amorphous, slightly slimy specimen living as an epizoan on a red alga. The consistency of the living specimen is soft, but it becomes crumbly when dried due to the great development of the aquiferous system. The ectosome is formed by paucispicular tracts forming roundish meshes. The choanosomal skeleton appears slightly different : rather confused, it illustrates more elongated meshes with paucispicular tracts and transverse spicules disorderly arranged. The spongin is scarce.

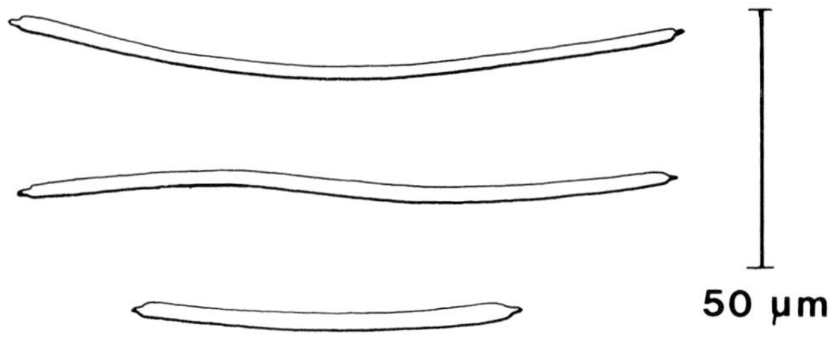

Figure 10 - Spicules of *Reniera* sp.

Spicules. (Fig. 10) - Oxea are curved or slightly flexuous with generally mucronated ends. They measure 65-145 by 2-4 µm.

Remarks. The closest species according to the skeletal organisation, appears to be *Reniera valliculata* Griessinger, whose spicules differ from those of the present specimen by being slightly longer and showing frequent centrotylotism. Further conclusions cannot be drawn on the present material.

CALLYSPONGIIDAE De Laubenfels, 1936

Siphonochalina coriacea Schmidt

Siphonochalina coriacea Schmidt, 1868

Material. BT 8 and BT 8 bis, st. 4, Terribile Bank, 40 m, rock, 18.7.81.

Description. Erect, cylindrical tubes, showing intermediate and terminal narrowings. The tubes may have a common base but they do not anastomose or ramify. The bigger specimen (BT 8 bis) is 25 cm high and 6 cm across (Pl. If). The thickness of the wall exceeds 1 cm. The surface is smooth and only a faint hispidation can be detected with a lens. The consistency is firm but elastic. The colour in vivo is light drab in BT 8 bis while BT 8 exhibits lilac shades. A tangential reticulation of single spicules forming a rather irregular ectosomal pattern is supported by the primary fibres of the choanosome. Spongin is not abundant and often limited to nodal points. The choanosome is a regular reticulation of primary paucispicular tracts and transversal, secondary fibres cored by one or two oxea. The meshes are more or less rectangular, sometimes lengthened (Fig. 11a).

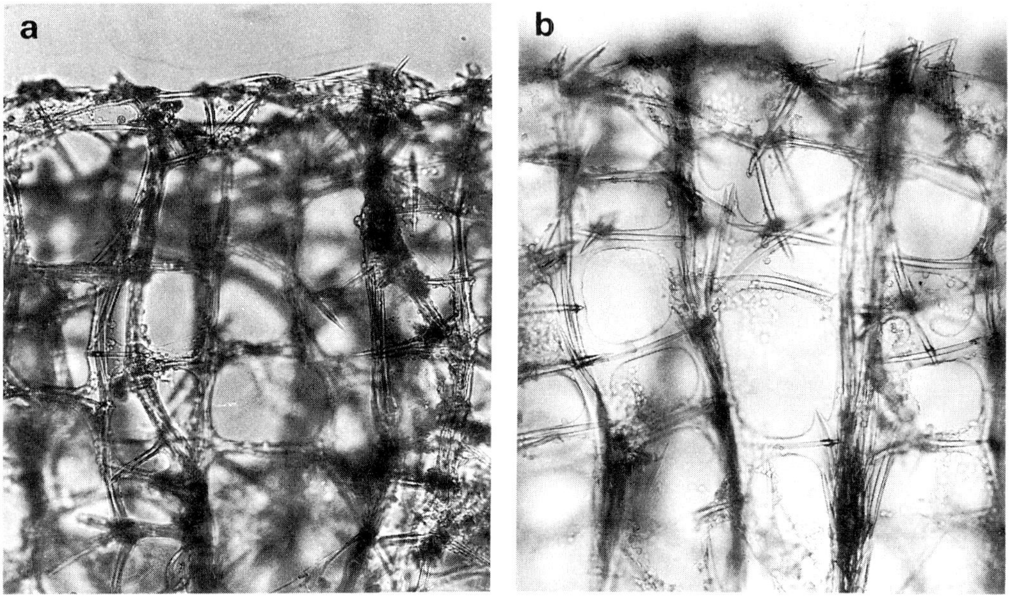

Figure 11 - a/ *Siphonochalina coriacea*, a cross section ; b/ *Siphonochalina balearica*, a cross section.

Spicules (Fig. 12) - Regular, well pointed, gently curved oxea, 125-140 by 5-6 μm.

Distribution. The distribution of the species seems to be restricted to the southern Mediterranean coasts and to the banks of the Strait of Sicily.

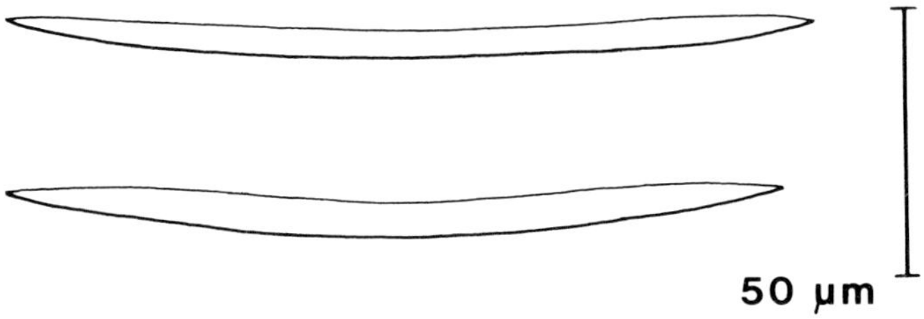

Figure 12 - Spicules of *Siphonochalina coriacea* Schmidt.

Siphonochalina balearica Ferrer-Hernandez

Siphonochalina balearica Ferrer-Hernandez, 1916

Material. BT 15, st. 4, Terribile Bank, rock, 42 m, 18.7.81 ; BN 3, BN 6, st. 7, off Alboran, rock, 32 m, 9.9.85.

Description. Erect, more or less cylindrical tubes, 1 to 3 cm in diameter, growing from a thin basal plate. They may be up to 25 cm high as in BT 15 and BN 3 (Pl. Ig) or become more stout and short as in BN 6. The tubes which ramify and anastomose may produce irregular processes often showing pointed extremities. The round, generally apical, oscula slightly restrict the diameter of the tubes. The sponge is soft but elastic and rather tenacious. The surface is smooth except in the processes where it becomes rather irregular. The colour in vivo is light drab with pink shades.

The ectosome is an isodictyal reticulation forming irregular meshes. The fibres are cored by one or two spicules. Spongin is always abundant and appears very thickly stratified at the nodes of the reticulation. The choanosomal skeleton is formed by ascending paucispiculated fibres and single or double transverse spicules. The resulting meshes are generally rectangular but become irregular and condensed close to the surface (Fig. 11b). No extraneous material has been noted in the abundant spongin, while the spherulous cells observed by Griessinger (1971) are clearly visible.

Spicules.(Fig. 13). Regular, slightly curved oxea with well pointed ends. Transformations into styles may be occasionally observed. The dimensions of the oxea appear slightly variable in the different specimens : BN 3 and BT 15 : 110-145 by 3-6.5 µm ; BN 6 (the specimen with short, stout tubes) : 130-160 by 3-7.5 µm.

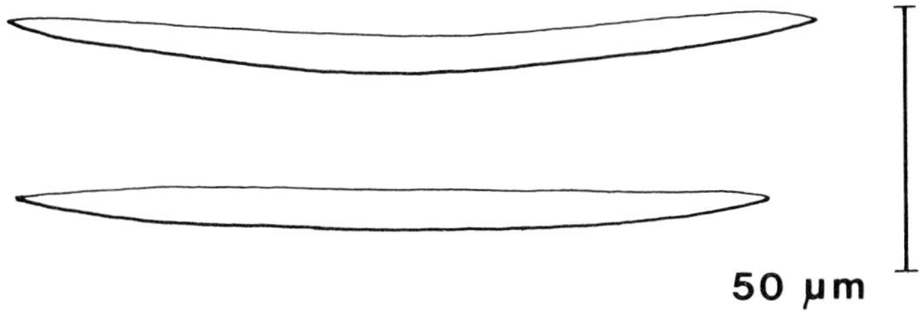

Figure 13 - Spicules of *Siphonochalina balearica* Ferrer-Hernandez.

Remarks. The characteristics of these specimens, except for the absence of extraneous bodies in the spongin fibres, perfectly fit with the hitherto known material (Ferrer-Hernandez, 1916 and Griessinger, 1971) so reinforcing the correct identification of this very conspicuous species. Its habitat appears to be limited to very clear waters such as those found offshore or around the islands, at medium depths (at least 25 m), where water movement is rather reduced. The known records, including a specimen from the Island of Ustica (Corriero, pers. comm.) restrict the present distribution of the south-western Mediterranean.

OCEANAPIIDAE van Soest, 1980

Pellina cf. semitubulosa (Lieberkühn)

Halichondria semitubulosa Lieberkühn, 1859

Material. BN 23, st. 7, off Alboran, rock, 32 m, 9.9.85.

Description. Small, amorphous, whitish specimen growing among the laminarians stems. Rather consistent in vivo but very fragile in the dry state. It displays only one stout, broken fistule, whose skeleton could not be observed.

<u>Spicules</u>. Oxea, regular, well pointed, gently curved, measuring 145-206 by 2-6.5 µm.

<u>Remarks</u>. Not withstanding the scarcity of the material the species is tentatively attributed to *Pellina semitubulosa* because of the skeletal characteristics, the dimensions of the oxeas and the presence of numerous spherulous cells.

Pellina sp.

<u>Material</u>. BN 23, st. 7, off Alboran, laminarians, 32 m, 9.9.85 ; BN 27, st. 5, off Alboran, detritic bottom, 70-80 m, 10.9.85.

<u>Description</u>. Small, about 1 cm in diameter, amorphous, whitish specimens living on laminarian stems (BN 23) and in a cavity of a roundish organic concretion (BN 27). No fistules are detectable. The consistency of the living specimens is soft, but they become very brittle when dried. The ectosome, easily detachable, is formed by a multilayered tangential reticulation of single oxea. The structure of the choanosomal skeleton appears rather confused and difficult to ascertain, but irregular multispicular tracts with abundant interposed spicules are detectable.

<u>Spicules</u>. Oxea regular, straight or gently curved, generally with well pointed ends : 170-210 by 1.5-6 µm.

<u>Remarks</u>. The specimens are here attributed to the genus *Pellina* according to the presence of the tangential layer of spicules and to the characteristics of the aquiferous system.

<u>PETROSIIDA</u> Bergquist, 1980

PETROSIIDAE, van Soest, 1980

Petrosia ficiformis (Poiret)

Spongia ficiformis Poiret, 1789

<u>Material</u>. BT 10, st. 4, Terribile Bank, rock, 42 m, 18.7.81.

Description. Cushion shaped, whitish in formalin, stony hard when alive and brittle in dry state. The ectosome is made by a tangential network of plurispicular tracts (15-40 µm thick) forming triangular or quadrangular meshes very variable in size. Many interstitial spicules are also present. This layer is supported by stouter spicular tracts (50-120 µm thick) forming roundish meshes (500-800 µm wide). The choanosome is a network of stout plurispicular tracts, up to 220 µm thick, forming series of roundish or elongated meshes arranged in planes parallel to the surface. Interstitial spicules are sparsely scattered.

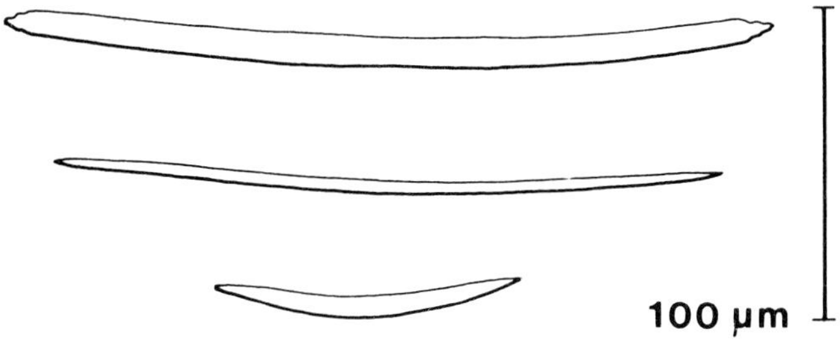

Figure 14 - Spicules of *Petrosia ficiformis* (Poiret).

Spicules (Fig. 14). Oxea which cannot be divided in size categories, but showing different forms. The stouter ones have stepped, rather blunt extremities, while the thinner ones show sharpened points. Both may be straight, bent or sinuous. They measure 43-280 by 2-9 µm.

Remarks. According to the considerations of de Weerdt & van Soest (1986) this specimen showing only oxea is here referred to *Petrosia ficiformis*. The opportunity to attribute specimens which lack thick oxea and strongyles to *Petrosia clavata* (Esper), as done by Pulitzer-Finali (1983) is not accepted.

DENDROCERATIDA Minchin, 1900

APLYSILLIDAE Vosmaer, 1883

Aplysilla sulfurea Schulze

Aplysilla sulfurea Schulze, 1878

Material. BN 20, st. 6, Alboran S-E coast, rock, 7-10 m, 10.9.85.

Description. Encrusting. Since the colour changes in different parts of this specimen from dull violet to yellow green it may not be regarded as a stable character in this family as suggested by Bergquist (1980) in order to maintain a specific distinction between *Aplysilla rosea* (Barrois) and *Aplysilla sulfurea* Schulze. Vacelet's (1959) opinion of considering as synonyms the two species is followed here.

Hexadella racovitzai Topsent

Hexadella racovitzai Topsent, 1896

Material. BT 6, st. 4, Terribile Bank, rock, 42 m, 18.7.81.

Description. Encrusting, with a beautiful net of turgid canals converging into round oscula. A series of strings, orderly arranged, raises the ectosome of the sponge. The colour in vivo is a pale brick red with orange shades. Dissociating a portion of the ectosome by a fine (20 µm meshes) cloth, even after preservation plenty of fusiform cells, which probably come from the strings, were obtained. The sponge is free of inclusions and debris.

Ecology. The species was abundant on the subvertical walls of big rocks scattered on the bottom.

DICTYOCERATIDA Minchin, 1900

THORECTIDAE Bergquist, 1978

Ircina variabilis (Schmidt)

Hircinia variabilis Schmidt, 1862

Material. BN 12, st. 7, off Alboran, rock and laminarians, 32 m, 9.9.85 ; BN 20, st. 6, Alboran S-E coast, rock, 7-10 m, 10.9.85.

Description. BN 12 is an encrusting specimen, whitish in colour. BN 16 is cushion shaped, with several rounded lobes, brown-green in colour.

Ecology. *Ircinia variabilis* is most common with massive and even erect forms in shallow water along the south-east coast of Alboran where it withstands high intensity water-movements. Greenish specimens, due to the presence of symbiotic algae, are very abundant.

DYSIDEIDAE Gray, 1867

Spongionella pulchella (Sowerby)

Spongia pulchella Sowerby, 1806

Material. PV 4, st. 3, Pantelleria Vecchia Bank, rock, 22 m, 21.7.81.

Description. The specimen is flabellate, with several elongated vertical fans (about 3 cm high and 0.5 cm thick) intercrossing one another. The oscula are round and apical. The colour in formalin is brown. All the characters fit with Vacelet's (1959) description.

LITERATURE CITED

Bergquist P.R., 1980 - A revision of the supraspecific classification of the orders Dictyoceratida, Dendroceratida and Verongida (Class Demospongiae). *N. Z. J. Zool.*, 7 : 443-503.

Boury-Esnault N. & Lopes M.T., 1985 - Les Démosponges littorales de l'archipel des Açores. *Ann. Inst. océanogr.*, Paris, 61(2) : 149-225.

Descatoire, A., 1967 - Les peuplements sessiles de l'infralittoral rocheux de l'archipel de Glénan : Spongiaires. Th. 3ème cycle, Univ. Paris, 190p.

Ferrer-Hernandez F., 1916 - Fauna del Mediterraneo occidental. Esponjas españolas. *Trab. Mus. Nac. Cienc. Nat. Zool.*, 27 : 1-52.

Griessinger J.M., 1971 - Etude des Reniérides de Méditerranée (Démosponges Haplosclérides). *Bull. Mus. natn. Hist. nat.*, 3(3) : 97-176.

Lévi C., 1960 - Les démosponges des côtes de France : les Clathriidae. *Cah. Biol. Mar.*, 1 : 47-87.

Pulitzer-Finali G., 1978 - Report on a collection of sponges from the Bay of Naples. III : Hadromerida, Axinellida, Poecilosclerida, Halichondrida, Haplosclerida. *Boll. Mus. Ist. Biol. Univ. Genova*, 45(1977) : 7-89.

Pulitzer-Finali G., 1983 - A collection of Mediterranean Demospongiae (Porifera) with, in appendix, a list of the Demospongiae hitherto recorded from the Mediterranean sea. *Ann. Mus. Civ. St. nat. Genova*, 84(1982) : 445-621.

Sarà M., 1958 - Contributo alla conoscenza dei Poriferi del Mar Ligure. *Ann. Mus. Civ. St. nat. Genova*, 70 : 207-244.

Sverdrup H.U., Johnson W.M. & Fleming R.H., 1942 - The oceans. Their physics, chemistry and general biology. Prentice-Hall, Inc., New York, 1087p.

Topsent E., 1894 - Etude monographique des Spongiaires de France. I. Tetractinellida. *Archs Zool. exp. gén.*, 2(3) : 259-400.

Topsent E., 1924 - Révision des *Mycale* de l'Europe occidentale. *Ann. Inst. océanogr. Monaco, N.S.*, 1 : 77-118.

Topsent E., 1936 - Eponges observées dans les parages de Monaco. 2ème partie. *Bull. inst. océanogr. Monaco*, 686 : 1-70.

Vacelet J., 1959 - Répartition générale des éponges et systématique des éponges cornées de la région de Marseille et de quelques stations méditerranéennes. *Recl Trav. St. mar. Endoume*, 16(26) : 39-101.

Vacelet J., 1980 - Les affinités du peuplement des Spongiaires de la Méditerranée. *Journées Etud. Systém. Biogéogr. Médit., Cagliari, C.I.E.S.M.* : 29-30.

Weerdt W.H., de & Van Soest R.W.M., 1986 - Marine shallow-water Haplosclerida (Porifera) from the south-eastern part of the North Atlantic ocean. *Zoologische Verhand. Leiden*, 225 : 1-49.

Plate I.
a/ *Stelletta mediterranea* (Topsent). **b/** *Pseudosuberites hyalinus* (Ridley & Dendy). **c/** *Stylostichon equiosculatus* sp.n., BN 1, holotype. **d/** *Acarnus tortilis* Topsent. **e/** *Reniera cratera* Schmidt. **f/** *Siphonochalina coriacea* Schmidt, specimen BT 8 bis. **g/** *Siphonochalina balearica* Ferrer-Hernandez, specimen BN 3.

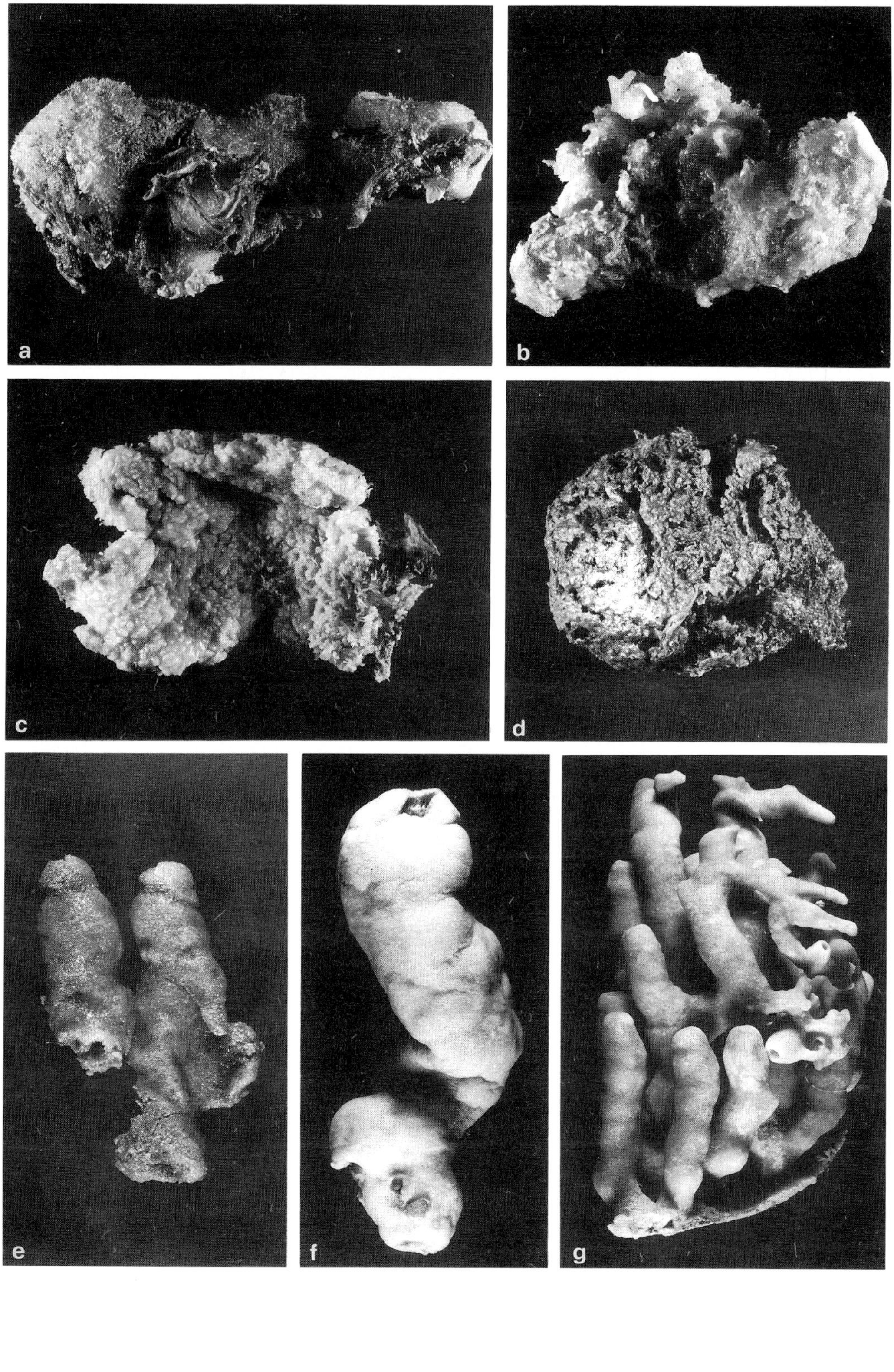
a
b
c
d
e
f
g

TETILLIDAE (*SPIROPHORIDA, PORIFERA*) : A TAXONOMIC REEVALUATION

K. RÜTZLER

Department of Invertebrate Zoology
National Museum of Natural History, Smithsonian Institution,
Washington, D.C. 20560, U.S.A.

SYNOPSIS

A large series of tetillid sponges was analysed using museum specimens and live material. The study led to a reevaluation of genera in that family, as well as of characters used to distinguish species. Properties of cortex, porocalices, and unusual accessory spicules are deemed adequate for distinguishing seven genera at the present state of our knowledge : *Tetilla, Craniella, Cinachyra, Paratetilla, Amphitethya, Cinachyrella,* (elevated from subgeneric status), and *Acanthotetilla*. Only few traditional species characteristics are considered stable enough to allow separating western Atlantic *Cinachyrella*. These apply to five species : *C. alloclada, C. kuekenthali, C. tarentina, C. apion,* and *C.* new species (van Soest & Stentoft, in press). It is recommended to avoid descriptions of new species without representative specimen series and observations of live populations, and to examine additional biological and molecular biological properties of the sponges for possible application in taxonomy.

INTRODUCTION

During the early 1980s a large collection of sponges, among other marine organisms, was made in the course of a continental shelf survey of the United States. The survey was sponsored by the Mineral Management Service (United States Department of Interior) and the collections were turned over to the Smithsonian Institution for deposit. Several hundred diverse looking specimens of the tetillid genus *Cinachyra* (as defined by all contemporary authors) are part of this material and stimulated the present study.

A survey of the literature showed that much thought has been given to

NATO ASI Series, Vol. G13
Taxonomy of Porifera
Edited by J. Vacelet and N. Boury-Esnault

the variability of characters used in genus and species diagnoses but that few workers actually examined large series of specimens, or studied live ones suited for evaluation of taxonomic criteria. The material on hand satisfies both needs because sizeable populations from different habitats and localities can be compared and because living specimens were observed and manipulated in the field.

MATERIAL AND METHODS

Most of my views expressed in this study were formed by examining some 200 specimens of *Cinachyra* (of authors) collected in the subtropical and tropical western Atlantic. Localities include the coasts of South Carolina and Georgia (20-40m depth, non-muddy "live" bottom); Bimini (Bahamas, 3-10m, coral reef) ; southwestern Florida (Gulf of Mexico, 24-77m, "live" bottom); Puerto Rico and Virginia Islands (8-16m, reef and rock bottoms) ; Jamaica (30m, reef) ; Belize (1m, mangrove ; 3-30m, reef) ; and Brazil (off mouth of Amazon River, 60-70m, rock bottom). Experiments with live specimens were conducted at the Carrie Bow Cay field laboratory, Belize, and included observations on current flow, using fluorescein dye, and on morphological changes during handling and fixation. Detailed spicule analyses were made for 64 specimens belonging to four species (Rützler & Smith, in press). Spicule preparations and epoxy resin sections were made the usual way (Rützler, 1978). Spicules were split into two size fractions (light and heavy) and mounted on separate microscope slides to avoid missing large forms and to be able to examine small microscleres under high power. Multiple preparations were made from different parts of a specimen if spicule variety appeared not to be representative.

DIAGNOSTIC CHARACTERS ON THE GENUS LEVEL

The family Tetillidae encompasses tetractinomorph sponges of massive and generally globular shape, with radial skeleton structure, with numerous monaxons and protriaenes as megascleres and spinispire-type microscleres (Dendy, 1922 ; Lévi, 1973). Ten genera are generally recognized belonging to this family. Criteria for separating them include differentiations of ectosome, aquiferous system, and spicule complement. These characteristics are certainly useful for establishing taxonomic hierarchies but they have not always been properly employed by a number of authors.

Presence or absence of a cortex, for example, is not easily determined unless special spicules reinforce this structure. Purely organic, fibrous (spongin) ectosomal reinforcements can range from thin layers, detectable only by microscope, to dense, cortex-like structures of 1 mm or more in thickness. Actually, all genera but one are described with some form of a cortex, *Tetilla* being the genus characterized by lacking one.

Characteristics of the aquiferous system concern the incurrent and excurrent openings and may have been subject to misinterpretation. A stucture called porocalyx is characteristic of the genus *Cinachyra* Sollas (*sensu stricto*, see definition below) and of the subgenus (*Tetilla*) *Cinachyrella* Wilson. Porocalices are circular, poriferous depressions in the ectosome, interrupting the cortex where present, and appearing as distinctive pits or even flask-shaped structures. At least one genus, *Fangophilina* Schmidt (1880:72), type species *F. submersa* Schmidt, has been defined by having two unlike poriferous pits, one inhalant, the other exhalant (Lendenfeld, 1907 ; Wilson, 1925). I doubt that this diagnosis can be upheld without being physiologically substantiated. Simple, typical oscula have been located in *Cinachyra barbata* Sollas, type species of the genus (Kirkpatrick, 1905 ; Boury-Esnault & Van Beveren, 1982 ; own observations on uncatalogued U.S.N.M. specimens) and in all specimens of four *Cinachyrella* species studied alive by myself, indicating that simple oscula may also occur in *Fangophilina*. Live specimens observed in Belize and then fixed in either absolute ethyl alcohol, 10% formalin-seawater, or by freezing demonstrate that contraction during preservation reduced the oscular openings to solid mounds or wartlike structures, whereas the porocalices remained practically unchanged. This test explains why preserved material usually does not diplay oscula clearly. Experiments with fluorescent dye administered to live *Cinachyrella* show also that not all porocalices are equally active drawing in water at all times. In a few observations it appeared that flow was reversed, water actually leaving the porocalyx, but closer examination revealed inhabiting polychaetes and amphipods to cause this effect.

Presence of certain spicule types in or below the ectosome, or of unusual (for the family) spicules anywhere in the body appears to be a very useful generic distinction. Absence of microscleres, on the other hand, is hardly a logical reason for establishing a separate genus, particularly since that spicule type (the sigmaspire) is part of the diagnosis of the family. With this view, already expressed by Dendy (1924:313), two genera should be dropped : *Tethyopsilla* Lendenfeld (1888:45), type species *T. stewartii* Lendenfeld, and *Craniellopsis* Topsent (1913:14), type species *Tethya zetlandica* Carter. These microscleres are often overlooked by authors, as could be demonstrated by van Soest (1977:2) for *Acanthotetilla* Burton (1959:201), thus placing *Acanthocinachyra* Lévi (1964:306) in synonymy with the former genus. However, single specimens or, possibly, whole populations of a species may indeed lose or not develop microscleres, as I could verify in specimens (USNM 31200, 31415) from the Gulf of Mexico, tentatively identified porocalices and special cortical oxea, should be assigned to the genus *Cinachyra* (*sensu stricto*). Others in a similar situation should also be assigned to more appropriate genera.

Another dubious genus is *Chrotella* Sollas (1886:181), type species *C. macellata* Sollas, which is defined by having a "cortex excavated by subdermal cavities and furnished with tangentially disposed spicules" (Sollas, 1888:17). The cavities seem hardly significant and the cortical spicules are neither mentioned in the text describing the type species, nor shown in the accompanying figures (Sollas, 1888 : Pl. IV, Figs 17-18) ; only uncharacteristic, broken spicules seem to occur in a criss-cross fashion. The type species is reported to have toxospires concentrated in the cortex, in addition to the usual sigmaspires of the choanosome, but even this feature is not considered important enough to justify a separate genus. Lendenfeld (1903:21) transferred *C. macellata* to *Tetilla*.

REVISED DIAGNOSES OF TETILLID GENERA

As a result of the above discussion the following seven genera of Tetillidae are now considered valid (see also table 1).

1. *Tetilla* Schmidt (1868:40)

   Type species : *T. euplocamos* Schmidt (1868:40)

   Diagnosis : Tetillidae without cortex, porocalices, or unusual spicule types.

2. *Craniella* Schmidt (1870:66).

   Type species : *C. tethyoides* Schmidt (1870:66)

   Diagnosis : Tetillidae with cortex traversed by special spicules, without porocalices or unusual spicule types.

Remarks : cortical oxeas are reported to be in radial orientation (Sollas, 1888:30). However, I have examined a specimen of *C. tethyoides* from Barbados, identified by Schmidt himself (USNM 984), in which the cortical oxeas occur criss-cross in the cortex, as well as, although less densely, in the choanosome.

Table 1 - Valid genera of Tetillidae and principal characteristics used to distinguish them (+ = present ; - = absent).

Genus	Cortex	Porocalices	Conspicuous accessory spicules	Nature of accessory spicules
Tetilla Schmidt, 1868	-	-	-	-
Craniella Schmidt, 1870	+	-	-	-
Cinachyra Sollas, 1886	+	+	-	-
Paratetilla Dendy, 1905	-	+	+	calthrop-like
Amphitethya Lendenfeld, 1907	+	-	+	amphiclads
Cinachyrella Wilson, 1925	-	+	-	-
Acanthotetilla Burton, 1959	-	-	+	acanthoxea

3. *Cinachyra* Sollas (1886:183)

Type species : *C. barbata* Sollas (1886:183)

Diagnosis : Tetillidae with pronounced cortex reinforced by special, robust oxeas, with flask-shaped porocalices, without unusual spicule types (Figs 1a, c, e ; 2e).

Remarks : cortical oxeas were originally described to traverse the cortex radiately (Sollas, 1888:23). Boury-Esnault & Van Beveren (1982) found their orientation rather confused except near the porocalices where they become radial. My own observations (uncatalogued specimens in U.S.N.M. collection) confirm this pattern for large specimens, but very small specimens (10-15 mm diameter) have the cortical oxea in almost perfect tangential orientation.

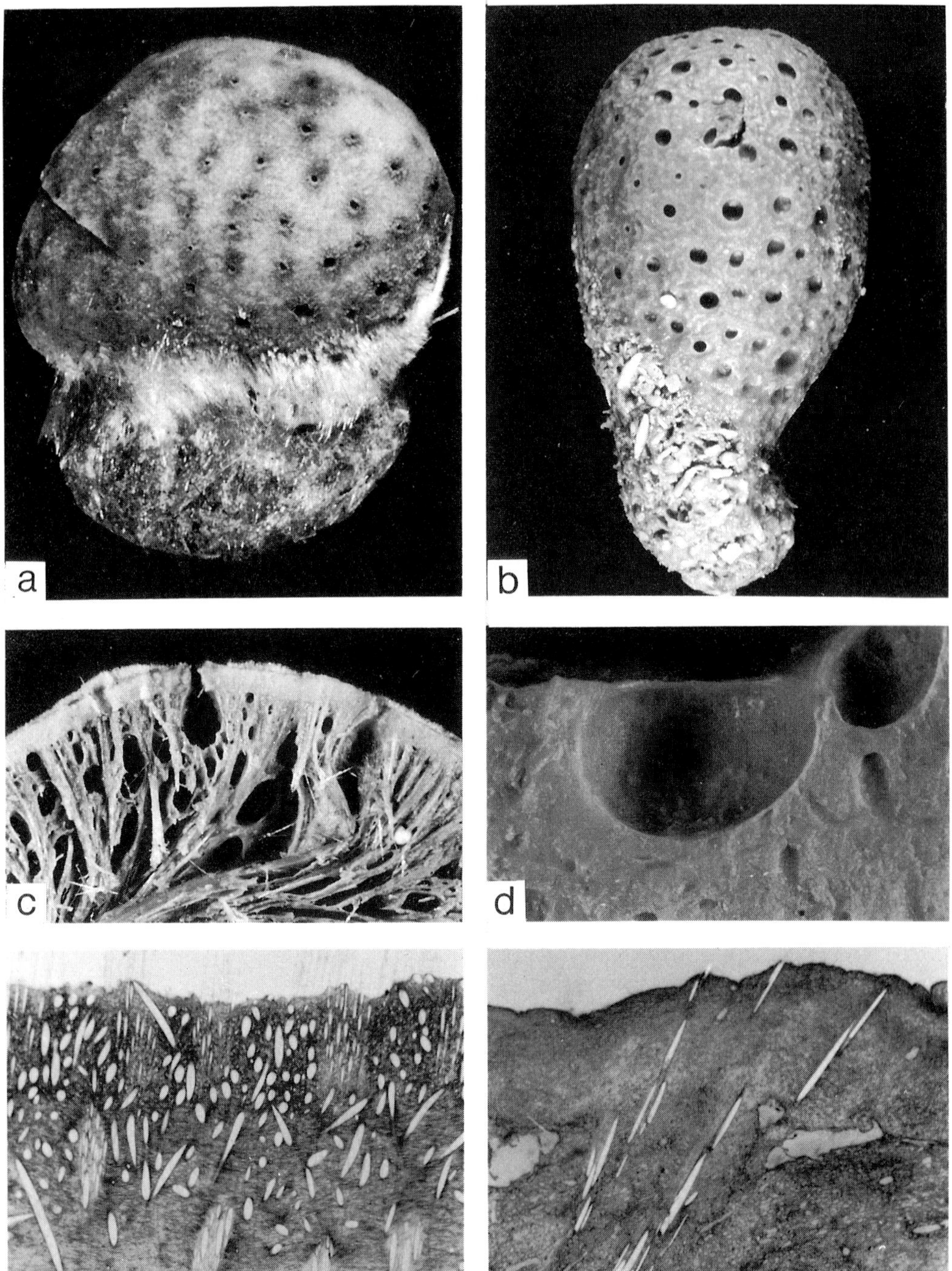
a
b
c
d
e
f

4. *Paratetilla* Dendy (1905:97)

Type species : *P. cineriformis* Dendy (1905:97)

Diagnosis : Tetillidae without cortex, with porocalices, and with a special layer of modified triaenes, resembling calthrops, lying in the ectosome or at the junction between ectosome and choanosome.

Remarks : Wilson (1925:380) reports that species without porocalices occur in this genus. The type species, however, is described as having poriferous "pocket-like depressions" (Dendy, 1905:97).

5. *Amphitethya* Lendenfeld (1907:126)

Type species : *A. microsigma* Lendenfeld (1907:126)

Diagnosis : Tetillidae with cortex, without porocalices, with a layer of cortical amphiclads (amphi-triaenes,-diaenes,-monaenes) and plagiotriaenes.

Remarks : in the type species, at least, the cortex is restricted to the stalk of the sponge (Lendenfeld, 1907:127).

6. *Cinachyrella* Wilson (1925:363)

Type species : *Tetilla hirsuta* Dendy (1889:75)

Diagnosis : Tetillidae without cortex, with poriferous pits (porocalices), without unusual spicule types (Figs 1b, d, f).

Remarks : Dendy (1922:11) already noted that the genus *Cinachyra* has been widely accepted and that many additional species have been assigned to it but that "none of these possesses the radially arranged cortical oxeas so characteristic of the type". With the same observation in mind Wilson (1925) established *Cinachyrella* as a subgenus of *Tetilla* but none of the later authors followed this practice. As a result of a recent study of *Cinachyra barbata* (Boury-Esnault & Van Beveren, 1982), N. Boury-Esnault (in lit., 1986) suggested to raise

---

Figure 1 - Comparison between *Cinachyra* and *Cinachyrella* : a/ *Cinachyra barbata* Sollas, habit, type species of the genus, Antarctica, x 0.9 ; b/ *Cinachyrella paterifera* Wilson, habit, originally described under *Tetilla (Cinachyrella)*, Philippines, x 0.9 ; c/ *Cinachyra barbata*, cortex with porocalyx in cross section, x 1.4 ; d/ *Cinachyrella alloclada*, ectosome with porocalyx in cross section, x 1.4 ; e/ *Cinachyra barbata*, spicule reinforced cortex in cross section, x 26 ; f/ *Cinachyrella alloclada*, ectosome and outer choanosome in cross section, x 26.

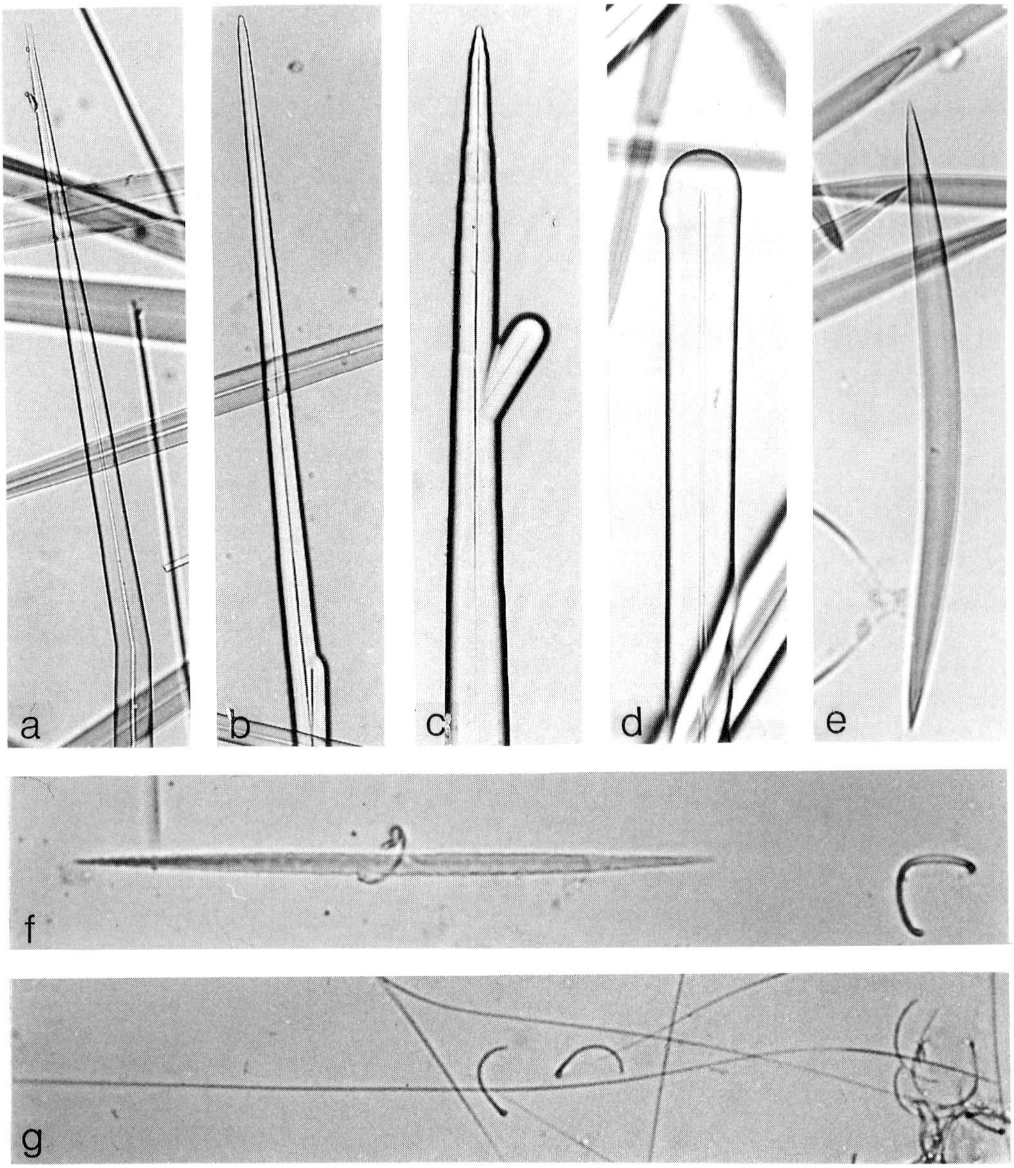

Figure 2 - Type and variability of monaxon spicules in *Cinachyrella* and *Cinachyra* : a/ simple kink near tip of large oxea, *Cinachyrella alloclada*, x 180 ; b/ kink near tip of large oxea showing branched axial canal, *C. alloclada*, x 180 ; c/ stepped tip and branching of large oxea, *C. alloclada*, x 180 ; d/ styloid modification of large oxea in *C. alloclada*, x 180 ; e/ cortical oxea, *Cinachyra barbata*, x 180 ; f/ crenulate accessory oxea and sigmaspires, *Cinachyrella kuekenthali*, x 720 ; g/ raphids and sigmaspires, *C. apion*, x 720.

Cinachyrella to the rank of genus to accomodate the many subtropical and tropical tetillids having poriferous pits but lacking a cortical armature of oxeas.

7. *Acanthotetilla* Burton (1959:201)

Type species : *A. hemisphaerica* Burton (1959:201)

Synonym : *Acanthocinachyra* Lévi (1964:386)

Diagnosis : Tetillidae without cortex, with porocalices, and with conspicuous midsized acanthoxeas as accessory megascleres.

Remarks : van Soest (1977:2) demonstrated that Burton (1959) overlooked the sigmaspires in his new sponge. This discovery eliminated the generic differences between *Acanthotetilla* Burton and *Acanthocinachyra* Lévi, placing the latter into synonymy with the former.

DIAGNOSTIC CHARACTERS ON THE SPECIES LEVEL

Based on the detailed analysis of 64 specimens of subtropical and tropical western Atlantic *Cinachyrella*, I tried to form an opinion about the importance and reliability of taxonomic characters used by previous authors to separate species within the family. The specimens were selected from a pool of more than 200 and represented 12 diverse morphological types, 3-10 individuals for each type. The morphological types display combinations of different body shapes (ball, erect pear, horizontal egg, presence of root tufts), size classes, porocalyx properties (diameter, shape, and distribution pattern), and surface conditions (smooth to the touch, bristly). Color, habitat conditions, and depth distribution were also evaluated.

The hypothesis was formed that anatomical details (spicule complement, shape and size, histology) should agree more between individuals whose morphological and environmental characteristics are identical than between those who display very different appearances, or come from different habitats or localities. With this perspective in mind I compared spicules and histological preparations with the following results.

Oxeas (Fig. 2) : they can appear in three distinct size classes ; raphides (some clearly arranged in trichodragmata) can also occur. Care has to be taken not to confuse small foreign oxeas (one of the most common forms in the sediments) with a proper size category. The largest class oxea can be modified to styloid, substyloid, and strongylote forms, a varying number can display stepped tips or a sharp kink close to one tip. The smallest category

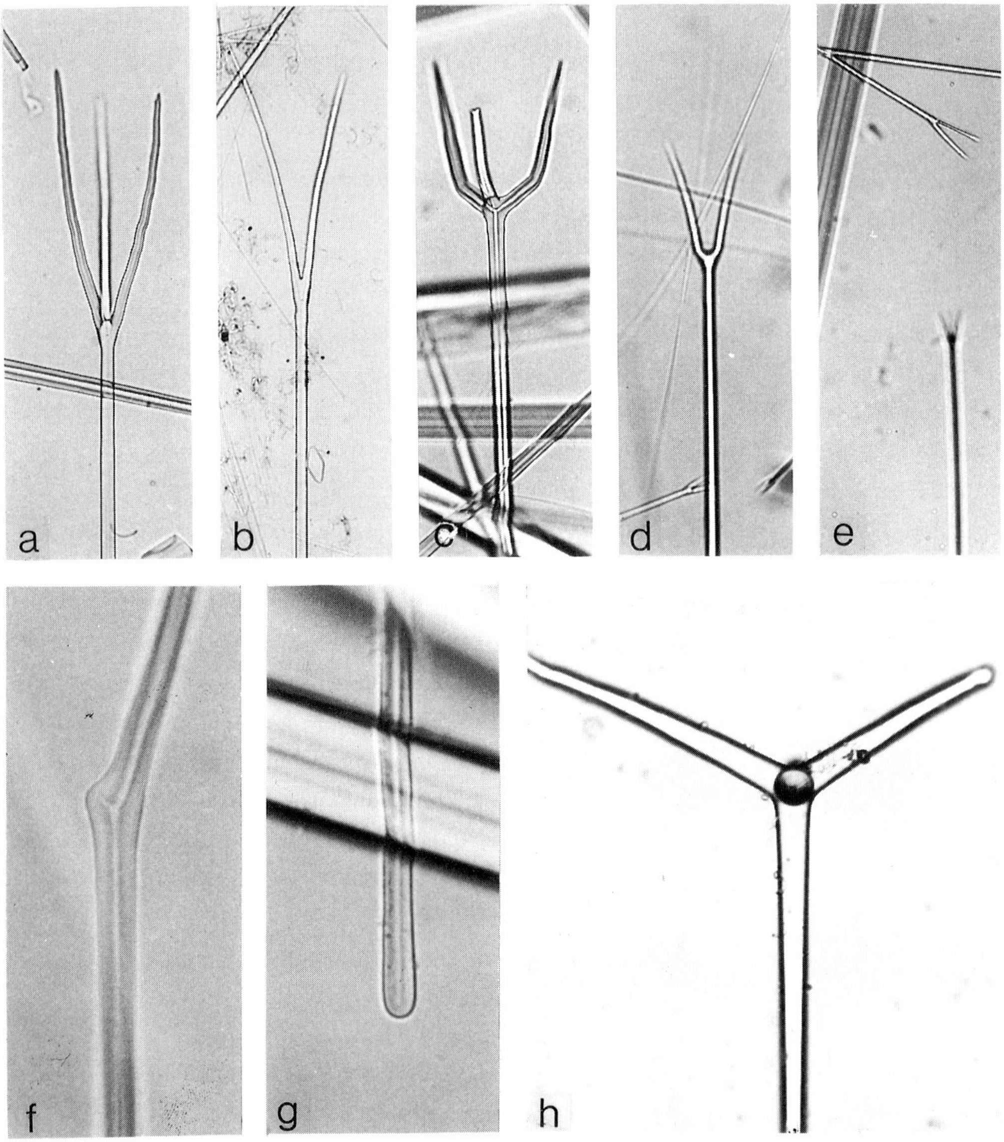

Figure 3 - Types and variability of protriaen-related spicules in *Cinachyrella* : a/ Protriaen, large size class, *C. alloclada*, x 180 ; b/ Prodiaen, *C. apion*, x 180 ; c/ Protriaen, crooked clads, *C. apion*, x 180 ; d/ Prodiaen, shaft diameter increasing toward midlength, *C. apion*, x 180; e/ Prodiaenes, small size class, *C. apion*, x 180 ; f/ Promonoaen showing forked central canal, *C. apion*, x 720 ; g/ rounded shaft point of small protriaen, *C. apion*, x 720 ; h/ Plagiotriaen, *C. tarentina*, x 180.

of oxea can be crenulate or roughened, also bent or curved sharply, or s-shaped.

Protriaenes (Figs 3a-g) : these occur in one or two distinct size classes and usually include reduced forms, such as diaenes (common) or monoaenes (rare). Rhabds are usually thickest just below the cladome, but some increase their width toward mid-length. Some rhabds end in sharp points, others in rounded ones. Clads can be long and slender, short and stout, crooked, bifurcate, or reduced to knobs ; some may diverge in angles as much as 120°.

Plagiotriaenes (Fig. 3h) : those observed in this material were common where they occurred and distinctly different from protriaenes because, in addition to the wide (120°) angles between their clads, they have considerably shorter, thicker shafts and longer clads.

Anatriaenes (Fig. 4) : occur only in one size class. Clads can be long, slender, and strongly curved, or short, stubby and almost straight ; some are bifurcate (rare) or reduced to knobs. Anatriaenes can be absent or extremely abundant in specimens that are otherwise undistinguishable and come from the same locality.

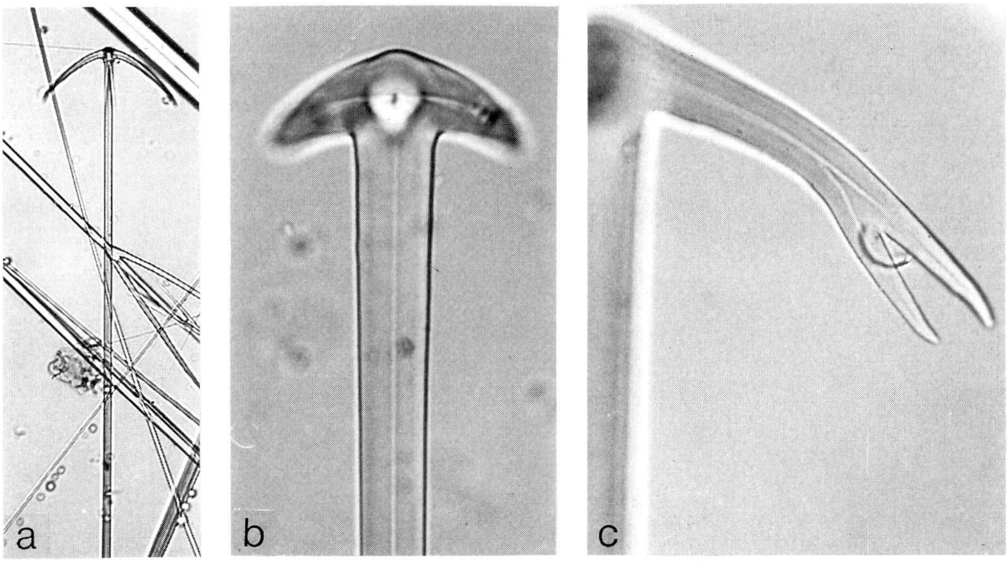

Figure 4 - Types and variability of anatriaenes in *Cinachyrella* : a/ typical anatriaen, *C. apion*, x 180 ; b/ stubby anatriaen, *C. tarentina*, x 720 ; c/ branched clad of anatriaen, *C. apion*, x 720.

Sigmaspires (Figs 2f, g ; 5) : these are c- or s-shaped, twisted to a varying degree, rarely with central swelling (centrotyl), and are beset by fine, thorn-shaped spines which are pointing away from the tips toward the center of the spicule.

Histology : this aspect is difficult to persue in material, such as most of the present one, fixed for museum purposes. Histological features are not very diverse. The ectosome has a stratified appearance due to layers of spongin, interrupted by the presence of numerous spherulous cells which also occur in the choanosome although they are less dense there. Choanocyte chambers are more or less spherical and measure 25-35 µm in maximum diameter. Symbiotic bacteria have only been noted in one species (*Cinachyrella kuekenthali*). Many incorporated foreign spicules and sediment particles (mostly calcareous, 20 µm and 150-300 µm) are noticeable.

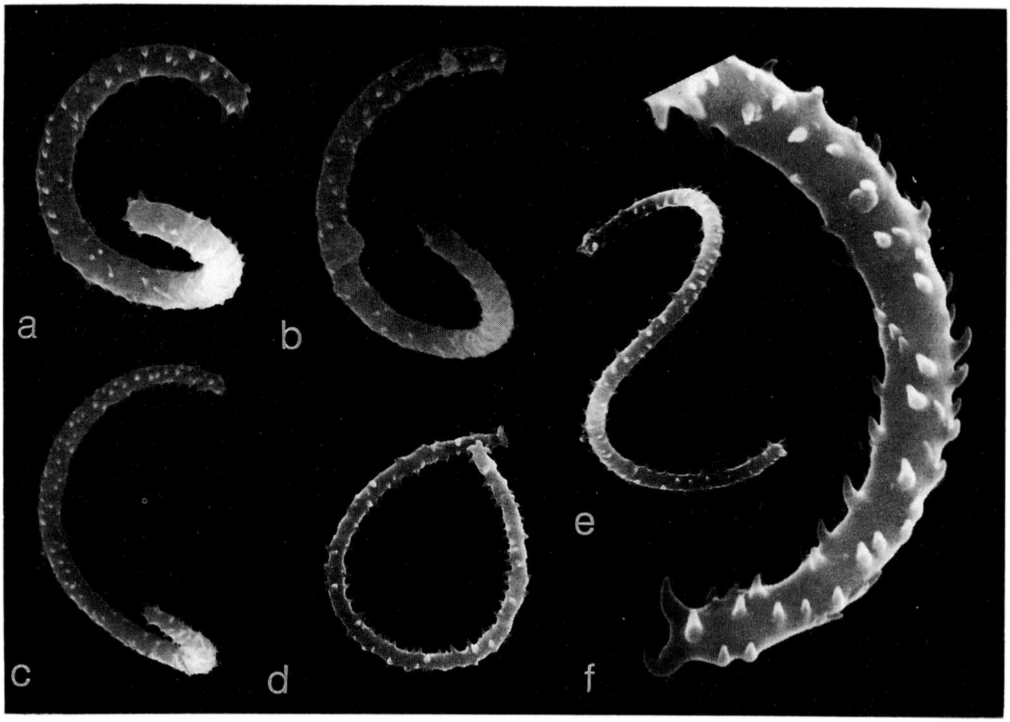

Figure 5 - Sigmaspires viewed by scanning electron microscope : a/ *C. alloclada*, x 3300 ; b/ centrotyl form, *C. alloclada*, x 3300 ; c/ *C. kuekenthali*, x 2100 ; d,e/ *C. apion*, x 3300 ; f/enlarged tip, *C. apion*, x 8200.

Comparing morphologically identical and diverse type groups, and groups from similar and dissimilar habitats, with anatomical characteristics of each specimen I find but few consistencies that allow objective (and, hence, generally applicable) separation of low level taxonomic units (species). Field observations of live material substantiated many theories suggested by study of museum specimens. I have therefore come to the following conclusions.

Specimen shape varies considerably with environment and age. Size is useful only if population averages can be obtained.

Porocalyx size, frequency, and location on the sponge varies much with environmental conditions (water turbulence, sediments) but little with specimen size or method of fixation. Deep (calm) water populations exposed to low sedimentation levels, for instance, develop large (relative to specimen size) porocalices.

Oscula contract easily and fully during fixation ; they are also easily formed new where needed.

Surface properties (bristly or smooth) can be species-characteristic but they are also dependent on environmental conditions (water movement and sediments), growth phase, and circumstances (in museum specimens) of collecting and handling.

Spicule variability is also considerable and only a few properties can be singled out as taxonomically useful. After lumping the 64 specimens of my analysed material into four species (based on at least two stable characters for each) I find that large oxeas with kinks, styloid, stongylate, and subtylostylote forms occur in three (but not all in the same three) species, while two size classes of oxeas occur in two. Protriaenes with rounded rhabd tips are found in all four species, those with rhabds widening at mid-length in three, with crooked and reduced clads in another combination of three. Prodiaenes are always present. Promonaenes (other than the large oxea with kink) were rarely counted because most of them looked distinctly crippled, with one or two knobs in addition to the single clad, or with indication of a forking of the axial canal. Anatriaenes are present in all species, even much more common than protriaenes in some specimens, but rare or absent in many other individuals. Stubby or stout clads occur regularly in two of the species.

A few remaining spicular characters are considered stable and were employed for diagnosing four species of *Cinachyrella* contained in the studied specimen series (Rützler & Smith, in press) : two or more size classes

Table 2 - Described western atlantic species of *Cinachyrella* and proposed taxonomic allocation.

Species named	New allocation	Remarks
Cinachyra rhizophyta Uliczka (1929:38)	*Cinachyrella apion* (Uliczka)	In agreement with van Soest & Sass (1981). The name *C. apion* is given preference because the type specimen of this species, presumed lost, is represented by a good quality photographic illustration.
Cinachyra alloclada Uliczka (1929:41)	*Cinachyrella alloclada* (Uliczka)	
Cinachyra apion Uliczka (1929:43)	*Cinachyrella apion* (Uliczka)	See comments under *C. rhizophyta*.
Cinachyra kuekenthali Uliczka (1929:43)	*Cinachyrella kuekenthali* (Uliczka)	
Cinachyra schistospiculosa Uliczka (1929:45)	*Cinachyrella kuekenthali* (Uliczka)	In agreement with van Soest & Sass (1981)
Trachygellius cinachyra Laubenfels (1936:158)	*Cinachyrella alloclada* (Uliczka)	A new spicule preparation from the holotype (U.S.N.M. 22433) revealed protriaenes, prodiaenes and anatriaenes which were obviously missed by the original describer.
Cinachyra subterranea van Soest & Sass (1981:337)	*Cinachyrella alloclada* (Uliczka)	After examination of a microscope preparation from the holotype (U.S.N.M. 32231, courtesy R.W.M. van Soest) I consider this an immature or stressed specimen of *C. alloclada*.
Cinachyra n. sp. van Soest & Stentoft (in press)	*Cinachyrella* n. sp. van Soest & Stentoft	Possesses rhaphides, but megascleres are different from *C. apion* (1 size class only).
Cinachyra tarentina Pulitzer-Finali (1983:477)	*Cinachyrella tarentina* (Pulitzer-Finali)	Possesses plagiotriaenes. This listing is the first record of the species for the western Atlantic.

of oxeas (*C. alloclada, C. kuekenthali*) ; smallest oxeas rough or crenulated (*C. kuekenthali*) ; presence of plagiotriaenes (*C. tarentina*) ; presence of two size classes protriaenes and of rhaphides (*C. apion*).

VALID SPECIES OF WESTERN ATLANTIC *CINACHYRELLA*

The four species emerging from our analysis, and a fifth species still unpublished (van Soest & Stentoff, in press), appear to be the only ones described for the western Atlantic region. Carrying further the evaluation presented by van Soest & Sass (1981:338, table 2) I am proposing the allocations summarized in table 2.

CONCLUSIONS AND RECOMMANDATIONS

One lesson learned by this study was already well expressed by Wilson (1925:356) who wrote that "*Tetilla* and its relatives offer excellent illustrations of the fact that sponge genera become more and more difficult to distinguish as the number of known species increases". The same seems to hold true for species and specimens. Chances are that the taxonomic criteria accepted here may turn out just as invalid as the ones rejected.

Perhaps a more biological approach than most museum collections allow could help to better classify families of so few and so variable characters as the tetillids. Until advanced molecular systematical methods have been tested it seems well advised to undertake careful regional studies with live material and to describe and illustrate morphological and anatomical characteristics before they become distorted by preservation, handling, and shipping. Without such data it will be impossible to ever attempt objective ecological or zoogeographical analyses.

ACKNOWLEDGMENTS

I wish to thank Kathleen P. Smith for her dedicated assistance in this project. Michael Carpenter and Heidi Wolf helped with photographic work and Cornelia Holley with typing the text. Specimens were made available by the Mineral Management Service (formerly, Bureau of Land Management ; U.S. Department of Interior) through Meredith Jones, Smithsonian Institution. I thank my colleagues Nicole Boury-Esnault and Jean Vacelet for constructive discussions, for co-hosting the September 1986 NATO workshop, and for financial support in Marseille. The Smithsonian Research Opportunities Fund covered most travel expenses for my attendence of the workshop. This is contribution number 201, Carribbean Coral Reef Ecosystems (CCRE) program, National Museum of Natural History, Washington, D.C., partly supported by the EXXON Corporation.

LITERATURE CITED

Boury-Esnault N. & Van Beveren M., 1982 - Les démosponges du plateau continental de Kerguelen-Heard. *Comité National Français des Recherches Antarctiques*, 52:1-175.

Burton M., 1959 - Sponges. *Sci. Rep. J. Murray Exped. (1933-1934)*, 10 (5):151-281.

Dendy A., 1889 - Report on a second collection of sponges from the Gulf of Manaar. *Ann. Mag. nat. Hist.*, ser. 6, 3:73-99.

Dendy A., 1905 - Report on the sponges collected by Professor Herdman at Ceylon in 1902. In: *Herdman Rep. Pearl Oyster Fisheries, London, suppl.*, 18:57-246.

Dendy A., 1922 - Report on the Sigmatotetraxonida collected by H.M.S. Sealark in the Indian Ocean. *Transactions, Linn. Soc. London*, ser. 2, 18:1-158.

Dendy A., 1924 - Porifera I : Non-Antarctic Sponges. In : *British Antarctic ("Terra Nova") Expedition, 1910. Zoology*, 6:269-392. London, British Museum (National History).

Kirkpatrick R., 1905 - On the oscules of *Cinachyra*. *Ann. Mag. nat. Hist.*, ser. 7, 16:662-667.

Lendenfeld R. von, 1888 - Descriptive catalogue of the sponges in the Australian Museum. Sydney, Australian Museum. 260 pages.

Lendenfeld R. von, 1903 - Tetraxonia. In : *Das Tierreich*, 19, Lfg. Friedländer und Sohn, Berlin. 168 pages.

Lendenfeld R. von, 1907 - Tetraxonia. *Deutsche Südpolar Exped. 1901-1903*, 11:57-373.

Lévi C., 1964 - Spongiaires du Canal de Mozambique. *Bull. Mus. natl Hist. nat.*, ser. 2, 36:384-395.

Lévi C., 1973 - Systématique de la classe des Demospongiaria (Démosponges): pp 577-631 in : *Traité de zoologie, (3), Spongiaires*, P.P. Grassé (ed.), Masson & Co, Paris.

Rützler K., 1978 - Sponges in coral reefs : pp 299-313 in : *Coral Reefs : research methods. Monographs on oceanographic methodology*, (5). D.R. Stoddart & R.E. Johannes (eds), Unesco, Paris.

Rützler K. & Smith K.P., in press - Guide to western Atlantic species of *Cinachyrella* (Porifera : Tetillidae). *Proc. biol. Soc. Wash.*

Schmidt O., 1868 - Die Spongien der Küste von Algier. Wilhelm Engelmann, Leipzig. 44 pages.

Schmidt O., 1870 - Grundzüge einer Spongienfauna des atlantischen Gebietes. Wilhelm Engelmann, Leipzig. 88 pages.

Schmidt O., 1880 - Die Spongien des Meerbusen von Mexico. II. Gustav Fisher, Jena : 33-90.

Soest R.W.M. van, 1977 - A revision of the Megacanthoxea-bearing tetillids (Porifera, Spirophorida), with a description of a new species.*Stud. Fauna Curaçao other caribb. Isl.*, 53(172):1-14.

Soest R.W.M. van & Sass D.B., 1981 - Marine sponges from an island cave on San Salvador Island, Bahamas. *Bijdragen tot de Dierkunde*, 51:332-344.

Soest R.W.M. van & Stentoff N.,in press - Barbados deep-water sponges. *Stud. Fauna Curaçao other caribb. Isl.*

Sollas W.J., 1886 - Preliminary account of the Tetractinellid sponges dregded by H.M.S. Challenger, 1872-1876, Part I, The Choristida. *Sci. Proc. R. Dublin Soc.*, 5:177-199.

Sollas W.J., 1888 - Report on the Tetractinellida collected by H.M.S. Challenger. In : *Rep. sci. Results of the voyage of H.M.S. Challenger during the years 1873-1876. Zoology*, 25:1-455. Challenger office, Edinburgh.

Topsent E., 1913 - Spongiaires provenant des campagnes scientifiques de la Princesse-Alice dans les mers du Nord. *Résult. Campagnes sci. ... par Albert Ier, Monaco*, 45:1-67.

Uliczka E., 1929 - Die Tetraxonen Schwämme Westindiens. *Zool. Jahrb. Suppl.*, 16:35-62.

Wilson H.V., 1925 - Silicious and horny sponges collected by the U.S. fisheries steamer "Albatross" during the Philippine expedition, 1907-10. *Smithson. Inst., U.S. natl Mus., Bull.*, 100:273-506.

A STUDY ON THE GENUS *TETHYA* (PORIFERA DEMOSPONGIAE) AND NEW PERSPECTIVES IN SPONGE SYSTEMATICS

M. SARA

Institute of Zoology, University of Genova
Via Balbi, 5, 16126 Genova, Italy.

SYNOPSIS

A general survey of the cosmopolitan genus *Tethya* (Porifera, Demospongiae), whose systematics is very difficult and intricate, in spite of the quantity of literature data available, is reported. After a critical examination of the status of its classification, an analysis of the different traits (morpho-functional, ecological, reproductive, biogeographical, etc.), which could be used in its systematics, has been given as a necessary introduction to its revision. In the Mediterranean and North-Eastern Atlantic the occurrence of two easily recognizable sympatric species of *Tethya*, *T. aurantium* (Pallas) and *T. citrina* Sarà & Melone, (long confused, under the name of *T. aurantium* and its synonymies), has offered good opportunities for research expanded to include the whole range of their morphological and biological differences. These species illustrate besides a great genetic distance, diverging morpho-functional features mainly connected to their different cortical structure and adaptatively related to their different ecological niches. This type of multivarious research, may also be useful for focusing general problems such as the mechanisms of speciation and the significance of taxonomic categories, primarily the "species", in sponges.

INTRODUCTION

The evolutionary field of systematics as a tool for renewal of taxonomy is also being applied increasingly to sponge research. The classical taxonomy of sponges (based on the spiculation) cannot deal with modern questions of speciation without the support of new trends in investigation and new methods recently introduced in sponge research. These new areas of research

NATO ASI Series, Vol. G13
Taxonomy of Porifera
Edited by J. Vacelet and N. Boury-Esnault

include : cytotaxonomy (Simpson, 1968) ; chemiotaxonomy (Bergquist & Wells, 1983 ; Lee & Gilchrist, 1985) ; reproductive biology (Reiswig, 1973 ; Chen, 1976 ; Fell & Lewandrowsky, 1983) ; cladistics (van Soest, 1984) ; immunology (Buss, 1976 ; Curtis *et al.*, 1982 ; Negel & Avise, 1983 ; Kaye & Reiswig, 1985) ; biochemical genetics through enzyme electrophoresis (Solé-Cava & Thorpe, this book).

The taxonomic problems of sponges, particularly for the revision of a genus, can be effectively approached utilizing a wide range of methods and characters, such as : the morphology ; cytology ; reproduction ; ecology ; biogeography ; biochemistry ; and genetics. This approach is particularly useful when the interest lies not only in the phylogenetic trees (i.e. cladists), but also in the evolutionary mechanisms. These mechanisms may lead to the establishment of adaptations that may result in structural and biological novelties. Very little is hitherto known on the adaptative significance of the morphological and biological characters of sponges. It may be noted that adaptation in growth form which until present has been the main object of study in this field, may be also ecophenotypically regulated. Study of the adaptative value of spicular and skeletal characters as well as the life histories of sponges, could possibly reveal important data not only from an evolutionary point of view but also for taxonomic purposes.

The genus *Tethya* displays many opportunities for a sound evolutionary study. It is sharply distinguished from other genera ; it has a world-wide distribution with several species, often widespread in shallow waters and easily recognizable in the field ; it exhibits a richness of characters (which may be taken into consideration for classification purposes) ; it illustrates both sexual and asexual reproduction ; ecophenotypic plasticity is relatively rare ; biochemical genetic methods through electrophoresis have been applied with success. Furthermore the occurrence in the Mediterranean Sea of two shallow water species, *T. aurantium* (Pallas) and *T. citrina* Sarà & Melone, which are easily recognizable and live sympatrically in two contiguous microhabitats, offers a remarkable opportunity to study the adaptative significance of their morpho-functional divergence.

A general description of the genus *Tethya* with a critical survey of the status of its classification follows. An analysis of the different features (i.e. morpho-functional, ecological, reproductive, biogeographical, etc.) which could be used to progress its systematics is also included. This report is an introduction to a next revision.

GENERAL REMARKS ON THE GENUS *TETHYA* AND ITS CLASSIFICATION

In spite of a considerable amount of literature dealing with the various aspects of spiculation, morphology, reproductive biology, systematics, etc., the taxonomy of *Tethya* remains confused.

Tethya is generally recognized as a member of the family Tethyidae in the order Hadromerida. However its position in classification should be examined more carefully. It has been recently suggested (van Soest, this book) that *Tethya* may be better phylogenetically classified with the Tetractinellida. The spiculation of the genus is characterized by : a type of monaxon megasclere, generally a strongyloxea but sometimes a style ; two categories of asters, a megaster (spheraster or spheroxyaster) and a micraster (oxyaster, chiaster or tylaster). Other types of micrasters can occur in the same species and are often located in different regions of the sponge. *Tethya* is also characterized by a radiate structure determined by the strongyloxeas which run in bundles from the center to the periphery of the sponge. It also contains a generally well developed ectosome or cortex. The cortex surrounds the endosome (choanosome) or medulla (which contains the choanocyte chambers) and may display varied structure and development.

Tethya has a dominant position, nearly exclusive, in the family Tethyidae. Only two other genera are attributed to this family according to Lévi (1963) : *Xenospongia* Gray 1867 and *Aaptos* Gray 1867. The inclusion of *Aaptos* among the Tethyidae however is questionable, considering the importance of the cortex in defining the general architecture of these sponges. Topsent (1928) states that *Aaptos* should more likely be included in the Polymastiidae. Burton (1924) includes *Tethyorraphis* Lendenfeld and *Halicometes* Topsent in the Tethyidae along with *Tethya* and *Xenospongia*. *Tethyorraphis laevis* Lendenfeld is distinguished by raphids and *Halicometes stellata* Schmidt is characterized by vestigial triaenes and a stipitate form. Their status is questionable. Perhaps the two single species, on which they are based, are aberrant species of the genus *Tethya*. Another genus is *Cryptotethya* Laubenfels (i.e. *C. crypta* Laubenfels). However *C. crypta*, in spite of its peculiar shape and dimensions, has been considered by Wiedenmayer (1977) as a true representative of the genus *Tethya*. His suggestion has been accepted by the latter authors. Its status should be reevaluated in the revision of the genus. For remarks on the history and synonymies of *Tethya* see Topsent (1920) and Sarà & Melone (1965).

THE SPECIES OF *TETHYA*

Authors who have applied general sponge classification principles to the genus *Tethya* disagreed on the number and identity of "good" species. As a result, there is presently much confusion on this point.

Lindgren (1898) distinguished three groups of species according to the characters of the micrasters. Topsent (1918) and Burton (1924) did not retain *Tethya* species in three distinct groups. However, they (in accordance with Lindgren) stated the micraster characteristics, together with other spicular characters, to be the main trait for distinguishing the species. Vosmaer (1933) criticized these classifications. He stated that many transitions may be observed between the different types of spiculation among the suggested species of *Tethya*. Therefore, Vosmaer excessively lumped the hitherto 22 recent species recorded for *Tethya* into one species, *T. aurantium* (Pallas) (according to the genotype).

Topsent (1918) had made the first careful revision of the genus. He retained only six fully valid species and two doubtful species (Tab. 1).

Table 1 - List of *Tethya* species (under the name *Donatia*) according to Topsent (1918).

1. *T. lyncurium* (Linné)	=	*T. aurantium* (Pallas)
		+ *T. citrina* Sarà & Melone
2. *T. diploderma* (Schmidt)	=	*T. seychellensis* (Wright)
3. *T. japonica* Sollas		
4. *T. fissurata* Lendenfeld		
5. *T. globostellata* Lendenfeld		
6. *T. arabica* (Topsent)		
Dubious species : *T. nux* Selenka ; *T. deformis* Thiele		

Six years later, Burton (1924) published the second and last revision of the genus. Topsent's revision had been practically discarded. Burton's classification leads to a very different list of species (Tab. 2).

Table 2 - List of *Tethya* species (under the name *Donatia*) according to Burton (1924).

1.	*T. peracuata* (Topsent)	[for *japonica* var. *peracuata* (Topsent)]
2.	*T. lyncurium* (L.)	[= *T. aurantium* (Pallas) + *T. citrina* Sarà & Melone]
3.	*T. repens* Schmidt	
4.	*T. deformis* Thiele	
5.	*T. robusta* (Bowerbank)	
6.	*T. magna* (Kirkpatrick)	
7.	*T. fissurata* Lendenfeld not Topsent	
8.	*T. multistella* Lendenfeld	
9.	*T. japonica* Sollas	
10.	*T. diploderma* (Schmidt)	(= *T. seychellensis* Wright)
11.	*T. maza* Selenka	
12.	*T. monstrosa* (Burton)	

Dubious species : *T. caudata* Dezso, *T. globum* (Duchassaing & Michelotti), *T. nux* Selenka, *T. innocens* Schmidt, *T. squamata* (Schmidt), *T. parasitica* Higgin.

His classification contains 12 "good" and six questionable species. It is interesting to note that only three species from Topsent's list [*T. lyncurium (= T. aurantium), T. diploderma (= T. seychellensis)* and *T. japonica*] are retained among the 12 fully valid species by Burton. *Tethya fissurata* Lendenfeld, common in both lists, conceals two separate entities. The *Tethya fissurata* described by Topsent, according to Burton is not the same species as his. On the contrary, it is likely that *T. robusta* Bowerbank of Burton's list and *T. arabica* listed by Topsent are the same species. The disagreement becomes more intriguing considering that both authors have based their lists on the same spicular traits, with particular emphasis on shape, size and localization of the micrasters.

Burton has tried to trace the evolutionary trend of the genus starting from *T. peracuata*. He considered *T. peracuata* as the more primitive species, based on the uniform oxeote type of the rays in the micrasters. *Tethya lyncurium, T. multistella, T. deformis* follow in which a reduction of the oxeote rays is observed. According to Burton , next group in succession is formed by *T. robusta*, *T. magna* and *T. fissurata*. In these species a differentiation of the micrasters into two distinct categories may be observed. This differentiation is stronger in *T. diploderma* and the allied species *T. maza* and *T. monstrosa*. Finally, *T. japonica*, characterized by tylasters only, is derived from the *T. diploderma* group because of the secondary disappearance of the long-rayed oxyasters. This had already been suggested by Sollas (1888) and Lindgren (1898). This evolutionary trend, until now the only advanced for *Tethya*, should be tested through a further, more rigorous investigation, which would not be based solely on a small number of spicular traits.

New data on *Tethya*, subsequent to Burton (1924), regards the description of few new species, the discussion of single species, and records of their ecology, biology and geographical distribution. A survey of the literature shows that only three new valid species of *Tethya* have been described subsequent to Burton (1924). They are : *T. crypta* Laubenfels and *T. actinia* Laubenfels both from the West Indies ; *T. citrina* Sarà & Melone from the Mediterranean Sea. *Tethya citrina* has been previously confused with *T. aurantium* (Pallas). *Tethya compacta* Bergquist (1961), from the Chatham Islands (New Zealand), is characteristic in its shape, but is similar in spiculation to *T. japonica* with which it should be provisionally synonymyzed. *Tethya limski* Müller & Zahn (1968), collected from Rovigno in the Mediterranean Sea, is clearly a synonym of *T. aurantium*. *Tethya crypta* and *T. citrina* have been studied in detail not only for their spiculation and morphology but also in their reproduction and ecology. *Tethya actinia* has been redescribed and recognized as a valid species by Hechtel (1965).

The list of species of *Tethya*, which may be tentatively advanced on the basis of a critical review of the literature, is given in tab.3. The list retains 11 of the 12 fully valid species of Burton (1924), with the substitution of *T. seychellensis* for *T. diploderma* (in accordance with Hechtel, 1965) and the rejection of *T. multistella*. This species was also doubtful for Burton and is probably a synonym of *T. aurantium*. The new list contains 14 species and includes the three new species. However, it may be modified considerably by further research.

Some bibliographic information needs to be clarified. Furthermore, a preliminary study of the *Tethya* collection in British Museum of Natural History reveals that many identifications are erroneous and that several undescribed or overlooked species of *Tethya* should be represented in the collections. These collections contain many type specimens as well as a large quantity of slides and specimens from all the oceans. Therefore, nearly all the described species of *Tethya* are gathered here.

Table 3 - List of *Tethya* species according a critical revision of the literature (until 1986).

1. *T. aurantium*	(Pallas, 1776)
2. *T. repens*	(Schmidt, 1870)
3. *T. robusta*	(Bowerbank, 1873)
4. *T. maza*	Selenka, 1879
5. *T. seychellensis*	(Wright, 1881)
6. *T. fissurata*	Lendenfeld, 1888
7. *T. japonica*	Sollas, 1888
8. *T. deformis*	Thiele, 1898
9. *T. magna*	(Kirkpatrick, 1903)
10. *T. peracuata*	(Topsent, 1918)
11. *T. monstrosa*	(Burton, 1924)
12. *T. crypta*	(Laubenfels, 1949)
13. *T. actinia*	(Laubenfels, 1950)
14. *T. citrina*	Sarà & Melone, 1965

Dubious species : *T. norvegica* Bowerbank ; *T. nux* Selenka ; *T. raphiroides* (Burton) ; *T. philippensis* Lendenfeld ; *T. ingalli* (Bowerbank) ; *T. aurantia* var. *californiana* Laubenfels.

MORPHOLOGICAL AND SPICULAR CHARACTERS

A preliminary analysis of *Tethya* characters is presented here as a basis for a more precise classification. In perspective, the evolutionary trends of the genus may be more understandable as a result.

General morphology

The improvement of field studies has brought about an increasing interest in the external features of *Tethya* as a mean for recognition and distinction of the species. Such features as : dimensions, shape, colour, ectosomic papillae and hardness may be considered important. Mean dimensions could also represent a specific character. An example of this are the species *T. aurantium* and *T. citrina*. Large numbers of specimens that have been studied in relation to sympatric populations have also exhibited clearcut dimensional differences (Sarà & Melone, 1965).

The shape is generally globular but may be irregularly rounded, hemispherical or sometimes encrusting. The specific value of the ecophenotypic or genotypic shape variations needs more investigation. Species of *Tethya* may be also stipitate but this character is sometimes related to reproductive or regenerative processes which allow the deplacement of the sponge. *Tethya crypta* has an irregular, massive shape which represents a distinctive characteristic of this species.

The two species *T. aurantium* and *T. citrina* illustrate that colour may also represent a specific distinctive character. In the Mediterranean, *T. aurantium* is orange-red while *T. citrina*, when uncovered by sediment, is yellow or yellowish-green. Some records of colour exist for other species, but the data are often unreliable because of confusions in nomenclature or because the colour is rarely recorded for the living specimens. Covering by sediment (another fact for consideration) may change the sponge's natural colour. For example, when sedimentation does not cover *T. citrina* and *T. crypta*, they are greenish-yellow. However when covered by sediment, they are brownish or grayish in colour. The specific value of colour, in the majority of the species, should be examined further. Differences in colour recorded for some species (i.e. *T. seychellensis* and *T. japonica*) can even be attributed to erroneous colour records or misidentifications of the species. The external characters of surface structure and papillae may also have a specific value (Hallmann, 1914), but require a careful analysis as they are variable in relation to the biological and reproductive conditions of the sponge.

Hardness or softness of the sponge body are traits related to the ectosomal structure and therefore are important distinctive characters. These are distinguishing features between *T. aurantium* and *T. citrina*. Too little information is presently available on specific differences in the oscula distribution and, in general, the structure of the aquiferous system, (in spite of its ecophysiological relevance).

Cortex

The different widths and structure of the cortex have not been fully evaluated until recently. The comparative study of *T. aurantium* and *T. citrina* reveals that these two species differ strongly with respect to those features. The differing degrees of corticalization is in turn, related to spicular traits such as the shape and distribution of the spherasters as well as the differentiation in categories of the micrasters. Both *T. citrina* and *T. crypta*, which display scarce development of the cortex, are characterized by a more slender oxyspheraster shape of the megasters and by a lack of separation into two distinct categories of the micrasters. The different degrees of corticalization in *Tethya* appears important from a phylogenetic as well as a physiological point of view. Phylogenetically, *Tethya* with its radiate stucture appears to have been derived from deep water ancestors without a cortex (Burton, 1927). Therefore, species with lesser developed cortices may be considered as the more primitive. However, the idea that *T. citrina* and *T. crypta*, possessing more reduced cortices, are specialized and secondarily derived species, cannot be disregarded.

From an eco-physiological standpoint, the cortex, which according to Burton developed in *Tethya* as a protective device for shallow water habitats, may protect the choanosome against high intensities of hydrodynamism and light. A preliminary comparison between the ecological requirements and distribution of *T. aurantium* (possessing a thick, well structured cortex) and *T. citrina* (characterized by a thinner cortex) in the Mediterranean waters reveals that, where the two species coexist in parapatry, *T. citrina* is on the whole, characterized by greater sheltering from hydrodynamism and light than *T. aurantium*. Furthermore, *T. citrina*, with a thinner and softer cortex, shows a greater capability of contraction and expansion. The same has been recorded for *T. crypta* (Reiswig, 1973) in which the cortex is nearly indistinct. This behaviour could be related to areas of high sedimentation and still water habitats in which these sponges, one from temperate and the other from tropical waters, live with the consequent need for quicker water exchange. The partially anaerobic metabolism of *T. crypta* has been related to its sandy habitat (Bergmann & Feeney, 1950).

However, the bibliographic data, even if incomplete on the subject of the cortex, illustrates at least three main types of structures :

1/ Cortex thin and nearly indistinct (in *T. crypta* nearly absent) with only one type of micrasters, as in *T. citrina* and *T. crypta*.

2/ Cortex thick, fibrous and compact with two types of micrasters, an ectosomic and an endosomic, slightly differentiated, as in *T. aurantium*.

3/ Cortex developed but subdivided in two regions, clearly distinguished, an external lacunar region and an internal more compact region, with types of micrasters strongly differentiated, as in *T. seychellensis*.

Megasters

Spicular traits, such as shape and distribution of the megasters, may be related to the structure of the cortex. The megaster may be a spheraster with a center diameter greater than the rays, (i.e. *T. aurantium*) or a slender oxyspheraster with a center diameter less than or equal to the length of the rays (eg. *T. citrina*). *Tethya crypta*, exhibiting an indistinct cortex, has spheroxyasters (Wiedenmayer, 1977). *Tethya seychellensis*, which has a well developed but lacunar cortex in its upper region, has a true spheraster, often with forked rays. The number of rays is another feature which should be taken into consideration. The distribution of the spherasters in the cortex also seems adaptative. *Tethya citrina*, with a thin cortex and slender oxyspherasters, has only one layer of these spicules around the choanosome, while *T. aurantium*, with a thick cortex and massive spherasters, has a multilayered envelope of these spicules around the choanosome. The same characteristic also occurs in *T. seychellensis*.

The shape, dimensions and localization of the megasters are characters whose diagnostic values have not been well explored until the present. The dimensions of the megasters are highly variable intraspecifically and interspecifically. A range in diameter between 20 um (*T. deformis* from Victoria, Australia) and 320 µm (*T. repens* from the Indian Ocean, Murray Expedition) has been observed in the slides from the BMNH. But Burton (1959) records megasters with a diameter of 400-600 µm for specimens of *T. repens* from the Indian Ocean. The location and function of these enormous asters are unknown. However, the dimensions most frequently observed are between 40 and 100 µm. The megasters are generally cortical but there are species that occur with both cortical and choanosomal megasters. Besides the location, their pattern is also an important characteristic. More research needs to be done to evaluate the diagnostic value of their shape which primarily depends on the expansion of the center leading to a spheraster or an oxyspheraster. The shape may be expressed easily by the index R/C : length of the ray/diameter center (Sarà & Melone, 1965). In addition, the shape of the rays, their spination, forking and number should be considered.

Micrasters

The number of rays, and shape of the micrasters, are characters on which a considerable amount of information has been accumulated. In spite of their variability, they may be of considerable systematic value. The basic number of the rays in the micrasters appears to be 12. This number may often be reduced but it has also been observed in greater numbers. In *T. aurantium* and *T. citrina* the most frequent averages are 8-12 with a range of 6 to 14, as in many other species of *Tethya*. However in *T. seychellensis* and related species, the number is often reduced to 6 and even to 4. *Tethya japonica* is normally fixed at 6.

The micrasters shape, which is determined by the forms and spination of the rays and the small or large expansion of the center, may also be very variable in the same region for a single specimen. The phenomenon is illustrated in the cortical micrasters of *T. aurantium* and *T. citrina*. However, these two species not only differ because *T. aurantium* has a special type of choanosomal micraster in addition to the cortical micraster but also because the shape of the cortical micraster of *T. aurantium* generally displays a more tylote shape while *T. citrina* is more oxeote. The shape of the micrasters has been considered very important for recognizing species. *Tethya japonica* is easily identified by its uniform tylasters. However, this character in itself is not sufficient to distinguish species. A relationship between the shape and the mechanical function inside the membranes coating the body surface, the lacunes and the channels may exist. This is displayed by the specialized large and few-rayed micrasters of *T. seychellensis* which occur in species with strongly developed subdermal cavities in its cortex. Dimensions of the same type of micrasters vary intraspecifically and are not distinctive between species. The set of micrasters'types and their location are more important. Their diameters generally range between 5 and 100 um.

Megascleres

The straight megascleres which give *Tethya* its radiate structure are monactines. They exhibit pointed extremity with the other end rounded after narrowing. Strongyloxeas possess a limited amount of dimensional variation between the species compared to the intraspecific variability. The shape is relatively constant. Some species, however, can be recognized from the more stylote or tylote form of its megascleres. The strongyloxeas are arranged in bundles whose consistency, number and disposition have never been analyzed in spite of their usefulness in the systematics of *Tethya*.

CYTOLOGICAL AND BIOCHEMICAL CHARACTERS

A comparative survey of the cytological characters which could be used in the systematics of *Tethya* is presently lacking. Simpson (1968) indicated that these characters may be important in distinguishing sponge species.

Biochemical characters, even if generally employed for the distinction of major taxa, may also eventually identify species. This mechanism is illustrated by differences in the presence of carotenoid pigments between *T. aurantium* and *T. citrina*. *Tethya aurantium* contains only echinenone while *T. citrina* contains both echinenone and ß-carotene (Liaci, 1964). Special nucleosides (Bergmann & Burke, 1965) and extra DNA (Nigrelli & Stempien, 1963) have been found in *T. crypta* and correlated to its ecology with facultative anaerobism (Stempien *et al.*, 1965).

Biochemistry is also used for genetic analysis by allozyme electrophoresis. Research on gene-enzyme systems has indicated a great genetic distance (D=1,469) between the sympatric species *T. aurantium* and *T. citrina*. Such a distance in other phyla separates genera (Sarà, in press).

REPRODUCTION AND LIFE HISTORIES

Sexual reproduction, studied mainly on *T. aurantium* from The Channel (Lévi, 1956) occurs through gonochorism and oviparity. Simple eggs, ovoid in shape and measuring 60 µm in diameter, are released and in 2-3 days give raise to a parenchymella which settles and produces the post-larval sponge. The morphogenetic period is long (Lévi, 1956) but the time for sponge maturation has not been evaluated. Mature eggs of *T. aurantium* are released at the end of July-beginning of August at Roscoff. Specimens found at Palese (Bari) located in the Southern Adriatic Sea spawn from July to October (with maximum egg release from August to September) (Scalera-Liaci *et al.*, 1971). Sperm release, observed only at Palese, occurs during the same time period.

Tethya citrina populations from Palese sympatric with those of *T. aurantium* have a somewhat more delayed cycle ; eggs are released from August to November and sperm from August to October (Scalera-Liaci *et al.*, 1971). The delay of *T. citrina* maturation in comparison with *T. aurantium* has a high level of statistic significance ($p<<0.05$) and is consistent with the hypothesis of a parapatric type speciation for the separation of *T. citrina* from *T. aurantium* (Sarà, in press). The maturation of eggs-cysts is synchronous and that of spermiocysts is asynchronous. Conversely, *T. crypta*, a tropical species, is sexually mature during the three months exhibiting high storm frequencies and decreasing water temperatures (November to January) (Reiswig, 1973). From a sexual reproduction point of view, *Tethya* species behave as K-strategists in both tropical and temperate waters. This is indicated by the short reproductive period which expends a moderate allocation of energy for sexual reproduction. This allocation has been evaluated for *T. crypta* as being only 1 % of that of parent. Other interesting points revealed from the relatively scarce data available are : the different timing of sexual maturity for the species living in the temperate versus tropical waters ; the slight shift of the time of maturation according to latitude (and therefore to water temperature) between the Mediterranean and North-European populations of *T. aurantium* ; the little but significant difference in the time of maturation between the sympatric Mediterranean populations of *T. aurantium* and *T. citrina* as previously mentioned. No comparative data on the egg morphology of different species is presently available, but the egg of *T. citrina* shows a very interesting ultrastructure with a well developed, multilayered cortical zone (Gaino *et al.*, in press). It would be interesting to compare the development of the embryo and the behaviour of the larva in relation to the different ecologies of the species.

Asexual reproduction through buds (sorites) has been observed in many species (*T. aurantium*, *T. citrina*, *T. seychellensis*, *T. japonica*, *T. actinia*, *T. maza*, *T. ingalli*). Budding occurs year round, but for the temperate species *T. aurantium* it is especially frequent during the winter months (Connes, 1978) when the energy allocation in sexual reproduction is lesser. The means of asexual reproduction occurring in *Tethya* are various because the buds may be produced in different ways and reproduction can also occur through peduncles and stolons. The capacity for regeneration and the production of filaments and protuberances is very high and may also be utilized for deplacements (Fishelson, 1981). A comparative study of the buds of *T. aurantium* and *T. citrina* inhabiting the same area shows specific differences. The buds of *T. citrina* are more numerous and smaller (Sarà & Melone, 1965). The red and yellow varieties of *T. aurantium* (the latter more likely *T. citrina*) studied by Connes at Sète (Mediterranean) also reveal differences in budding. Sponge development from the bud is very slow, in some cases lasting more than ten months.

Tethya species have a characteristic behaviour which depends primarily on the strong contractibility of the cortex. Rhythmic pumping behaviour has been observed in *T. crypta* (Reiswig, 1971) : a circadian rhythm in relation to illumination and a 16-21 days rhythm which is endogenous and related to the clearance of the water system. During contraction, the sponge suspends its pumping activity and can also undergo histological changes. No data for these rhythmic activities are available in other species, though oscular and body contractility have been observed in *Tethya* following incidental contact and water movement before the contact. The contractility in *T. citrina* is accompanied by a strong reduction in body volume. The capacity for greater contractility of *T. citrina* is another distinguishing feature from *T. aurantium* and may be related to the thinner and softer structure of the cortex.

ECOLOGICAL CHARACTERS

The ecology of different species of *Tethya* is badly known although three species have been studied in detail : *T. aurantium* (Crozier, 1918 ; Burton, 1948) ; *T. citrina* (Sarà & Melone, 1965) ; *T. crypta* (Reiswig, 1973). Auto-ecological and physiological data may be applied to future studies as demonstrated by Reiswig (1971, 1973, 1974) for *T. crypta*.

Reiswig (1973) has analyzed, in detail, the life history of *T. crypta* in the tropical waters of Jamaica. The species appears restricted to nutrient-rich waters, in habitats free from deposition of fine sediment and marginal habitats of the tropical region. Lack of data prohibits the assumption of whether the observations by Reiswig on *T. crypta* life-history as a K-strategist can be extended to other species of *Tethya* or if these species differ greatly in life histories. However, it is probable that *Tethya* species, in comparison with other sponges, may be considered as specialists : all reproduce through oviparity and are characterized by slow growth ; definitive shape ; complexity of structure and adaptations. Their existence is bounded to definite ecological conditions. Although they are widespread in the littoral zone of all the temperate and tropical seas, their distribution is patchy and linked to special habitats. This patchiness may also be observed on a smaller scale in areas where they are abundant. However, considerable ecological differences among the species can be detected. *Tethya crypta* and *T. citrina*, with their specialized structure and peculiar ecological requirements, may by considered among the more stenotopic and specialized species of the genus.

Tethya aurantium and *T. seychellensis*, widespread in larger geographic areas, are also more eurytopic. *Tethya seychellensis* is characteristic of lagoons and atolls (including low tide stations) but has been also found in dark galleries and holes in the platform of the coral reef (Vacelet & Vasseur, 1971). In some stations where different species of *Tethya* have been found, they appear to occupy different niches with different ecological requirements. In the Mediterranean Sea, *T. aurantium* and *T. citrina* ranges may overlap in some stations, but the general distribution is that of parapatric species with repartition along contiguous but different bathymetric belts. *Tethya citrina* occurs mainly between depth of 0 to 1 m and below 40 m while *T. aurantium* is generally found between 1 and 40 m and only occasionally below this depth. Moreover, *T. aurantium* inhabits areas more exposed to light and hydrodynamism while *T. citrina* is restricted to more sheltered places with higher sedimentation. As previously reported, the constancy of these ecological preferences with the respective morphological features, represents an interesting example of morpho-functional adaptations in sponges.

In Cuba, Alcolado (1985) has found that *T. diploderma* (= *T. seychellensis*), *T. aurantium* and *T. crypta* populations dominate different stations. When they occasionally share the same station, *T. aurantium* is the more ubiquitous : living on the surface of rocks ; shell fragments ; pebbles and any other hard surface ; in horizontal positions or under roofs. Elsewhere it is also apparently tolerant to muddy or sandy substrata (Topsent, 1928 ; Burton, 1948). It has been found on the rhizomes of *Posidonia* in different localities, and is widespread in the cold shallow waters of the Arctic (Koltun, 1970). However, the data from the Arctic should be critically reevaluated, considering the confusion of *T. aurantium* with *T. citrina* and the uncertain status of *T. norvegica*. *Tethya aurantium* has also been recorded from great depths, (up to 805 m) but it needs to be ascertained whether or not deep populations pertain to *T. citrina*. Among the other species of *Tethya* only *T. repens* has been recorded from greater depths (878 m, at Maldives) (Burton, 1959). A sound analysis of the fine distribution and ecological requirements of the different species of *Tethya* would be very interesting for their characterization.

BIOGEOGRAPHICAL CHARACTERS

The genus *Tethya* is cosmopolitan. Its center of origin is of course unknown but the major part of its species seems localized in the Indo-Pacific area. The European waters are inhabited by two valid species and perhaps a third if *T. norvegica* proves to be a "good" species. In the Atlantic Ocean no more than 7-8 species (out of 30 suggested species) can be presently recognized. Tropical waters seem more rich in species than temperate, and subpolar waters are only exceptionally inhabited (probably only by *T. aurantium*). In an hypothetical way an Indo-Pacific, tethysian origin of *Tethya* can therefore be assumed.

Paleontological data on Tethyidae are lacking. Only from references to the other families of Demospongiae can the origin be dated around the Jurassic-Cretaceous. The genus *Tethya* may be very old and the origin of some extant species seems remote, as indicated by the great genetic distance between *T. aurantium* and *T. citrina* (Sarà, in press).

It is very difficult to indicate, on the basis of the present knowledge and the intricated status of the systematics of *Tethya*, the geographic range of the different species. Some species show a very large, nearly cosmopolitan, range. Whether the very distant populations from different oceans pertain to the same species or to a group of related species needs to be ascertained. This control is needed especially for *T. aurantium*, *T. seychellensis* and *T. japonica*. In addition to the European waters (Mediterranean and North-European) *T. aurantium* has been recorded : in the Western Atlantic coast between Florida and the Caribbean Sea ; in the Eastern Atlantic coast until the Islands of Cabo Verde ; in the Indian Ocean (Red Sea, Seychelles); and questionably in the Pacific Ocean (Japan, Australia, New Zealand, and California). The range of *T. seychellensis*, if conspecific with *T. diploderma*, extends throughout all circumtropical seas. Outside the Pacific and Indian Oceans *T. japonica* has been found in the Atlantic Ocean (along the coasts of South America). Other species have a more restricted range. *Tethya citrina* occurs in the Mediterranean, Eastern Atlantic including Canaries and Azores. *Tethya crypta* has been found only in the Caribbean area. The range of different species often overlaps but a different area of major diffusion is generally recognizable for each of them. For example : *T. aurantium* is more frequent in the European seas ; *T. japonica* in the Japanese and Indonesian seas ; *T. seychellensis* in the Caribbean Sea and in the Indian Ocean ; *T. magna* in the South African seas ; *T. robusta* along the Australian coasts.

The only hypothesis hitherto suggested on the evolutionary trend of the genus and phylogenetic relationships between species is that of Burton (1924). As previously stated, a more precise effort in the direction requires a considerable amount of taxonomic work.

A NEW APPROACH TO THE SYSTEMATICS OF *TETHYA*

A new approach is required to clarify the confused status of the systematics of *Tethya*. The first objective is to elucidate some key characters of the cortical structures with its skeletal architecture and the distribution of micrasters and megasters in the different species. This may be done with the existing collected materials as the basis. New data from field investigations are also required for a better understanding of the ecological requirements and distribution, external appearance in life, abundance and coexistence of the different species in diverse biotopes. Important data for a biological characterization of the species may be obtained through a study of life histories, population dynamics, biochemistry and cytology. The role of sexual and asexual reproduction in their life histories, and their relationship to the different ecological requirements of the species, should also be investigated. The enzyme electrophoresis, applied to *T. aurantium* and *T. citrina* shows that these species are not only separated by a great genetic distance but that they also differ in their intraspecific variability pattern. This fact may be related to their different ecological niches (Sarà, in press). An extension of these studies may give valuable information not only on the population structure, reproductive and survival strategy of single species, but also on the mechanisms of speciation which have operated in *Tethya*. A splitting of the genus into subgenera may be tried on the basis of the different types of cortical structure. In this case *T. aurantium* and *T. citrina* could be considered as representative of different subgenera as supported by their high genetic distance. A scenario in which the subgenera are the well differentiated species, and in which many diagnostic characters change in a correlated way, in relation to their different ecological niches, may therefore be envisaged. Their divergence may have arisen following mutations in the regulatory genetic system which determined heterochronic changes in morphogenesis, in accordance with the theory of punctuated equilibria (Gould & Eldredge, 1977). Adaptation of the new forms to suitable open niches would then follow through disruptive selection. On the other hand widely distributed species such as *T. aurantium, T. seychellensis* and *T. japonica* could represent groups of cryptic allopatric species separated by geographic isolation and speciated in allopatry mainly through stochastic processes.

Enlightenment of these problematic questions may interest not only the taxonomy of *Tethya*, but also the general systematics of sponges.

ACKNOWLEDGMENTS

I thank Miss Shirley Stone for assistance in my study of the BMNH collections of *Tethya*. The research has been supported by a grant M.P.I.

LITERATURE CITED

Bergmann W. & Burke D.C., 1955 - Contributions to the study of marine products. XXXIX. The nucleosides of sponges. III. Spongothymidine and Spongouridine. *J. Org. Chem.*, 20 : 1501-1507.

Bergman W. & Feeney C.J., 1950 - The isolation of a new thymine pentoside from sponges. *J. Amer. Chem. Soc.*, 72 : 2809.

Bergquist P.R., 1961 - Demospongiae (Porifera) of the Chatham Islands and Chatham Rise, collected by the Chatham Islands 1954 Expedition. *Bull. N.Z. Dpt Sci. Industr. Res.*, 139 : 169-206.

Bergquist P.R. & Wells R.J., 1983 - Chemotaxonomy of the Porifera. The development and current status of the field : pp 1-50 in *Marine Natural Products, 5*. Scheuer P.J. (ed.), Academic Press.

Bowerbank J.S., 1864-1882 - A monograph of the British Spongiadae. Vol.I. IV, Ray Society, London.

Burton M., 1924 - A revision of the sponge family *Donatiidae*. *Proc. Zool. Soc. London*, 4 : 1033-1046.

Burton M., 1928 - Comparative study of the shallow water and deep sea sponges, with notes on their external form and reproduction. *J. Queckett Microsc. Cl. London*, 16 : 49-70.

Burton M., 1948 - The ecology and natural history of *Tethya aurantium*. *Ann. Mag. Nat. Hist.*, (12)1 : 122-130.

Burton M., 1959 - Sponges. *Sci. Rep. Murray Exped. 1933-34.*, 10 (5): 151-281.

Buss L.W., 1976 - Better living through chemistry : the relationship between allelochemical interactions and competitive networks : pp 315-348 in *Aspects of sponge biology*. F.W. Harrison & R.R. Cowden (eds), Academic Press, New York.

Chen W.T., 1976 - Reproduction and speciation in *Halisarca* : pp 113-137 in *Aspects of sponge biology*. F.W. Harrison & R.R. Cowden (eds), Academic Press, New York.

Connes R., 1971 - Variations du bourgeonnement de *Tethya lyncurium* Lamarck, dans la nature. *Vie Milieu*, 22 : 133-142.

Crozier B.B., 1918 - Bermudan species of *Donatia (Tethya)*. *Ann. Mag. Nat. Hist.*, 9(1) : 11-18.

Curtis A.S.G., Kerr J. & Knowlton N., 1982 - Graft rejection in sponges. Genetic structure of accepting and rejecting populations. *Transplantation*, 33 : 127-133.

Fell P.E. & Lewandrowski K.B., 1983 - Seasonal fluctuations in standing stock biomass of *Halichondria* sp. (Porifera Demospongia) within a New England keelgrass community. *J. exp. mar. Biol. Ecol.*, 66 : 11-24.

Fishelson L., 1981 - Observations on the moving colonies of the genus *Tethya* (Demospongia, Porifera). 1. Behaviour and cytology. *Zoomorphology*, 98 (1): 89-100.

Gaino E., Burlando B., Buffa P. & Sarà M., in press - Ultrastructural study on the mature egg of *Tethya citrina* Sarà & Melone (Porifera, Demospongiae). *Gamete Research*.

Gould S.J. & Eldredge N., 1977 - Punctuated equilibria : the tempo and mode of evolution reconsidered. *Paleobiology*, 3 : 115-151.

Gray J.E., 1867 - Notes on the arrangement of sponges, with the description of some new genera. *Proc. Zool. Soc. London*, 492-558.

Hallmann E.F., 1914 - A revision of the monaxonid species described as new in Lendenfeld's catalogue. Part I. *Proc. Linnean Soc. New South Wales*, 39 : 263-315.

Hechtel G.J., 1965 - A systematic study of the Demospongiae of Port-Royal, Jamaica. *Peabody Mus. nat. Hist. Yale Univ. Bull.*, 20 : 1-94.

Kaye H.R. & Reiswig H.M., 1985 - Histocompatibility responses in *Verongia* species (Demospongiae). Implications of immunological studies. *Biol. Bull. (Woods Hole)*, 168(1) : 183-188.

Kirkpatrick R., 1903 - Description of South African Sponges. Part III. *Marine Invest. South Africa*, 2 : 233-264.

Koltun N.M., 1970 - Sponges of the Arctic and Antarctic; a faunistic review: pp 285-297 in *The Biology of the Porifera, Symp. Zool. Soc. London, 25*. W.G. Fry (ed.), Academic Press, London.

Laubenfels M.W. de, 1949 - Sponges of the Western Bahamas. *Am. Mus. Novit.*, 1431 : 1-25.

Laubenfels M.W. de, 1949 - The sponges of Woods Hole and adjacent waters. *Bull. Mus. comp. Zool. Harvard,* 163(1) : 1-55.

Laubenfels M.W. de, 1950 - The Porifera of the Bermuda Archipelago. *Trans. Zool. Soc. Lond.*, 27 : 1-154.

Lee W.L. & Gilchrist M., 1985 - Carotenoid patterns in 29 species of sponges in the order Poecilosclerida (Porifera, Demospongiae) : a possible tool for chemiosystematics. *Mar Biol.*, 86(1) : 21-36.

Lendenfeld R. von, 1888 - Descriptive catalogue of the sponges in the Australian Museum, Sydney. *Pub. Austr. Mus.*, London. 260 p.

Lévi C., 1956 - Etude des *Halisarca* de Roscoff. Embryologie et systématique des Démosponges. *Archs Zool. exp. gén.*, 93 : 1-184.

Lévi C., 1973 - Systématique de la classe des Demospongiaria (Démosponges) : pp 577-631 in *Traité de Zoologie. III, Spongiaires.* P.P. Grassé (ed.), Masson.

Liaci L., 1964 - Differenze sulla pigmentazione di due forme di *Tethya aurantium* (Pallas) Gray (Porifera). *Atti Soc. Peloritana Sci. fis. mat. nat.*, 10(4) : 569-572.

Lindgren N.G., 1898 - Beiträg zur kenntniss der Spongien Fauna des malayischen Archipels und der chinesischen Meere. *Zool. Jahrb. Abt. Syst.*, 11 : 283-378.

Müller W. & Zahn L.K., 1968 - *Tethya limski* n.sp., eine Tethyide aus der Adria (Porifera : Homosclerophorida, Tethyidae). *Senckenb. Biol.*, 49(6) : 469-478.

Neigel J.E. & Avise J.C., 1985 - The precision of histocompatibility response in clonal recognition in tropical marine sponges. *Evolution*, 39 : 713-723.

Nigrelli R.F. & Stempien M.F., 1963 - The possible presence of an abnormal nucleic acid in the sponge *Cryptotethya crypta*. *J. Histochem. Cytochem.*, 11 : 395-400.

Pallas P.S., 1766 - Elenchus Zoophytorum. *Hagae-Comitum apud Petrum van Cleff.*

Reiswig H.M., 1971 - In situ pumping activities of tropical Demospongiae. *Mar. Biol.*, 9 : 38-50.

Reiswig H.M., 1973 - Population dynamics of three Jamaican Demospongiae. *Bull. mar. Sci.*, 23 : 191-226.

Reiswig H.M., 1974 - Water transport, respiration and energetics of three tropical marine sponges. *J. exp. mar. Biol. Ecol.*, 14 : 231-249.

Sarà M., in press - Divergence between the sympatric species *Tethya aurantium* and *T. citrina* and speciation in sponges. *Proc. 3rd int. Conf. on the Biology of Sponges, Woods Hole.*

Sarà M. & Melone N., 1965 - Una nuova specie del genere *Tethya*, *T. citrina* sp.n. del Mediterraneo (Porifera Demospongiae). *Atti Soc. Peloritana Sci. Fis. Mat. Nat.* Suppl., 11 : 123-138.

Scalera-Liaci L., Sciscioli M., Papa O. & Lepore E., 1971 - Raffronto fra i cicli sessuali di *Tethya aurantium* (Pallas) Gray e *Tethya citrina* Sarà & Melone (Porifera, Hadromerina). *Atti Soc. Peloritana Sci. Fis. Mat. Nat.*, 17 : 287-298.

Schmidt O., 1870 - Grundzüge einer Spongien Fauna des atlantischen Gabietes. Leipzig, 88 p.

Selenka E., 1879 - Ueber einen kieselschwamm von achtstrahaligen Bau, und uber Entwicklung der Schwammknosper. *Z. Wiss. Zool.*, 33 : 467-476.

Simpson T.L., 1968 - The structure and function of sponge cells : new criteria for the taxonomy of Poecilosclerid sponges (Demospongiae). *Peabody Mus. Nat. Hist. Yale Univ. Bull.*, 25 : 5-141.

Soest van R.W.M., 1984 - Deficient *Merlia normani* Kirkpatrick, 1908, from the Curaçäo Reefs, with a discussion on the phylogenetic interpretation of Sclerosponges. *Bijdr. Dierk.*, 54(2) : 211-219.

Sollas W.J., 1888 - Report on the Tetractinellida collected by H.M.S. Challenger, during the years 1873-1876. *Reports Challenger, Zool.*, 25 : 1-458.

Stempien M.F., Nigrelli R.F. & Pulitzer G., 1965 - Further observations on the Caribbean sponge *Cryptotethya crypta* (de Laubenfels). *Nature*, 207, (4993), 217.

Thiele J., 1898 - Studien uber pazifische Spongien. *Zoologica*, 24 : 1-72.

Topsent E., 1918 - Eponges de S. Thomé. Essai sur les genres *Spirastrella, Donatia*, et *Chondrilla*. *Archs Zool. exp. gén.*, 57 : 535-618.

Topsent E., 1920 - *Tethya aurantium* (Pallas) et les *Tethya* de Lamark. *Bull. Mus. natl Hist. nat.*, 7 : 640-649.

Topsent E., 1928 - Spongiaires de l'Atlantique et de la Méditerranée, provenant des croisières du Prince Albert Ier de Monaco. *Res. Camp. Sci. Albert Ier Monaco*, 74 : 1-376.

Vacelet J. & Vasseur P., 1971 - Eponges des récifs coralliens de Tuléar (Madagascar). *Tethys*, suppl. 1 : 51-126.

Vosmaer G.C.J., 1933-35 - The sponges of the Bay of Naples. Porifera Incalcaria, I-III. The Hague, Martinus Nijhoff.

Wiedenmayer F., 1977 - Shallow-water sponges of the Western Bahamas. Birkhauser Verlag, Basel, 287 p.

Wright E.P., 1881 - On a new genus and species of sponge (*Alemo seychellensis*) with supposed heteromorphic zooids. *Trans. Irish. Acad.*, 28 : 13-20.

PHYLOGENETIC EXERCISES WITH MONOPHYLETIC GROUPS OF SPONGES

R.W.M. van SOEST

Institute of Taxonomic Zoology, University of Amsterdam
P.O. Box 20125, 1000 HC Amsterdam, The Netherlands.

SYNOPSIS

The classification of the Porifera is critically evaluated using cladistic principles. The (syn-)apomorphic characters and character states are established for 34 groups of Porifera which are generally accepted as monophyletic. These characters/states are compared with the currently accepted classification of Hartman (1982), resulting in the conclusion that there are many non-parsimonious homoplasies included. Two consecutive attempts to amend this are presented. The first consists of a self-constructed strictly cladistic classification characterized by the assumption of many assumed losses. The second is a computer-based analysis characterized by a neutral construction of a classification with the least number of assumed homoplasies.

INTRODUCTION

In a paper in press (van Soest), some general classification principles to be used in the phylogenetic classification of sponges were outlined. Apart from examples, it demonstrated how current classifications may be violating these principles (notably the universal principles of out-group comparison and parsimony). However, no attempt was made to test the currently generally accepted classification of the Porifera on these principles of phylogenetic classification. The purpose of the present paper is to do so using Hartman's classification as the most recent and complete reflexion of current ideas. Next to a critical evaluation of this classification two attempts are presented to remedy some of the more serious inconsistencies in Hartman's classification. The first consists of a self-constructed classification based on the principle that the independent loss of a character

NATO ASI Series, Vol. G13
Taxonomy of Porifera
Edited by J. Vacelet and N. Boury-Esnault

during an evolutionary process happens much more readily than the independent parallel development of a character. The second exercise consists of an analysis of characters and character-states of groups of Porifera by a computer program using numerically cladistic principles, but without discrimination between true homoplasies and assumed independent losses.

METHODS

Thirty-four groups of sponges comprising a representative proportion of the Porifera, have been selected. These groups are selected on the basis of their presumed firm monophyletic status, but their rank is often quite different. For example : the order Spirophorida is treated as one group, because all members manifest characteristics such that there is no disagreement among spongologists over their common ancestry. On the other hand, within the Poecilosclerida five groups are distinguished, because no general agreement seems to exist over their common ancestry. Desmacellidae lack the most important synapomorphy (i.e. chelae) and they have been associated with Axinellidae in the past. Mycalidae (*sensu lato* including Cladorhizidae, etc.) because they show some similarities in microsclere complement with Desmacellids are also treated separately. Desmacidids (with which Spongillids are also associated : see the latest conclusions of Volkmer-Ribeiro, in press) have been associated with Haplosclerida in the past. Clathriids (or better Microcionids) are associated with Raspailiid Axinellids by Hooper (in press), Myxillids (*sensu lato* including Hymedesmiids, Phorbasiids, Coelosphaerids and Cornulids) are left over. Likewise, Astrophorids, Lithistids, Hadromerids, Halichondrids (which include microsclere bearing genera such as *Didiscus*), Axinellids, Haplosclerids, Dictyoceratids and Dendroceratids have been split apart. Some of these subdivisions may appear unnecessary, but they may serve as a test for the efficiency of the phylogenetic analysis presented below. A few difficult groups (e.g. Sigmaxinellidae) and most groups known only as fossils have been ignored because of the author's lack of knowledge of their characters.

The next step is finding possible synapomorphies for any two or more groups. As each of the chosen monophyletic groups contains good synapomorphies (otherwise they could have been split up), these characters have been ignored because, for the purpose of a further attempt to classify the groups, they must be considered autapomorphies, which have no value.

Table 1 - List of characters and character-states used for the phylogenetic analysis.

1.Skeleton of siliceous megascleres
a.present - b.absent

2.Calcareous spicules
a.absent - b.present

3.Spongin fibre skeleton
a.absent - b.present

4.Hexactine spicules
a.present - b.absent

5.Rhaphides a.absent - b.present

6.Microrhabds a.absent - b.present

7.Sigmata a.absent - b.present

8.Toxa a.absent - b.present

9.Acanthostyles
a.absent - b.present -c.verticillated

10.Desmata a.absent - b.present

11.Sclerosponge basal skeleton
a. absent - b.present

12. Larval strategy
a.oviparous - b.viviparous

13. Larval morphology
a.parenchymella fully ciliated
b.parenchymella bare post.pole
c.coeloblastula
d.amphiblastula

14. Ectosomal crust of microscleres
a.absent - b.present

15.Spongocytes a.absent - b.present

16.Excavating properties
a.absent - b.present

17.Choanocyte size a.small b.large

18.Choanocyte chambers
a. large ascon-type
b. large (rhagon-)type
c. small (rhagon-)type

19.Gemmule formation
a.absent - b.present

20. Sigmaspires a.absent - b.present

21.Taurin synthesis
a.through hypotaurin - b.cystic acid.

22. Reticulate skeletal architecture
a. absent - b.present

23.Incorporation of foreign material
a.absent - b.present

24.Calthrops-type tetractines
a.absent - b.present

25.Triaenes a.absent - b.present

26.Huge radiating oxeotes
a. absent - b.present

27.Small oxeotes a. absent - b.present

28.Tylostyles a.absent - b.present

29.Styles a.absent - b.present

30. Euasters a.absent - b.present

31.Amphiasters a.absent - b.present

32.Spirasters a. absent - b.present

33.Chelae a.absent - b.present

34.Radiate architecture
a. absent - b.present

35.Plumoreticulate architecture
a. absent - b.present

36.Confused architecture
a. absent - b.present

37.Subradiate architecture
a. absent - b.present

38.Tangential ectosomal skeleton
a.absent - present

39.Sphaerasters a.absent - b.present

40.Special ectosomal megascleres
a.absent - b. present

41.Regular ectosomal unispicular reticulum
a.absent - b.present

42.Two or more categories of sigmata
a.absent - b.present

43.Axial condensation of skeleton
a. absent - b.present

Table 1 lists the characters selected to illustrate synapomorphies. Forty-three were chosen. They cover a variety of morphological, skeletal, spicular histological, and life history features. These characters are selected by simply scanning the literature for any character that is shared by more than one of the chosen groups. Each character is checked against the groups in

Table II Matrix of characters (horizontal axis) and monophyletic sponge groups (vertical axis). Numbers refer to characters given in Table I. Abbreviations used for the groups: HEXACT.=Hexactinellida, CALCIN.=Calcinea, CALCAR.=Calcaronea (these two groups include Pharetronids, Murrayonids, etc.),HOMOSC.=Homoscleromorpha, VACELET=Vaceletia (and Sphinctozoa),ASTR.ss= Astrophorida excluding Pachastrellidae, TETHYA= Tethyidae, PACHAST= Pachastrellidae, SPIROPH=Spirophorida, LITH.I=Desmophorida, LITH.II=Scleritodermidae and remaining true Lithistids, HADR.AST.=Aster-bearing Hadromerida, HADR.ss=Non-astrose Hadromerida, CHONDR.=Chondrosida, HAL.Tops.=Halichondrid genera with huge oxeotes (Topsentia et al.),HAL.ss=Halichondria,Hymeniacidon, HEMIAST=Hemiasterellidae, DESMOX=Desmoxyidae, AXINELL=Axinellidae s.l. (incl.Bubaridae), RASPAIL=Raspailiidae s.l.(incl. Euryponidae), DES+SPON=Desmacididae+Spongillidae, MYCAL.=Mycalidae s.l.(incl.Cladorhizidae), MYXILL.=Myxillidae s.l.(incl. Hymedesmiidae+Phorbasidae+Crellidae), CLATHR.=Clathriidae, PETROS=Petrosiidae+Oceanapiidae, HAPLO=Haplosclerida s.s., VERONG=Verongida, DICTYO=Dictyoceratida -Dysideidae, DENDRO= Dendroceratida-Halisarcidae, HALISAR=Halisarcidae,DYSIDEA= Dysideidae.

	1	2	3	4	5	6	7	8	9	10	11	12	13	14	15	16	17	18	19	20	21	22	23	24	25	26	27	28	29	30	31	32	33	34	35	36	37	38	39	40	41	42	43
HEXACT.	a	a	a	a	a	a	a	a	a	a	a	a	a	a	a	a	a	a	a	a	a	a	a	a	a	a	a	a	a	a	a	a	a	a	a	a	a	a	a	a	a	a	a
CALCIN.	b	b	a	b	a	a	a	a	a	a	b	b	c	a	a	a	b	a	a	a	a	a	a	b	a	a	a	a	a	a	a	a	a	a	a	a	a	a	a	a	a	a	a
CALCAR.	b	b	a	b	a	a	a	a	a	a	b	b	d	a	a	a	b	a	a	a	a	a	a	b	a	a	a	a	a	a	a	a	a	a	a	a	a	a	a	a	a	a	a
HOMOSC.	a	a	a	b	a	a	a	a	a	a	a	b	c	a	a	a	b	b	a	a	a	a	a	b	a	a	a	a	a	b	a	a	a	a	a	a	a	a	a	a	a	a	a
VACELET	b	a	a	b	a	a	a	a	a	a	b	b	c	a	?	a	a	c	a	a	a	a	a	a	a	a	a	a	a	a	a	a	a	a	a	a	a	a	a	a	a	a	a
ASTR.ss	a	a	a	b	a	b	a	a	a	a	a	a	?	b	b	a	a	c	a	a	a	a	a	a	b	b	a	a	a	b	a	a	a	b	a	a	a	a	b	a	a	a	a
TETHYA	a	a	a	b	a	a	a	a	a	a	a	a	c	b	b	a	a	c	a	a	a	a	a	a	a	b	a	a	a	b	a	a	a	b	a	a	a	a	b	a	a	a	a
PACHAST	a	a	a	b	a	b	a	a	a	a	a	a	?	b	b	a	a	c	a	a	a	a	a	b	b	b	a	a	a	b	b	a	a	b	a	a	a	a	a	a	a	a	a
SPIROPH	a	a	a	b	b	b	a	a	a	a	a	a	?	a	b	a	a	c	a	b	a	a	a	a	b	b	a	a	a	a	a	a	a	b	a	a	a	a	a	a	a	a	a
LITH.I	a	a	a	b	a	b	a	a	a	b	a	?	?	b	b	a	a	c	a	a	a	a	a	a	b	b	a	a	a	a	b	b	a	b	a	a	a	a	a	a	a	a	a
LITH.II	a	a	a	b	a	b	a	a	a	b	a	?	?	b	b	a	a	c	a	b	a	a	a	a	a	b	a	a	a	a	a	a	a	b	a	a	a	a	a	a	a	a	a
HADR.AST	a	a	a	b	a	a	a	a	a	a	b	a	c	b	b	b	a	c	a	a	a	a	a	a	a	a	a	a	b	a	b	a	b	a	a	a	a	b	a	b	a	a	a
HADR.ss	a	a	a	b	a	a	a	a	a	a	a	a	c	a	b	a	a	c	b	a	a	a	b	a	a	a	a	b	a	a	a	a	a	a	a	a	a	b	a	a	a	a	a
CHONDR.	b	a	a	b	a	a	a	a	a	a	a	a	c	a	?	a	a	c	a	a	a	a	a	a	a	a	a	a	a	b	a	a	a	a	a	a	a	a	b	a	a	a	a
HAL.Tops	a	a	a	b	b	a	a	a	a	b	a	?	?	a	b	b	a	c	a	a	a	a	a	a	a	a	b	b	a	a	a	a	a	a	a	b	a	b	a	a	a	a	a
HAL.ss	a	a	a	b	a	a	a	a	a	a	a	b	a	a	b	a	a	c	a	a	a	a	a	a	a	a	b	a	b	a	a	a	a	a	a	b	a	b	a	a	a	a	a
HEMIAST	a	a	a	b	a	a	a	a	a	a	a	a	a	b	b	a	a	c	a	a	a	a	a	a	a	a	b	b	b	a	a	a	a	a	a	a	a	a	a	a	a	a	b
DESMOX	a	a	a	b	b	a	a	a	a	a	a	a	?	a	b	a	a	c	a	a	a	a	a	a	a	a	b	b	a	a	a	a	a	b	a	a	a	a	a	a	a	a	b
AXINELL	a	a	a	b	b	a	a	a	a	b	a	a	?	a	b	a	a	c	a	a	a	a	a	a	a	b	b	b	a	a	a	a	a	b	a	a	a	a	a	a	a	a	b
RASPAIL	a	a	a	b	b	a	a	a	b	a	a	a	?	a	b	a	a	c	a	a	a	a	a	a	a	a	b	b	a	a	a	a	a	b	a	a	a	a	a	a	a	a	b
CERATOP	a	a	a	b	a	a	a	a	c	a	b	?	?	a	b	a	a	c	a	a	a	a	a	a	a	a	a	a	b	a	a	a	a	a	?	a	a	a	a	a	a	a	a
AGELAS	a	a	b	b	a	a	a	a	c	a	a	a	?	a	b	a	a	c	a	a	a	b	a	a	a	a	a	a	b	a	a	a	a	a	a	a	a	a	a	a	a	a	a
DESMACE	a	a	a	b	b	a	b	b	a	a	b	a	a	a	b	a	a	c	a	a	?	b	a	a	a	a	b	b	b	a	a	a	a	a	b	a	a	b	a	a	a	b	a
DES+SPON	a	a	a	b	a	a	b	b	a	a	a	b	b	a	b	a	a	c	b	a	b	b	b	a	a	a	b	b	b	a	a	a	b	a	a	a	a	a	a	a	a	a	a
MYCAL.	a	a	a	b	b	a	b	b	a	a	a	b	b	a	b	a	a	c	a	a	b	a	b	a	a	a	b	b	b	a	a	a	b	a	b	a	a	b	a	a	a	b	a
MYXILL.	a	a	a	b	b	a	b	b	b	a	a	b	b	a	b	a	a	c	a	a	b	b	b	a	a	a	b	a	b	a	a	a	b	a	a	a	a	b	a	b	a	a	a
CLATHR.	a	a	a	b	b	a	b	b	b	a	a	b	b	a	b	a	a	c	a	a	b	a	b	a	a	a	a	b	b	a	a	a	b	a	b	a	a	b	a	b	a	a	a
PETROS	a	a	a	b	a	a	b	b	a	a	a	a	?	a	b	b	a	c	a	a	?	b	a	a	a	a	b	a	a	a	a	a	a	a	a	a	a	b	a	a	b	a	a
HAPLO	a	a	a	b	b	a	b	b	a	a	b	b	b	a	b	a	a	c	b	a	?	b	b	a	a	a	b	a	a	a	a	a	a	a	a	a	a	b	a	a	b	a	a
VERONG	b	a	b	b	a	a	a	a	a	a	a	a	?	a	b	a	a	c	a	a	b	b	a	a	a	a	a	a	a	a	a	a	a	a	a	a	a	a	a	a	a	a	a
DICTYO	b	a	b	b	a	a	a	a	a	a	a	b	a	a	b	a	a	c	a	a	b	b	b	a	a	a	a	a	a	a	a	a	a	a	a	a	a	a	a	a	a	a	a
DENDRO	b	a	b	b	a	a	a	a	a	a	a	b	a	a	b	a	a	b	a	a	b	a	b	a	a	a	a	a	a	a	a	a	a	a	a	a	a	a	a	a	a	a	a
HALISAR	b	a	a	b	a	a	a	a	a	a	a	b	a	a	a	a	a	b	a	a	b	a	a	a	a	a	a	a	a	a	a	a	a	a	a	a	a	a	a	a	a	a	a
DYSIDEA	b	a	b	b	a	a	a	a	a	a	a	b	a	a	b	a	a	b	a	a	b	b	b	a	a	a	a	a	a	a	a	a	a	a	a	a	a	a	a	a	a	a	a

its state, primarily either present or absent. However, in some cases more possibilities were considered, e.g. larval morphology, or acanthostyles. By applying this method a matrix table (table 2) arose which could be used in finding the distribution of the characters, and their usefulness as synapomorphies at any given level in the desired cladogram.

Using the taxa-character matrix, the following consecutive exercices are made : first, the distribution of characters is compared with the most recent generally accepted classification of the Porifera as presented by Hartman (1982), which reflects the older classification starting with Lévi (1956) (later amended by other authors). Secondly, a different cladogram is constructed using author's ideas and hypotheses still under discussion. Finally, the matrix is exposed to a computer program called "Tree Tools" developed by Ellis (1986), which makes use of numerical cladistic techniques.

A proper cladistic analysis demands the recognition of an out-group in order to be able to establish the primitive vs. the advanced state of characters. Arbitrarily, the Hexactinellids have been chosen, based on Reiswig & Mackie's (1983) and Bergquist's (1985) evidence. However, a dissident opinion on this has been expressed by Mohn (1984), who assumes that Hexactinellids and the Demospongiae are more closely related to each other, than either is to the Calcarea.

RESULTS

I. Hartman's(1982) classification :

Hartman's classification has been translated into the tree of Fig. 1 as justly as possible. The next step is to find synapomorphies for the groups that are distinguished. These are given on the stems of the different parts of the cladogram. The numbers refer to the list of characters given in the table 2.

The following characteristics are apparent when analysing this tree :

— Many important unsolved synapomorphies exist (e.g. no reasonable characters are apparent for grouping the non-Calcarea and the non-Sclerosponge Demosponges) ; other characters are quite weak.

— A large number of assumed homoplasies (parallel developments that account for the fact that the same character(state) occurs in paraphyletic

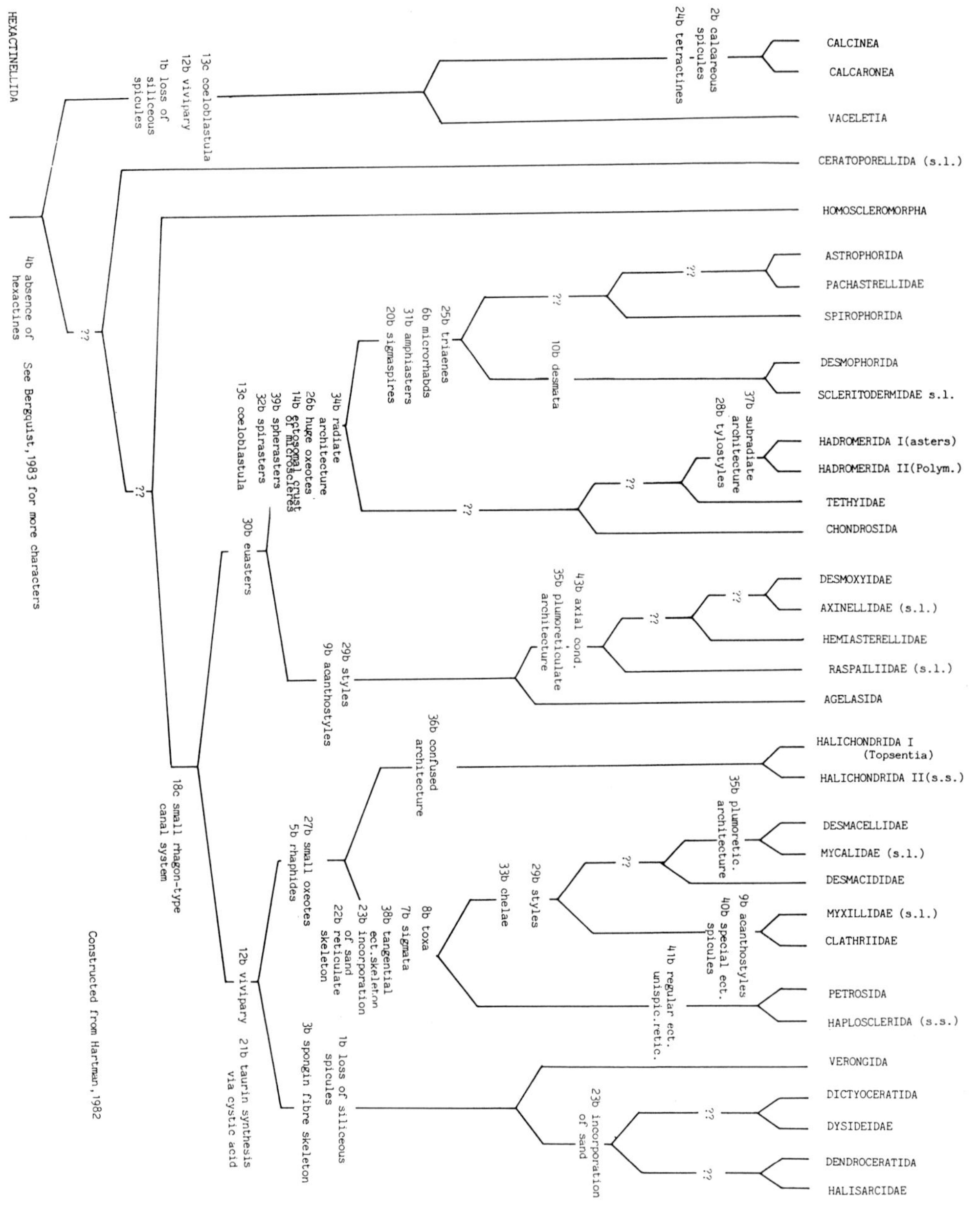

Figure 1 - Cladogram constructed from the classification of the Porifera published by Hartman (1982), representing the most recent and complete, more or less generally accepted, sponge classification. Numbers in the tree refer to the list of characters and character/states given in Table 1. Taxonomic groups are enumerated in the legend of Table 2.

groups). The most important homoplasies are given in table 3.

— A low number of assumed losses ; only those losses have been counted which lack concrete evidence (table 3).

Table 3- Analysis of the three consecutive cladograms presented in Figs 1-3. Abbreviations of taxon names are explained in the legend of Table 2 (in addition : cera = ceractinomorpha).

Hartman's(1982) classification	own tree	computree
Unsolved synapomorphies (total 11)	(7)	(10)
Non-Calcarea	Astrophorida+Tethyidae	Astrophorida+Tethyidae
Non-Sclerosponge Demospongiae	Astroph.+Tethy.+Pachastr.	Astroph.+Tethy.+Pachastr.
Astrophorida s.l.	Hadromerida+Chondrosida	Astroph.+Tethy.+Pachastr.+Spiroph.
Tetractinellida s.s.	Haplo+Desmacid.+Desmacel.+Mycal.	Desmox.+Axinell.
Hadromerida s.l.	Desmox.+Axinell.	Desmox.+Axinell.+Raspail.
Desmox.+Axinellid.	Desmox.+Axinell.+Raspail.	Cerac.+Axinellida
Hadromer.+Chondros.	Dendrocer.+Halisarc.	Vac.+Chondr.+Cerac.
Hemiast.+Desmox.+Axinell.		Halich.II+Poec.+Haplo.
Desmace.+Desmacid.+Mycal.		Desmacid.+Myxill.+Clathr.
Dictyo.+Dysid.		Dendroc.+Halisarc.
Dendroc.+Halisarc.		
Number of homoplasies (total 61)	(17)	(33)
Calcarea as a separate class	Homosclerom.+Calcarea	Homosclerom.+Calcarea
Sclerosponges as a separate class	Tetractinellida+Hadrom.+Halich.I+II	Tetractinellida+Hadrom.
Homoscleromorpha as separate subclass	Haplo.+Poecilo.separate from Dictyo.	Lith.I+II separated independently
Tetractinomorpha as separate subclass		Spirophorida separate from Lith.II
Haplo+Poecilo.separate from Dictyo.		Haplo.+Poecilo.separate from Dictyo
Dysideidae included in Dictyo		Petros.separate from Haplo.
		Vaceletia+Chondrosida
		Sclerosponges polyphyletic
		Agelasida separate from Ceratop.& Cerac.
		Halich.I separate from Halich.II
Number of characters involved: 35	13	21
Assumed losses (total 46)	(105)	(59)
3b(1x),5b(2x),9b(3x),12b(3x),14b(3x) 20b(3x),21b(2x),23b(3x),25b(1x),26b(3x) 30b(8x),31b(2x),32b(7x),33b(1x),34b(3x)	3b(1x),5b(4x),9b(13x),9c(5x),10b(8x), 11b(23x),12b(2x),14b(3x),19b(1x),21b(6x) 23b(2x),24b(2x),25b(2x),30b(22x),31b(1x) 32b(7x),33b(3x),6b(1x).	3b(1x),5b(12x),6b(1x),12b(2x),14b(3x) 22b(2x),23b(3x),27b(2x),30b(22x),32b(6x), 33b(6x),39b(4x).

II. Author's tree :

Two objectives are the basis for this phylogenetic tree ; the first is to bring down the number of unsolved synapomorphies (without synapomorphies there is no stability : the assumed phylogenetic relationships may be arbitrarily changed) ; the second is to reduce the number of homoplasies. The latter needs further explanation.

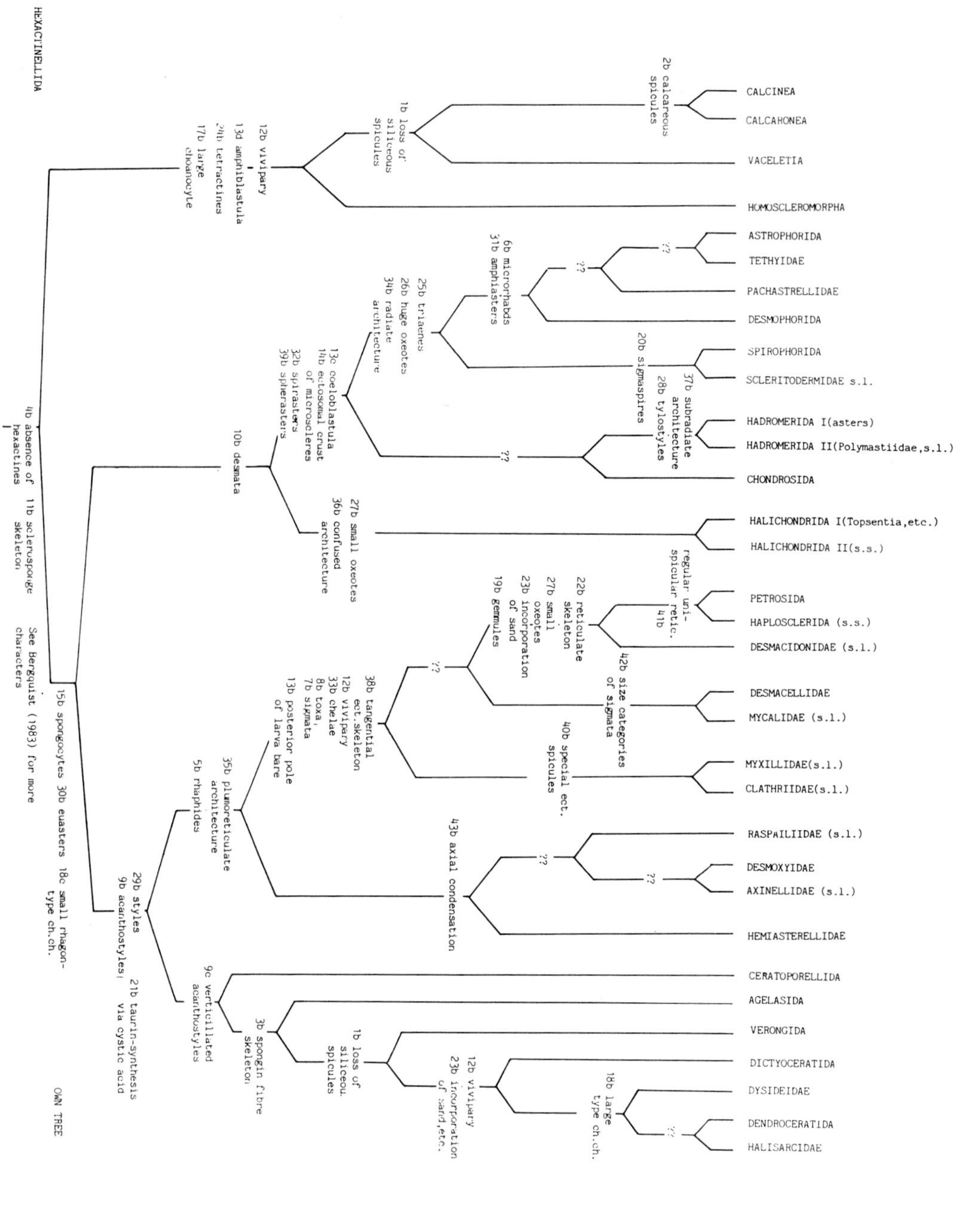

Figure 2 - Cladogram constructed by the author using strictly applied cladistic principles, with in addition the argument that loss of a character/ state occurs much more easily than independent parallel development of a character/state

It is established that parallel developments have occurred during the evolution of the Porifera. It is the opinion of the author that it is not possible to construct a cladogram of the Porifera in which no homoplasies are assumed. On the other hand, many characters appear to be unique and undoubtedly evolved only once judging from their limited distribution over the taxa of the Porifera. Many of the apomorphous characters of the groups considered here are quite limited. It is therefore assumed that unless evidence to the contrary exists or weighty arguments apply, a character is considered to have evolved only once, and its distribution is assumed to reflect common ancestry at some level. Lack of such a character in any given taxon otherwise related to taxa possessing it may easily be explained by its loss or non-expression. In other words, losses occur more easily than new developments, and often have no phylogenetic significance.

The following characteristics can be observed from this new cladogram (table 3) :

— Only a few minor unsolved synapomorphies are left, many of which reflect close relationship of the groups concerned ; a few solved synapomorphies are weak.

— Relatively few assumed homoplasies.

— A great number of inferred "losses" (twice as high as in Hartman's classification). Arguments in favour of those losses can be presented in most cases but remain speculations (e.g. loss of sclerosponge skeletons, desma-reticulation).

III. Computer-based classification (computree) :

Hartman's tree and the new phylogenetic tree illustrated in Fig.3, are based on analyses of characters using the following subjective criteria : judgments concerning the weight or significance of a character ; the probability of its being developed two or more times independently ; the probability that a character will disappear in the course of evolution ; possibly inadmissible criteria such as 'gut feeling' or 'instinct' which is difficult to translate into characters(states). An abstract statistical treatment of the taxa-character matrix appears to be an objective alternative. A new development is the use of numerical techniques originally developed for purely phenetic treatment of data, but now adapted for phylogenetic purposes. A colleague W.N. Ellis (an entomologist), has developed his own version of a computer program (written specifically for the MacIntosh PC)

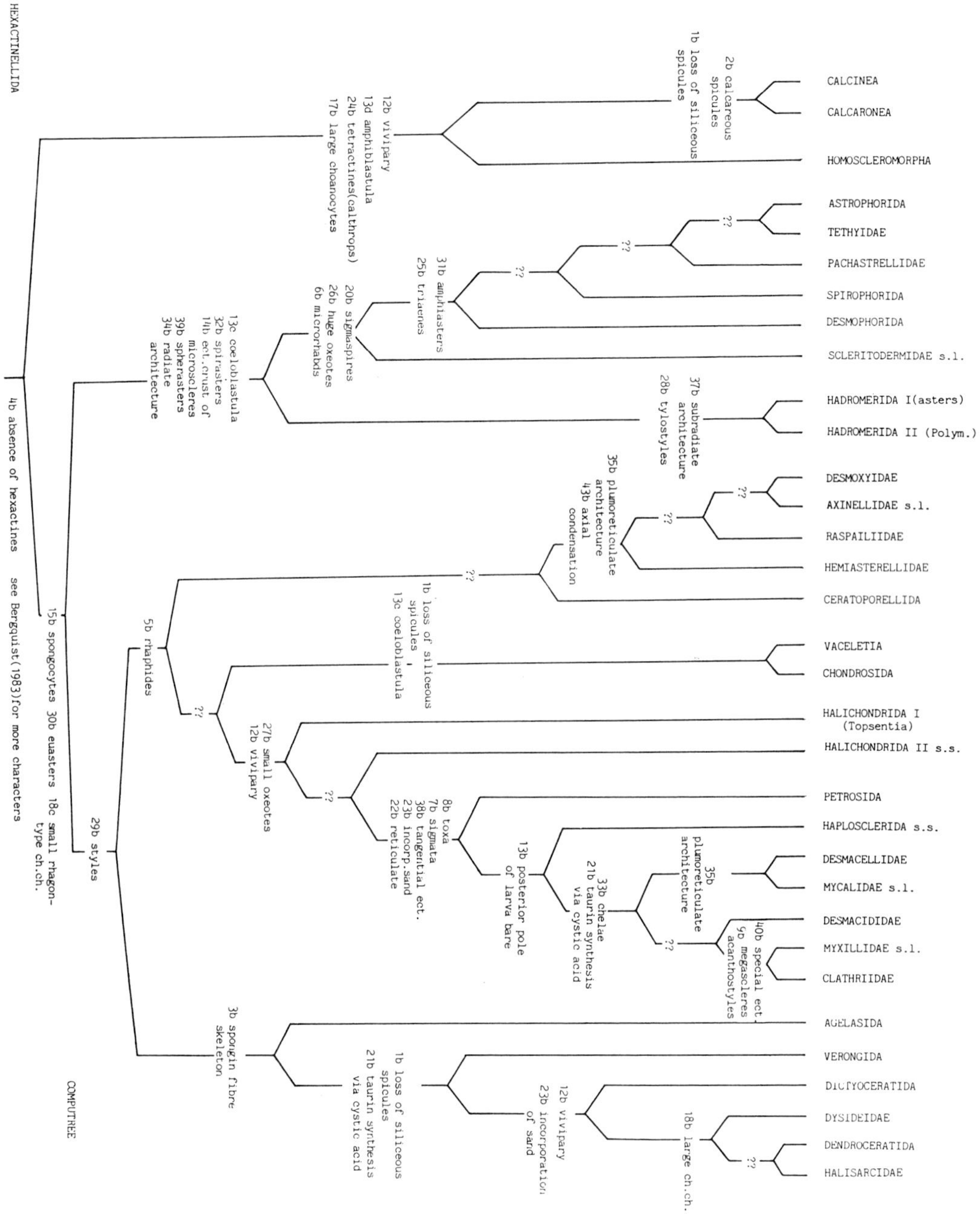

Figure 3 - Cladogram constructed by computerprogram 'Tree Tools' using numerical cladistic techniques ; the tree is one of many posible trees, but it shows the highest consistency with the taxa-character matrix of Table 2.

constructing so called "Wagner Trees' which comprises networks of taxa and hypothetical ancestors. These networks have no starting point but may be converted into a phylogenetic tree (such as the ones already presented) by indicating one of the taxa as the out-group illustrating the characters in their primitive state. In contrast to both previous tree constructions, the computer program does not weigh and cannot theorize over the possibilities of the subsequent loss of a character unless otherwise instructed. The program constructs a tree that shows the highest consistency with the taxa-character matrix, (i.e. tries to hold the homoplasies to the lowest possible number). It does not discriminate between parallel new developments and parallel losses. It is an interesting exercise to observe how the program deals with the big discrepancies between Hartman's and the author's cladogram. The computer program lists the exact character states of all hypothetical ancestors (diverging points in the cladograms), but in a few cases other character states are invented in order to produce solved synapomorphies. Therefore the cladogram is the computer's, but not all of the synapomorphies are ones indicated by the computer.

The following characteristics are apparent (table 3) :

— The tree contains a fair amount of unsolved synapomorphies (the taxa are then grouped on overall similarity), but these are for the most part minor.

— There is a fair amount of homoplasies observed, some of which appear untenable.

— A reasonable number of losses are illustrated.

IV Analysis :

▪ **Similarities consistent within the cladograms :**

a/ Tetractinellida + Hadromerida : within this group there are minor discrepancies, but all three trees agree on the monophyly, based on a number of strong synapomorphies.

b/ Keratosa : again a minor disagreement within the group but as a whole there is no problem. Great uncertainty exists, as to whether the Agelasida also belong here.

c/ Axinellida : despite the 'problems' of aster-bearing Hemiasterellids and acanthostyle-bearing Raspailiids, all three trees accept the group as monophyletic based on the axial condensation of the skeleton and the stylote nature of the majority of the megascleres.

d/ Poecilosclerida + Haplosclerida : united by a number of strong synapomorphies ; obvious disagreements are apparent over the precise relationships within this large group.

■ **Major disagreements :**

a/ Calcarea + Homoscleromorpha : although there are several good synapomorphies, their homology is still doubtful (van Soest, 1984); it is clear that any common ancestor of the groups, if there has been one at all, lived in the very distant past.

b/ Sclerosponges : Hartman's view that these constitute a separate development from Demosponges is difficult to defend. One of the consequences would be an independent development of a large number of complicated structures with a very limited distribution. Moreover, there is evidence of the instability in the sclerosponge basal skeleton of recent Sclerosponges (Vacelet, 1980 ; van Soest, 1984). The computree agrees with the author's tree (Fig. 3) in the denunciation of the Sclerosponges. However, the chosen solution, (independent development in at least 5 different lines in accordance with Vacelet, 1985), differs from the author's perspective (primitive character lost in most recent taxa). The segregation of Agelasida and Ceratoporellida is difficult to defend.

c/ The computree agrees with the author's opinion that Tethyids are related to Astrophorids *sensu stricto* because the only apparent difference is the absence of triaenes in the former. No synapomorphies are apparent however.

d/ Desmophorids and Scleritodermids (*sensu lato*) are apparently not monophyletic (also the view of the author), but the characters shared with Pachastrellidae and Spirophorida respectively are considered homoplasic. More study needs to be done on this subject along with an analysis of desma-bearing sponges in general.

e/ *Vaceletia* (and Sphinctozoa) : the problems are caused by the absence of spicules which is the basis for computer association with Chondrosida, (newly erected by Boury-Esnault & Lopez, 1985). Solutions such as this and alternative ideas of the author are very difficult to defend. More data are needed. This is an illustration of how computer analysis focuses attention on problematic parts of a system.

f/ Positions of Halichondrida and Axinellida : most authors at present are convinced of the problematic status of these two groups. Inclusion of the Axinellids into the 'Tetractinomorpha' by Hartman is presumably based on

Hemiasterellid asters and on the oviparity of the group. The latter character represents a primitive trait and is thus inadmissible. Further problems with this solution are the plumoreticulate architecture and the possession of (small) styles in all and acanthostyles in Raspailiids which have to be assumed to have developed at least twice. The inclusion of the Halichondrids in the 'Ceractinomorpha' by Hartman (1982) is based on the viviparity of the Halichondridae, which is hardly a strong character, particularly as the larval morphology differs from that of the remaining 'Ceractinomorpha' (e.g. Wapstra & van Soest, in this volume). The occurrence of sublithistids is difficult to explain in this group unless one assumes the possession of desmata to be homoplasic, too.

A proposed solution (reversing the positions of these groups) also leaves a few problems, such as the homoplasy in the small oxeotes, and the rather difficult assumption that asters are a primitive character for all the Demosponges.

For the time being the computree seems to have found the solution for this problem by including both Axinellids and Halichondrids in the same larger group. The synapomorphy (rhaphides) is rather weak, since these also occur isolatedly in *Cinachyra*. Hemiasterellids and Raspailiids remain exasperating groups.

g/ Haplosclerida-Poecilosclerida : the computree supports Hartman's classification of Poecilosclerida as a monophyletic group (including the achelate Desmacellidae). It leaves the position of the Desmacididae undecided and denies the monophyly of the Haplosclerida s.l. (in accordance with Bergquist, 1980).

h/ Dysideidae : as already intimated by Boury-Esnault *et al.* (in press) this group is probably more related to the Dendroceratids than to the Dictyoceratids.

DISCUSSION

The computer-generated tree seems an objective method in helping solve differences of opinion on the derived vs. advanced states of characters, as well as the consequences of such decisions for hypotheses on the phylogeny of the taxa concerned. Most important, however, is the taxa-character matrix with which it is fed. It is obvious that the presented matrix is still in need of amendment ; more and often stronger characters are needed (e.g. biochemical). Discussions over the homologous status of characters and character states are also very necessary. The present exercises are intended to generate these discussions. They should eventually lead to a 'COMPUTREE' which is generally accepted as a reflection of the phylogenetic relationships. Any formal proposals for a change of currently accepted classes, subclasses, etc... should wait until then.

LITERATURE CITED

Bergquist P.R., 1980 - The ordinal and subclass classification of the Demospongiae (Porifera) ; appraisal of the present arrangement and proposal of a new order. *N. Z. J. Zool.*, 7 : 1-6.

Bergquist P.R., 1985 - Poriferan relationships : pp 14-27 in *The origin and relationships of lower invertebrates*. S.C. Morris, J.D. George, R. Gibson & H.M. Platt (eds), The systematics Association special vol. 28.

Boury-Esnault N. & Lopes T., 1985 - Les démosponges littorales de l'archipel des Açores. *Ann. Inst. océanogr.*, 61 :149-225.

Boury-Esnault N., De Vos L., Donadey C. & Vacelet J, in press - Ultrastructure of choanosome and sponge classification. *Proc. 3rd int. Conf. Biol. sponges*, Woods Hole.

Ellis W.N., 1986 - Tree Tools. A set of programs for the construction of Wagner Networks and trees for the MacIntosh 512 computer, 1-44.

Hartman W.D., 1982 - Porifera : pp 640-666 in *Synopsis and classification of living organisms (1)*. S.P. Parker, (ed), MacGraw- Hill, New York.

Hooper J.N.A., in press - Investigation of character stability, systematics, and affinities between Clathriids (Poecilosclerida) and Axinellida. *Proc. 3rd int. Conf. Biol. Sponges*, Woods Hole.

Lévi C., 1956 - Etude des Halisarca de Roscoff. Embryologie et systématique des Démosponges. *Archs Zool. exp. gén.*, 93 : 1-181.

Mohn E., 1984 - System und Phylogenie der Lebewesen. Band 1. Schweizerbart, Stuttgard, XXXV+884 pp.

Reiswig H.M. & Mackie G.O. 1983 - Studies on hexactinellid sponges. III. The taxonomic status of Hexactinellida within the Porifera. *Philos. Trans. r. Soc. Lond. B biol. Sci.*, 301 : 419-428.

Soest R.W.M. van, 1984 - Deficient *Merlia normani* Kirkpatrick, 1908, from the Curaçao reefs, with a discussion on the phylogenetic interpretation of sclerosponges. *Bijdr. Dierk.*, 54 : 211-219.

Soest R.W.M. van, in press - Towards a phylogenetic classification of sponges. *Proc. 3rd int. Conf. Biol. sponges*, Woods Hole.

Vacelet J., 1980 - Squelette calcaire facultatif et corps de regénération dans le genre *Merlia*, éponges apparentées aux Chaetétides fossiles. *C. R. hebd. Seanc. Acad. Sci.*, Paris, 290 : 227-230.

Vacelet J., 1985 - Coralline sponges and the evolution of Porifera : pp 1-13 in *The origin and relationships of lower invertebrates*. S.C. Morris, J.D. George, R. Gibson & H.M. Platt, (eds), The systematics Association special vol. 28.

Volkmer-Ribeiro C., in press - A new insight into the systematics, evolution and taxonomy of freshwater sponges. *Proc. 3rd int. Conf. Biol. Sponges*, Woods Hole.

Wapstra M. & van Soest R.W.M., in this volume - Sexual reproduction, larval morphology and behaviour of Demosponges from the South West of the Netherlands. *Proc. 2nd Workshop Taxon. Porifera N.E. Atlantic Med.*

THE USE OF ELECTROPHORESIS IN SPONGE TAXONOMY

A.M. SOLE-CAVA & J.P. THORPE

Department of Marine Biology
The University of Liverpool, Marine Biological Station
Port Erin - Isle of Man - U.K.

SYNOPSIS

Enzyme electrophoresis detects products of individual gene loci and allows the calculation of gene frequencies and the estimation of exchange rates between populations. Speciation and subsequent evolution result in genetic divergence between populations and can therefore be studied. Electrophoresis has been used to distinguish sibling species and to establish the taxonomic status of dubious sub-species and colourmorphs of a wide range of organisms. The Porifera are a taxonomically difficult group, offering many possible applications for such methods. Preliminary results of electrophoretic studies on sponges are discussed and used to evaluate the potential for future work.

INTRODUCTION

The advancement of knowledge in science is achieved, principally, by the use of well established techniques to solve new problems or by the use of new techniques to solve old problems. Originally, the biological and genetic uses of enzyme electrophoresis were typical examples of the second case : after the association between gel electrophoresis and histochemical stains (Hunter & Markert, 1957), geneticists realised the potential of the technique to study the genetic structure of natural populations and the first studies, mainly on man and *Drosophila*, were published in the mid-sixties (e.g. Hubby, 1963 ; Hubby & Lewontin, 1966 ; Johnson *et al.*, 1966 ; Lewontin & Hubby, 1966 ; Harris, 1966). These works, which gave the first really useful estimates of levels of genetic variation in natural popula-

NATO ASI Series, Vol. G13
Taxonomy of Porifera
Edited by J. Vacelet and N. Boury-Esnault

tions, had a major impact on the evolutionary ideas of the time (for reviews see e.g. Dobzhansky, 1970 ; Lewontin, 1974 ; Dobzhansky *et al.*, 1977). Since then, biochemical genetics has been successfully used in a variety of fields such as the identification of juvenile stages of fish (e.g. Mork *et al.*, 1983), the identification of stocks for aquaculture (e.g. Moav *et al.*, 1976 ; Cruz *et al.*, 1982), physiological ecology (e.g. Zouros *et al.*, 1980 ; Beaumont *et al.*, 1985 ; Mallet *et al.*, 1986) and the use of genetic markers for fisheries (e.g. Simonarsen & Watts, 1969 ; Child, 1984).

Thus, it can safely be said that enzyme electrophoresis has revolutionised evolutionary and ecological genetics. Furthermore, this technique has transcended the field of pure genetics and also become a very important tool for the study of taxonomy and population dynamics. In taxonomy it has helped to identify and separate dubious or sibling species in a variety of groups (reviewed in Avise, 1974 ; Gotlieb, 1977 ; Ferguson, 1980 ; Ayala, 1983 ; Berlocher, 1984 ; Innes, 1984), and also provided the data for phylogenetic reconstructions of taxonomic groups, the later based on the "molecular clock hypothesis" (see e.g. Fitch, 1973, 1976 ; Wilson *et al.*, 1977 ; Vawter *et al.*, 1980 ; Thorpe, 1982, 1983 ; Avise, 1983).

THE TECHNIQUE

Electrophoresis is the migration, in response to an electric field, of electrically charged molecules through a solid or liquid medium. The rate of migration of these molecules is influenced mainly by the net charge and molecular weight. This property is used to separate proteins, the separation being regulated by the use of buffers with specific pH and of different support media, such as cellulose acetate, starch and polyacrylamide gels. The proteins (including enzymes) thus separated can then be stained by specific staining mixtures, and the banding patterns obtained interpreted genetically. For a full review of the techniques, see Brewer (1970), Harris and Hopkinson (1978), Ferguson (1980) and Gaal *et al.* (1980).

The interpretation of the banding patterns depends partially upon the knowledge of how different enzymes, with different tertiary structures, may be expected to appear in the gel for samples from different homozygous and heterozygous individuals. Monomeric enzymes, for example, will appear as two bands in heterozygotes, whereas dimeric enzymes will present three bands

(Fig.1). For further explanation of banding patterns see for example Harris & Hopkinson (1978), Ferguson (1980).

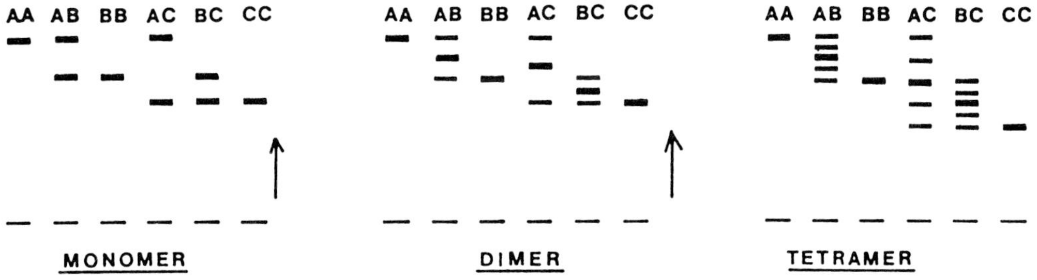

Figure 1 - Typical banding patterns for monomeric, dimeric and tetrameric enzymes. *AA, AB, BB,* etc. are the genotypes of the individuals for the given enzymes. The arrows indicate the direction of migration.

After genotype frequencies have been obtained from the banding patterns, gene frequencies can be calculated for each locus (Fig. 2) , and the populations can be compared by the use of genetic identity (Nei, 1972) or similarity (Thorpe, 1979) indices. Other parameters, such as the mean levels of polymorphism and heterozygosity can also be calculated, providing an insight into the population structure and biology of the species.

APPLICATIONS

Electrophoresis can be employed for various genetic, zoological and ecological purposes (Fig. 3). The majority of the biochemical genetic studies have been carried out on terrestrial animals, principally vertebrates (see Ward, 1978 and references therein) and insects (Berlocher, 1984). This is understandable since we are terrestrial animals, and our first interests lie in understanding phylogenetically close or economically important animals. For the same reasons, the marine animals most studied genetically are fish (reviewed by Smith & Fujio, 1982) and commercially important crustaceans (reviewed by Nelson & Hedgecock, 1980) and molluscs (e.g. Koehn & Gaffney, 1984 ; Beaumont *et al.*, 1985). Other marine invertebrates are less well studied, however, and only now is a significant amount of work starting to be produced on such organisms.

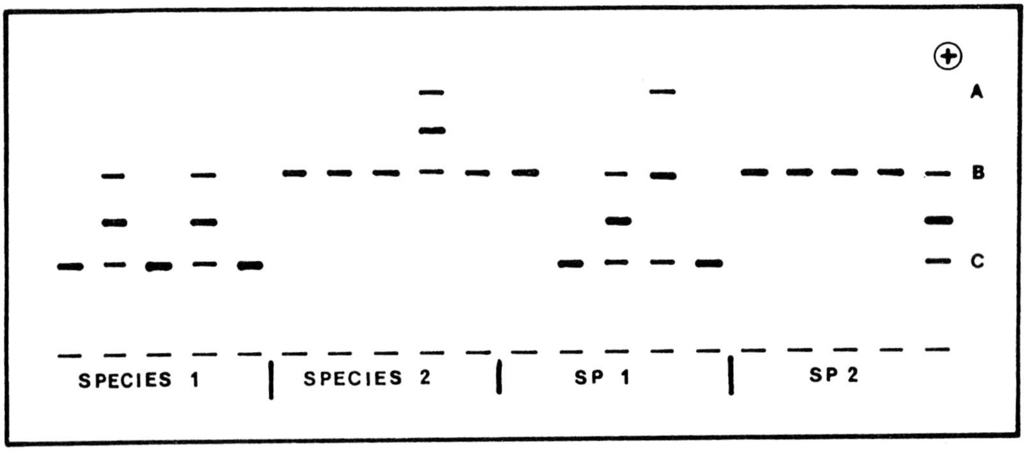

	Species 1	Species 2
AA	0	0
AB	0	1
BB	1	8
AC	1	0
BC	3	1
CC	5	0
<u>Total</u>	10	10

	Species 1	Species 2
fA	0.05	0.05
fB	0.25	0.90
fC	0.70	0.05
H_o	0.40	0.20
H_e	0.45	0.19

Figure 2 - A putative gel for the analysis of 10 individuals from two populations, belonging to two different species analysed for a dimeric enzyme. *A*, *B* and *C* are the three alleles studied. The genotype and allele frequencies for both populations are presented.

ELECTROPHORESIS

- Many enzymes / Few samples
  - *Taxonomy*
    - *Diagnostic loci*
    - *Reproductive biology*
    - *Hybridization*
  - *Evolution*
    - *Phylogeny*
    - *Ecological strategies*
    - *Evolutionary rates*
- Few enzymes / Many samples
  - *Pop. dynamics*
    - *Gene flow*
    - *Stock identification*
    - *Population history*
  - *Ecology*
    - *Physiological adaptations*
    - *Clinal structures*
    - *Reproductive biology*

Figure 3 - Different possible uses of enzyme electrophoresis in biology.

Among marine invertebrates, benthic species have received a moderate amount of attention. These are of interest because they include many "primitive" organisms, which frequently show unusual lifestyles and very often also present rather complicated taxonomic problems. Studies published so far include organisms such as sipunculids (Balakirev & Manchenko, 1983), tropical bivalve molluscs (Ayala *et al.*, 1973 ; Campbell *et al.*, 1975), gastropods (Snyder & Gooch, 1973 ; Gresham & Tracey, 1975 ; Johnson & Black, 1984) nudibranchs (Havenhand *et al.*, 1986), phoronids (Ayala *et al.*, 1974), brachiopods (Valentine & Ayala, 1974 ; Hammond & Poiner, 1984), pycnogonids (King *et al.*, 1986), bryozoans (Thorpe *et al.*, 1978 ; Thorpe & Ryland, 1979), polychaetes (Beckitt, 1980 ; Guérin & Kerambrun, 1983), echinoderms (Ayala *et al.*, 1975 ; Marcus, 1977 ; Bisol *et al.*, 1984), tunicates (Schimdtke *et al.*, 1977 ; Fisher *et al.*, 1980) and coelenterates (McCommas & Lester, 1980 ; Carter & Thorpe, 1981 ; Solé-Cava *et al.*, 1985).

Despite their great ecological importance as major components of the benthic fauna of temperate and tropical seas, sponges have been largely overlooked in biochemical genetics studies. This is particularly surprising given the highly problematic nature of sponge taxonomy. One of the major problems for taxonomists is to find characters which are at the same time conservative enough to be stable in the species, and variable enough to diverge following speciation events as shared derived (synapomorphic) characters. In animals like sponges, the number of different morphological characters which can be analysed is rather restricted, and taxonomists are often faced with the problem of having to choose not the characters which seem to be the most significant biologically, but, instead, characters which are the only ones available. In the case of sponges the skeletal structures are the most commonly used taxonomic features because these are often the only pratically useful part which remains in the common, "dry" collections in museums. This situation is, obviously, far from ideal, and several new techniques such as chemical (Bergquist *et al.*, 1980 ; Bergquist & Wells, 1983) and immunological (Van de Vyver, 1971 ; Connes *et al.*, 1974 ; Neigel & Avise, 1983) taxonomy have been tried with varying degrees of success. These techniques produce valuable information about the overall similarity between species or higher taxonomic ranks, and should be used, together with morphological, reproductive, cytological and ecological data to try to achieve more biologically meaningful schemes of classification. However, both chemical taxonomy and immunological taxonomy suffer from the same major drawback : they do not lead to any clear genetic interpretation of differences between taxa. It is obvious that, ultimately, almost all characters will have a genetic origin, which is modified in varying degrees by the environment. For example, new secondary metabolites, produced *de novo* by the sponge require specific enzymes (and hence specific genes) to cause their synthesis. Similarly, the strength of an immunological reaction is related to the molecular divergence in the primary structure of the antigen (again, genetically determined). However, because with such results it is generally completely unknown what proportions of the observed differences are genetically or environmentally determined, such results must be interpreted with extreme caution.

Enzyme electrophoresis produces results which can be genetically interpreted and, therefore, allows a high level of objectivity in the study of natural populations. The biological definition of species (Mayr, 1963 ; Genermont & Lamotte, 1980) is operationally useful, especially for the comparison of sympatric populations. When biochemical taxonomic methods are associated with the biological species concept, they become very powerful, for they produce results which are generally less open to arguments and which contribute, consequently, to a more stable classification than that obtained by more conventional morphological criteria. In other cases, however, electrophoretic studies can contribute little to the solution of specific problems. It is important, therefore, that the limitations and advantages of the technique be understood, in order to avoid the creation of false hopes and, with them, of possible frustrations for the sponge taxonomist.

Limitations

1/ **Preservation of the material** — Enzymes are proteins, and as such are very sensitive to the conditions of preservation. Therefore, studies of biochemical genetics have to be restricted to fresh or well (and freshly) frozen samples. This is the major limitation of the technique, since this means that alcohol fixed material, museum collections and fossils cannot be studied by enzyme electrophoresis. Even with frozen samples, special care must be taken. Artifact bands have been shown to be produced by the repeated freezing and thawing of samples (Scozzani *et al.*, 1980). In any case, all samples to be analysed electrophoretically should be preserved in the same way and it is always preferable that they should be fresh. By using fresh samples potentially misleading artifacts can be avoided or easily detected ; also levels of enzyme activity are likely to be generally higher in fresh material.

2/ **Cost** — The price of setting up an electrophoresis laboratory can be prohibitive to some zoology departments. However, the running costs are not so high if the quantity and discriminatory power of the data which can be obtained are taken into account. Typically, an electrophoresis laboratory should be able to run four gels simultaneously, which would mean that up to sixty to eighty individuals could be analysed for about twenty different enzymes (for slices per gel) in one week (five working days). Considering that sample sizes of twenty are normally sufficient for most work in biochemical taxonomy (Gorman & Renzi, 1979), electrophoresis can actually be highly cost-effective when compared to the many samples and large amounts of time required to carry out a comprehensive morphological study.

3/ **Small sample of the genome** — Commonly in biochemical taxonomy about ten to thirty enzyme loci are analysed. This is a very small number if compared with the total number of structural gene loci in one organism. This means that, although the technique is decisive about differences between populations, it is not so for similarities between them. In other words, if two sympatric populations show significant differences in allele frequencies at one or more of the loci analysed it can be positively said that they are reproductively isolated (i.e. they belong to different species), whereas if those populations show no differences at any of the enzyme loci analysed this does not provide conclusive evidence that they are interbreeding and therefore conspecific, since the two populations concerned may simply not have diverged at any of the loci studied.

4/ Small amount of enzymes in the tissues — Many sponges have tissues containing large amounts of siliceous or calcareous spicules. This means that the actual amount of biomass per unit of weight is effectively much smaller than in most of the other soft-bodied organisms ; also sponges generally are not highly active and therefore will probably have lower metabolic rates than many higher animals. An obvious consequence is that the enzymes will be present at low concentrations, and many will be difficult to stain after electrophoresis. There is no simple solution to this problem. Using fresh samples, homogenizing with a minimum amount of buffer and increasing the concentration of the staining reagents can be effective sometimes, and experimenting is the only way of finding the most appropriate methodology for each particular case.

Advantages

Some advantages of the technique, such as the objectivity and the production of direct genetic information have already been stated. However, three other important advantages of electrophoresis should be mentioned :

1/ Unbiased results — The major criterion governing the choice of enzymes for electrophoretic work is availability, that is, whether the reagents for the staining of the enzymes are available in the laboratory or whether the animals show any detectable activity for the enzymes assayed. This means that the results obtained are likely to be unbiased and therefore can be analysed in terms of statistical probability. For example, if two populations do not show any differences at twenty enzyme loci it can be said that these populations are probably similar over more than 95 % of their structural genome.

2/ Analogy implies homology — Most enzyme loci studied belong to key biochemical pathways. Therefore, the possibility that a given enzyme locus has been replaced by another, different locus in the course of evolution is extremely unlikely (Avise, 1974).

3/ Easy detection of unusual reproductive strategies — Sponges can present unusual life strategies, such as asexual reproduction (Korotkova, 1979 ; Bergquist, 1980) and larval fusion (Fry, 1971). By means of electrophoretic methods, the genetic structure of populations can be easily analysed, and the extend (if any) of asexual reproduction or fusion estimated. In sea-anemones, for example, electrophoresis has been used to show that the small "brooding" anemones found in the coelenteron of their parents had been produced asexually (Ottaway & Kirby, 1975 ; Orr *et al.*, 1982). In three species of *Suberites* enzyme electrophoretic data indicated that the incidences of both allogenetic fusion and asexual reproduction were likely to be negligible or zero (Solé-Cava & Thorpe, 1986).

Biochemical polymorphism and sponge taxonomy

The first published studies on the enzyme electrophoresis of sponges were basically concerned with the description of banding patterns, without any attempt to interpret the results genetically (Pakhomov *et al.*, 1974 ; Schoots *et al.*, 1977 ; Baden & Corbett, 1979 ; Urbaneja & Lin, 1981). The first work to use elctrophoresis as a tool in sponge taxonomy was that of Connes *et al.* (1974), who used stains only for total (soluble) proteins. These give banding patterns which are pratically impossible to interpret genetically. Enzyme biochemical genetic studies have started only very recently in sponges (Balakirev & Manchenko, 1985 ; Sarà, in press ; Solé-Cava & Thorpe, 1986, in press). The overall pattern apparently emerging from these few published studies (table 1) is that sponges seem to have very high levels of genetic variation (about three times larger than the average for vertebrates), in a similar way to coelenterates (see e.g. Bucklin, 1985 ; Solé-Cava *et al.*, 1985 ; Solé-Cava, 1986). The reasons for this are still unclear, but it is not unreasonable to suppose that these results are probably related to the particular, sessile lifestyle of these organisms or to their very old evolutionary history (Solé-Cava, 1986 ; Solé-Cava & Thorpe, in press).

Table 1 - Levels of genetic variation in sponges. ni = number of individuals analysed ; nl = number of enzyme loci analysed ; P(0.95) = proportion of polymorphic loci ; Ho = mean observed heterozygosity ; He = mean expected hetrozygosity. References : 1 - Solé-Cava & Thorpe (in press). 2 - Balakirev & Manchenko (1985). 3 - Solé-Cava & Thorpe (1986). 4 - Sarà (in press).

Species	ni	nl	$P_{(0.95)}$	H_0	H_e	Ref
Halichondria panicea	18	16	0.688	0.227	0.234	1
Mycale macilenta	7	18	0.500	0.246	0.189	1
Suberites domuncula	12	28	0.393	0.146	0.137	2
Suberites luridus	13	18	0.667	0.179	0.168	1,3
Suberites pagurorum	11	18	0.611	0.215	0.190	1,3
Suberites rubrus	16	16	0.750	0.365	0.335	1,3
Tethya aurantium	30	8	0.125	?	0.050	4
Tethya citrina	30	8	0.125	?	0.020	4

The great power and usefulness of the technique for taxonomic work in sponges has been shown in the separation of very close, sibling species of *Suberites* from the Irish Sea (Solé-Cava & Thorpe, 1986), and in the identification of sympatric species of *Tethya* (Sarà, in press).

It can be concluded, thus, that electrophoresis can be used successfully in resolving taxonomic problems in sponges, at least for well defined problems, such as the detection of sibling species amongst sympatric populations. It can be used as well for the analysis of the divergence between allopatric populations ; it has been shown (Thorpe, 1982, 1983) that values of genetic identity between different taxonomic levels have a tendency to be similar. For example, more than 98% of intra-specific population comparisons (out of 7000 vertebrate populations) give values of genetic identity (Nei, 1972) above 0.90, whereas only 2% of inter-specific (congeneric) population comparisons (out of 900 vertebrate populations) gave values above 0.85

(Thorpe, 1983). The results for the smaller amount of published data on invertebrates suggest that they follow the same trend (Thorpe, 1982, 1983). This means that if two populations show a genetic identity of 0.80, for example, they are very likely to belong to different species.

Electrophoresis must be used in association with classical taxonomy, particularly for the selection of reliable taxonomic characters and determination of taxonomic status of "morphs", "sub-species" and "species". Taxonomy is an indispensable science, and systems of classification are operationally important structures. Electrophoresis, together with ecology, chemistry and reproductive biology is helping taxonomy to become biologically more meaningful, and this will hopefully lead to a more stable classification of a traditionally problematic group like the Porifera.

LITERATURE CITED

Avise J.C., 1974 - The systematic value of electrophoresis data. Syst. Zool., 23 : 465-481.

Ayala F.J., 1983 - Enzymes as taxonomic characters : pp 3-26 in Protein polymorphism : adaptative and taxonomic significance. G.S. Oxford & D. Rollinson (eds), Academic Press, London.

Ayala F.J., Hedgecock D., Zumwalt G.S. & Ehrenfeld J.C., 1973 - Genetic variation in Tridacna maxima, an ecological analogue of some unsuccessful evolutionary lineages. Evolution, 27 : 177-191.

Ayala F.J., Valentine J.W., Barr E.G. & Zumwalt G.S., 1974 - Genetic variability in a temperate intertidal phoronid Phoronopsis viridis. Biochem. Genet., 11 : 413-427.

Ayala F.J., Valentine J.W., Hedgecock d. & Barr E.G., 1975 - Deep-sea asteroids : high genetic variability in a stable environment. Evolution, 29 : 203-212.

Baden D.G. & Corbett M.D., 1979 - Peroxidases produced by the marine sponge Iotrochota birotulata. Comp. Biochem. Physiol., 64B : 279-283.

Balakirev E.S. & Manchenko G.P., 1983 - High level of genetic variability in the sipunculan Phascolosoma japonicum. Genetika, 19 : 1638-1643 (in Russian).

Balakirev E.S. & Manchenko G.P., 1985 - Intraspecific genetic variation in the sponge Suberites domuncula . Biol. Moyra., 5 : 36-40.

Beaumont A.R., Gosling E.M., Beveridge C.M., Budd M.D. & Burnell G.M., 1985 - Studies on heterozygosity and size in the scallop, Pecten maximus : pp 443-454 in Proceedings of the 19th European marine biology symposium. P.E. Gibbs (ed), Cambridge University press, Cambridge.

Beckitt R., 1980 - Genetic structure of *Pileolaria pseudomilitaris*. *Genetics*, 96 : 711-721.

Bergquist P.R., 1978 - Sponges. Hutchinson Press, London.

Bergquist P.R., Hofheinz W. & Oesterhelt G., 1980 - Sterol composition and the classification of the Demospongiae. *Biochem. Syst. Ecol.* 8 : 423-435.

Bergquist P.R. & Wells R.J., 1983 - Chemotaxonomy of the Porifera : the developments and current status of the field :in *Marine Natural Products* vol.5. P.J. Scheuer (ed), Academic Press, London.

Berlocher S.H., 1984 - Insect molecular systematics. *Ann. Rev. Entomol.*, 29: 403-433.

Bisol P.M., Costa R. & Sibuet M., 1984 - Ecological and genetical survey of two deep-sea holothurians : *Benthogone rosea* and *Benthodytes typica*. *Mar. Ecol. Prog. Ser.*, 15 : 275-281.

Brewer G.C., 1970 - An introduction to isozyme techniques. Academic Press, New York.

Bucklin A., 1985 - Biochemical genetic variation, growth and regeneration of the sea anemone, *Metridium*, of British shores. *J. Mar. Biol. Assoc. U.K.*, 65 : 141-157.

Campbell C.A., Valentine J.W. & Ayala F.J., 1975 - High genetic variability in a population of *Tridacna maxima* from the great barrier reef. *Mar. Biol.*, 33 : 341-345.

Child A.R., 1984 - Biochemical polymorphism in charr (*Salvinus alpinus*) from three Cumbrian lakes. *Heredity*, 53 : 249-257.

Connes R., Diaz J.-P., Nègre G. & Paris J., 1974 - Etude morphologique, cytologique et sérologique de deux formes de *Suberites massa* de l'étang de Thau. *Vie Milieu*, 24 : 213-224.

Cruz T.A., Thorpe J.P. & Pullin R.S.V., 1982 - Enzyme electophoresis in *Tilapia zilii* : a pattern for determining bichemical genetic markers for use in *Tilapia* stock identification. *Aquaculture*, 29 : 311-329.

Dobzhansky T., 1970 - Genetics of the evolutionary process. Columbia University Press, New York.

Dobzhansky T., Ayala F.J., Stebbins G.L. & Valentine J.W., 1977 - Evolution. W.H. Freeman, San Francisco.

Ferguson A., 1980 - Biochemical systematics and evolution. Blackie, Glasgow.

Fisher S.E., Shaklee J.B., Ferris S.D. & Whitt G.S., 1980 - Evolution of five multilocus enzyme systems in the chordates. *Genetica*, 52 : 73-85.

Fitch W.M., 1973 - Aspects of molecular evolution. *Ann. Rev. Genet.*, 7 : 343-380.

Fitch W.M., 1976 - Molecular evolutionary clocks : pp 160-178 in *Molecular evolution*. F.J. Ayala (ed), Sinauer Associates, Sunderland, Massachusetts.

Fry W.G., 1971 - The biology of larvae of *Ophlitaspongia seriata* from two North Wales populations : pp 155-178 in *Fourth European Marine Biology Symposium*. Ed by D.J. Crisp. Cambridge University Press, London.

Gaal D., Medgyesi G.A. & Vereczkei A., 1980 - Electrophoresis in the separation of biological macromolecules. John Wiley & Sons, Toronto.

Génermont J. & Lamotte M., 1980 - Le concept biologique de l'espèce dans la zoologie contemporaine. *Mém. Soc. Zool. France*, 40 : 427-452.

Gorman G.C. & Renzi J., 1979 - Genetic distance and heterozygosity estimates in electrophoretic studies : effects of sample size. *Copeia*, 1979 : 242-249.

Gotlieb L.D., 1977 - Gel electrophoresis : a new approach to the study of evolution. *Bioscience*, 21 : 939-944.

Gresham M.L. & Tracey M.L., 1975 - Genetic variation in an intertidal gastropod, *Collisella digitalis*. *Genetics*, 80 : 37.

Guérin J.-P. & Kerambrun P., 1983 - Mise en évidence de différentes formes dans le complexe *Malacoceros fuliginosus* Claparède (annélide, polychète). *Rapp. Proc. Verb. Réun. Comm. Internat. Explorat. Scient. mer Méditerranée*, 28 : 287-288.

Hammond L.S. & Poiner I.R., 1984 - Genetic structure of three populations of the "living fossil" brachiopod *Lingula* from Queensland, Australia. *Lethaia* 17 : 139-143.8

Harris H., 1966 - Enzyme polymorphism in man. *Proc. R. Soc. London* 164B : 298-310.

Harris H. & Hopkinson D.A., 1978 - Handbook of enzyme electrophoresis in human genetics. North Holland, Armsterdam.

Havenhand J.N., Thorpe J.P. & Todd P.D., 1986 - Estimates of biochemical genetic diversity within and between the nudibranch molluscs *Adalaria proxima* (Alder & Hancock) and *Onchidoris muricata* (Muller) (Doridacea : Onchidorididae). *J. Exp. Mar. Biol. Ecol.* 95 : 105-111.

Hubby J.L., 1963 - Protein differences in *Drosophila*. I. *Drosophila melanogaster*. *Genetics*, 48 : 871-879.

Hubby J.L. & Lewontin R.C., 1966 - A molecular approach to the study of genetic heterozygosity in natural populations. I. The number of alleles at different loci in *Drosophila pseudobscura*. *Genetics*, 54 : 577-594.

Hunter R.L. & Markert C.L., 1957 - Histochemical demonstration of enzymes separated by zone electrophoresis in starch gels. *Science*, 125 : 1294-1295.

Innes D.J., 1984 - Genetic differentiation among populations of marine algae. *Helgol. Meeresunt.*, 38 : 401-417.

Johnson F.M., Kanapi C.G., Richardson R.H., Wheeler M.R. & Stone W.S. 1966 - An operational classification of *Drosophila* esterases for species comparisons. *Univ. Texas Publ.*, 6615 : 517-537.

Jonhson M.S. & Black R., 1984 - Pattern beneath the chaos : the effect of recruitment on genetic patchiness in an intertidal limpet. *Evolution*, 38 : 1371-1383.

King P.E. Thorpe J.P. & Wallis G.P., 1986 - A biochemical genetic and morphological investigation of the species within the genus *Endeis* Philippe (Pycnogonida ; Endeidae) in Britain. *J. Exp. Mar. biol. Ecol.*, 98 : 115-128.

Koehn R.K. & Gaffney P.M., 1977 - Genetic heterozygosity and growth rate in *Mytilus edulis*. *Mar biol.*, 82 : 1-7.

Korotkova A., 1979 - Peculiarities of somatic embryogenesis in Porifera : pp 53-58 in *Sponge Biology*. C. Lévi & N. Boury-Esnault (eds), CNRS, 291, Paris.

Lewontin R.C., 1974 - The genetic basis of evolutionary change. Columbia University Press, New York.

Lewontin R.C. & Hubby J.L., 1966 - A molecular approach to the study of genetic heterozygosity in natural populations. II. Amount of variation and degree of heterozygosity in natural populations of *Drosophila pseudobscura*. *Genetics*, 54 : 595-609.

Mallet A.L., Zouros E., Gartner-Kepay K.R. & Freeman K.R., 1986 - Genetics of growth in blue mussels : family and enzyme-heterozygosity effects. *Mar. Biol.*, 92 : 475-482.

Marcus N.H., 1977 - Genetic variation within and between geographically separated populations of the sea urchin, *Arbacia punctulata*. *Biol. Bull.*, 153 : 560-576.

Mayr E., 1963 - Animal species and evolution. Harvard University Press, New York.

McCommas S.A. & Lester L.J., 1980 - Electrophoretic evaluation of the taxonomic status of two species of sea anemone. *Biochem. Syst. Ecol.*, 8 : 289-292.

Moav R., Brody T., Wohlfarth G. & Hulata G., 1976 - Applications of electrophoretic genetic markers to fish breeding. I. Advantages and methods. *Aquaculture*, 9 : 217-228.

Mork J., Solemdal P. & Sundnes G., 1983 - Identification of marine fish eggs : a biochemical genetics approach. *Can. J. Fish. Aquat. Sci.*, 40 : 361-369.

Nei M., 1972 - Genetic distance between populations. *Amer. Natur.*, 106 : 283-292.

Neigel J.E. & Avise J.C., 1983 - Histocompatibility bioassays of population structure in marine sponges. *J. Heredity*, 74 : 134-140.

Nelson K. & Hedgecock D., 1980 - Enzyme polymorphism and adaptative strategy in the decapod crustacea. *Amer. Natur.*, 116 : 238-280.

Orr J., Thorpe J.P. & Carter M.A., 1982 - Biochemical genetic confirmation of the asexual production of brooded offspring in the sea anemone *Actinia equina*. *Mar. Ecol. Prog. Ser.*, 7 : 227-229.

Ottaway J.R. & Kirby G.C., 1975 - Genetic relationships between brooding and brooded *Actinia tenebrosa*. *Nature*, 225 : 221-223.

Pakhomov A.N., Sukhodol'skaya A.N. & Mukhlenkov A.G., 1974 - Electrophoretic investigation of esterase spectrum in sponge tissues. *Vestnik Leningr. gos. Univ.*, 1974 : 137-138 (in Russian).

Sarà M., in press - Divergence between the sympatric species *Tethya aurantium* and *T. citrina*, an example of speciation in sponges.*Proceedings of the third international Conference on the Biology of Sponges*. K. Rützler & W.D. Hartman (eds), Smithsonian Institution Press, Washington.

Schmidtke J., Weiler C., Kunz B. & Engel W., 1977 - Isozymes of a tunicate and a cephalochordate as a test of polyploidisation in chordate evolution. *Nature*, 266 : 532-533.

Schoots A.F.M., Schrijen J.J. & Denuce J.M., 1977 - Some characteristics of esterases in extracts of the freshwater sponge *Spongilla* sp. *Comp. Biochem. Physiol.* 56C : 49-51.

Scozzani R., Ulizzi L., Ferracini A. & Lunadei M., 1980 - An electrophoretic polymorphism that mimics a true genetic polymorphism in *Triturus cristatus carnifex* (Amphibia, Urodela). *Experientia*, 36 : 645-646.

Smith P.J. & Fujio Y., 1982 - Genetic variation in marine teleosts : high variability in habitat specialists and low variability in habitat generalists. *Mar. Biol.*, 69 : 7-20.

Simonarsen B. & Watts D.C., 1969 - Some fish muscle esterases and their variation in stocks of the herring (*Clupea harengus* L.). The nature of esterase variation. *Comp. Biochem. Physiol.*, 31 : 309-318.

Snyder J.P. & Gooch J.L., 1973 - Genetic differentiation in *Littorina saxatilis* (Gastropoda). *Mar. Biol.*, 22 : 177-182.

Solé-Cava A.M., 1986 - Studies of biochemical genetics and taxonomy in coelenterates and sponges. PhD Thesis, University of Liverpool.

Solé-Cava A.M. & Thorpe J.P., 1986 - Genetic differentiation between morphotypes of the marine sponge *Suberites ficus* (Demospongiae : Hadromerida). *Mar. Biol.*, 93 : 247-253.

Solé-Cava A.M. & Thorpe J.P., in press - High levels of genetic variation in marine sponges. *Proceedings of the third international Conference on Biology of Sponges*. K. Rützler & W.D. Hartman (eds), Smithsonian Institution Press, Washington.

Solé-Cava A.M., Thorpe J.P. & Kaye J.G., 1985 - Reproductive isolation with little genetic divergence between *Urticina* (= *Tealia*) *felina* and *U. eques* (Anthozoa : Actiniaria). *Mar. Biol.*, 85 : 279-284.

Thorpe J.P., 1979 - Enzyme variation and taxonomy : the estimation of sampling errors in measurements of interspecific genetic similarity. *Biol. J. Linn. Soc.*, 11 : 369-386.

Thorpe J.P., 1982 - The molecular clock hypothesis : biochemical evolution, genetic differentiation and systematics. *Ann. Rev. Ecol. Syst.*, 13 : 139-168.

Thorpe J.P., 1983 - Enzyme variation, genetic distance and evolutionary divergence in relation to levels of taxonomic separation : pp 131-152 in *Protein polymorphism : adaptative and taxonomic significance*. G.S. Oxford & D. Rollinson (eds), Academic Press, London.

Thorpe J.P., Beardmore A. & Ryland J.S., 1978 - Genetic evidence for cryptic speciation in the marine bryozoan *Alcyonidium gelatinosum*. *Mar. Biol.*, 49 : 27-32.

Thorpe J.P. & Ryland J.S., 1979 - Cryptic speciation detected by biochemical genetics in three ecologically important species of intertidal bryozoans. *Estuar. Coast. Mar. Sci.*, 8 : 395-398.

Urbaneja M. & Lin A.L., 1981 - A preliminary study on the isozyme patterns and taxonomy of tropical sponges. *Comp. Biochem. Physiol.*, 70B : 367-373.

Valentine J.W. & Ayala F.J., 1974 - Genetic variation in *Frieleia halli*, a deep-sea brachiopod. *Deep-sea Res.*, 22 : 37-44.

Van de Vyver G., 1971 - Analyse de quelques phénomènes d'histoincompatibilité intraspécifique chez l'éponge *Ephydatia fluviatilis*. *Archs Zool. exp. gén.*, 112 : 55-62.

Vawter A.T., Rosenblatt R. & Gorman G.C., 1980 - Genetic divergence among fishes of eastern Pacific and Caribbean : support for the molecular clock. *Evolution*, 34 : 705-711.

Ward R.D., 1978 - Subunit size of enzymes and genetic heterozygosity in vertebrates. *Bioch. Genet.*, 16 : 779-810.

Wilson A.L., Carlson S.S. & White T.J., 1977 - Biochemical evolution. *Ann. Rev. Biochem*, 46 : 573-639.

Zouros E., Singh S.M. & Miles H.E., 1980 - Growth rate in oysters : an overdominant phenotype and its possible explanations. *Evolution*, 34 : 856-867.

THE CALCIUM CARBONATE SPHERULES OF *HEMIMYCALE COLUMELLA* (DEMOSPONGES, POECILOSCLERIDA) AND THEIR TAXONOMIC VALUE

J. VACELET, C. DONADEY & C. FROGET

Centre d'Océanologie de Marseille, CNRS UA 41,
Station Marine d'Endoume, 13007 Marseille, FRANCE

SYNOPSIS

The Demosponge *Hemimycale columella* contains both a siliceous, spicular skeleton and unusual calcareous spherules. In formalin or glutaraldehyde fixed specimens, the spherules are observed as extracellular, irregular concretions, 2 to 55 µm in diameter. They are made of a non-magnesian calcite. Their occurence, size and morphology are variable according to fixation procedures, and they were not observed in living specimens. Although they were mentionned only in *H. columella*, preliminary observations indicate that nearly similar spherules exist in an undescribed Red Sea sponge which may belong to the same genus.

INTRODUCTION

Although in the phylum Porifera the skeleton may be composed of a large variety of minerals (Jones, 1979), calcium carbonate and siliceous structures very rarely occur together in the same sponge. The only two examples are : i) Coralline, or calcified sponges (= "Sclerosponges"), which possess a solid calcareous skeleton and siliceous spicules ; ii) *Hemimycale columella*, Class Demospongiae, in which calcium carbonate spherules occur alongside siliceous spicules. In this second example, no taxonomic value has been given to the presence of spherules, which were not mentionned in the diagnosis of the genus *Hemimycale* (Burton, 1934). In contrast, the taxonomic importance of the occurence of a calcareous skeleton in Coralline sponges has been emphasized so much that they are considered to be the basis for the distinction of a new Class of the Porifera, the Sclerospongiae (Hartman & Goreau, 1970). However, this new Class is challenged by Vacelet (1985), who

NATO ASI Series, Vol. G13
Taxonomy of Porifera
Edited by J. Vacelet and N. Boury-Esnault

considers that calcareous skeletons are plesiomorphies and have a value only at lower taxonomic levels.

The calcareous skeletons of calcified sponges, which are very important both in phylogeny and in paleontology, have been well studied recently. In contrast, very little is known on the spherules of *H. columella*. They have only been very vaguely described by Bowerbank (1874) as "gemmules". These were further identified by Forster (1955) as the calcareous spherules described by Topsent (1891 & 1934). According to Topsent (1891), the calcium carbonate material would occur within spherulous cells and would be responsible of the creamy colour of *H. columella*.

We have conducted some preliminary observations on these concretions in order to elucidate their nature and significance, and to investigate if such unusual structures may have an interest for taxonomic purposes.

MATERIAL AND METHODS

Numerous specimens of *Hemimycale columella* (Bowerbank, 1874) were collected in the Mediterranean Sea and one specimen in The Channel. Another sponge which may belong to the same genus and which will be soon described by Nobbe (personnal communication), was collected in the Red Sea (Jeddah and Eilat) and in the Gulf of Aden near Djibouti.

Most of the specimens were fixed in neutral formalin, 10 % in sea water, or in 95 % ethanol. For transmission electron microscope (TEM) studies, the sponges were fixed for 20 hr in 2.5 % glutaraldehyde in a buffer composed of 0.4 M sodium cacodylate and sea water 1/1. They were then rinsed in sea water and postfixed in osmium tetroxide 2% in sea water for 1 hr and embedded in "Araldite". Thin sections were cut using a diamond knife, stained in uranyl acetate and lead citrate, and examined with a Phillips EM 300 electron microscope.

For light microscope and scanning electron microscope (SEM) observations and for X ray diffraction analysis, the spherules were isolated from the tissue by boiling in a solution of sodium hypochlorite in water, followed by several rinses in distilled water and in ethanol. They were observed with a Hitachi S 570 SEM after being coated with about 150 Å of gold.

X-ray diffraction analysis was conducted according to the powder diffraction method : spherules were dried, crushed and X-rayed after being set in powder on a glass slide. A C.G.R. Thêta 60 diffractometer (copper K alpha radiation focused by a quartz carved monochromator) was used at scan speed of 1°20 / min.

OBSERVATIONS

General description of *Hemimycale columella*

The morphology of this sponge has been well described by several authors (Bowerbank, 1874 ; Topsent, 1891, 1925, 1934 ; Forster, 1955). It is an encrusting, somewhat massive sponge. Living, healthy specimens have a distinctive appearance, due to very conspicuous pore areas, or "cribs", which have a darker colour than the other parts of the surface. Two different colour forms occur side by side on the littoral of Provence (France) in the North Western Mediterranean : one which has the usual pink, creamy colour, and another one which is more brownish. No significant difference was found between these two forms.

The siliceous spicules are mainly strongyles, most of them possessing unequal ends, and a few styles of the same size or a little smaller. In Mediterranean specimens, strongyles are thinner than in the Atlantic specimens (Table I). Their axial canal is large (e.g. 1.2 µm for a spicule 3 µm wide, or 2 µm for 4 µm). When they are boiled several minutes in a solution of sodium hypochlorite in water, these thin spicules with a large axial canal are often deformed.

Table I : Strongyle measurements in µm in *H. columella* : ▪ minimum and maximum (mean and standard deviation), respectively for length and width, based upon 20 measurements. * minimum and maximum for 4 specimens. # after a drawing of one spicule.

Marseille (Port-Miou 1)▪	225-310 (285 ± 20.8)	/ 2.0-4.0 (3.0 ± 0.4)
Marseille (Port-Miou 2)▪	320-410 (369 ± 14.2)	/ 2.5-3.8 (3.1 ± 0.2)
Marseille (Pomègues)▪	220-320 (273 ± 16.7)	/ 2.0-4.0 (2.7 ± 0.3)
Topsent, 1925 (Naples)*	270-470	/ 4.0-6.0
Roscoff▪	290-465 (394 ± 24.6)	/ 4.0-7.0 (5.1 ± 0.5)
Topsent, 1891 (Roscoff)	400	/ 6.0
Foster, 1955 (? Plymouth)	330-420 (373)	/ 5.0-6.0 (5.85)
Bowerbank, 1874 (Exmouth)#	376	/ 7.0

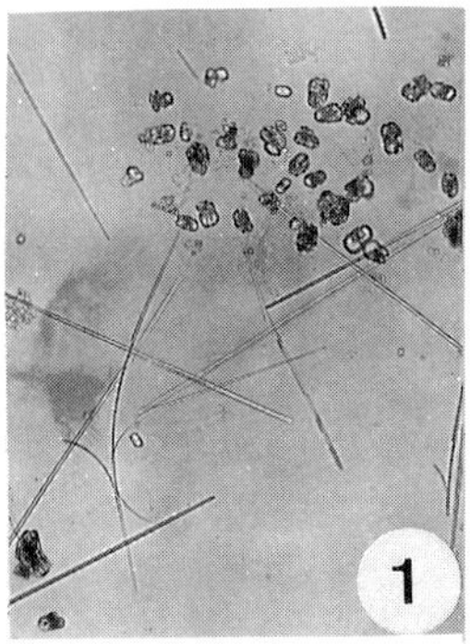
1

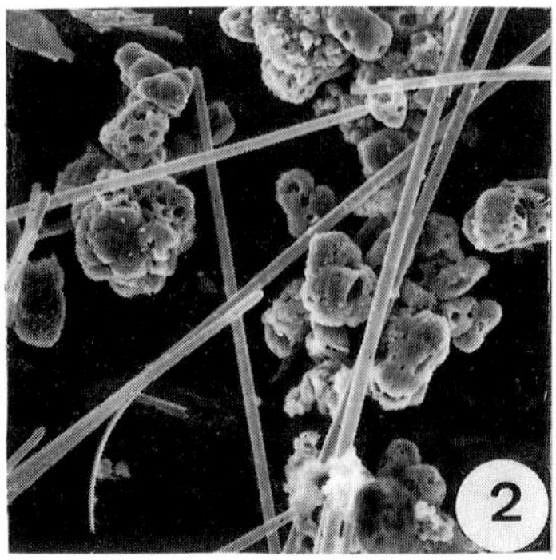
2

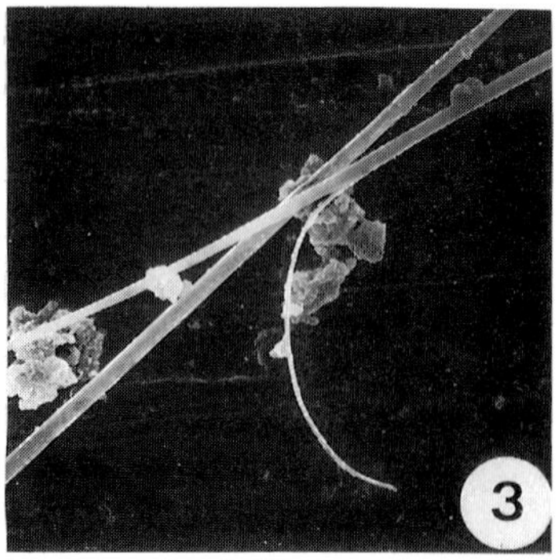
3

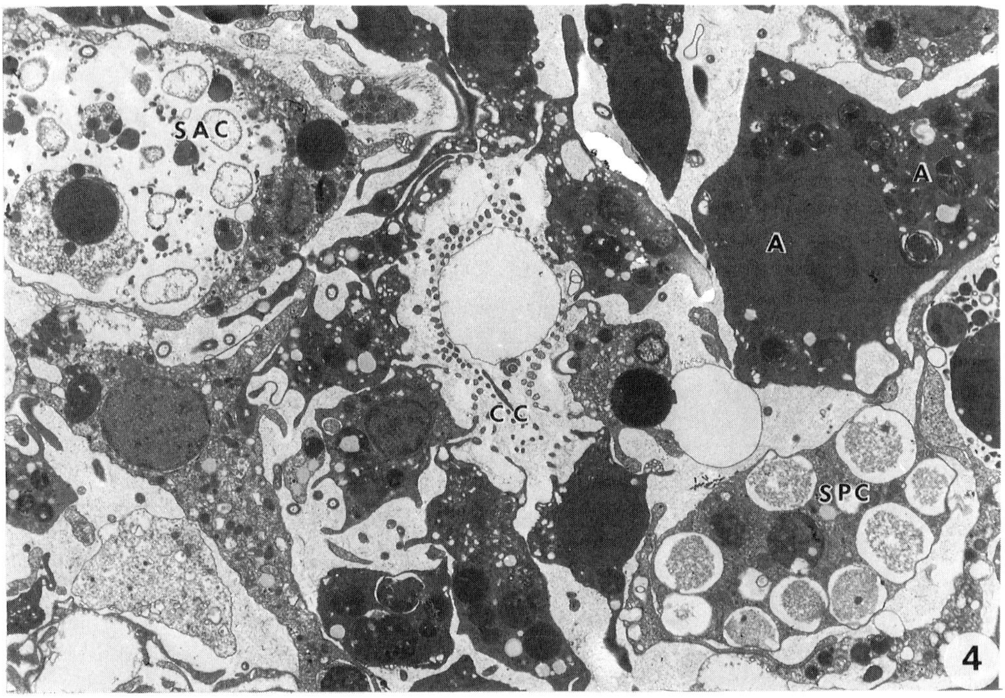
SAC
A
A
C C
SPC
4

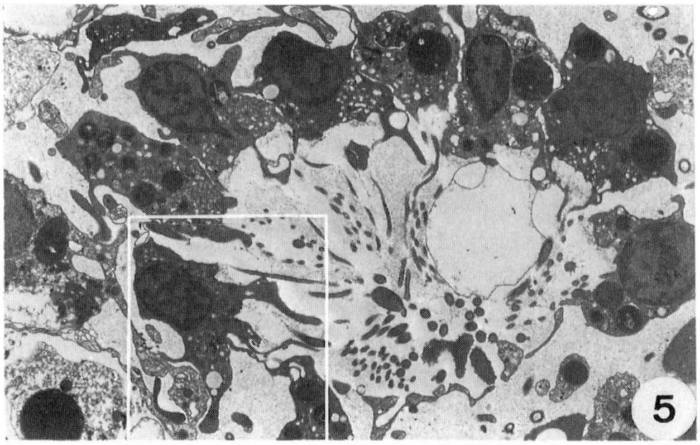
5

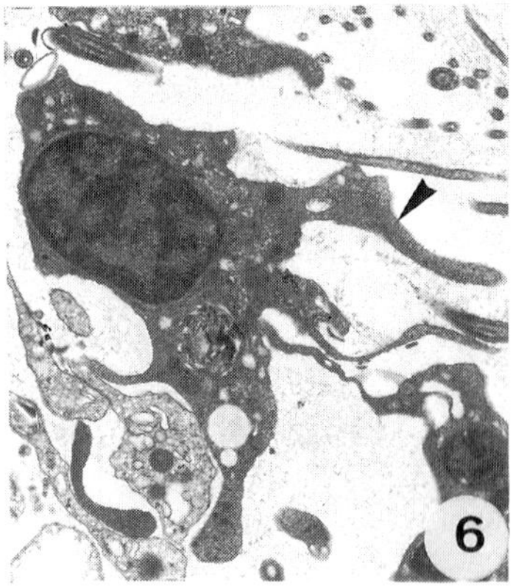
6

The walls of the canal stand apart and recoil up on themselves, as if they were set free at one end of the spicule and were softening (Figs 1-3). These deformations never occur with the conventionnal mode of siliceous spicules cleaning, i.e. boiling in concentrated nitric acid.

In the specimen from The Channel, the spicules are thicker, with a smaller axial canal. Subtylostrongyles are more frequent. No deformation occurs when boiling in sodium hypochlorite.

These spicules are linked in loose tracts by a few spongin. The few styles do not have a definite position.

The cytology of this sponge has been described by Boury-Esnault (1972) and by Willenz (1982). A very distinctive cuticle has been found by Willenz on the surface between the pore areas, which may be related to nutritive particles capture.

Cells with inclusions are very numerous (Fig. 4). As Willenz pointed out, they belong to two main types. Spherulous cells contain inclusions 2 to 3 µm in diameter, the content of which looks fluffy and is of a variable density. Sacculiferous cells contain annular inclusions, 1.5 to 2 µm in diameter, with an empty central area enclosed by a fibrillar, thin sheet. The latter have been frequently observed releasing their inclusions into the intercellular matrix. No calcium carbonate mineralizations are evident in these cells with inclusions, neither in sections for microscopy nor in living cells, in contradiction with Topsent's statement that the calcareous spherules are contained in spherulous cells.

The choanocyte chambers (Figs 4-6) are about 35-40 µm in maximum diameter. Their choanocytes often display a periflagellar sleeve. Symbiotic bacteria are absent or very rare.

Embryos and stages of spermatogenesis have been observed at the end of April and in October in Mediterranean specimens. Eggs and embryos were previously recorded in August and September in Roscoff (Brittany) by Topsent (1891) and Lévi (1956).

Fig. 1 - Siliceous spicules and calcium carbonate spherules. x 80. **Figs 2 & 3** - SEM image of siliceous spicules and calcium carbonate spherules. Note the deformation occuring on some spicules after boiling in sodium hypochlorite. 2 : x 250 ; 3 : x 320. **Fig 4** - A choanocyte chamber (CC) surrounded by diverse cell types, including an archeocyte (A), a spherulous cell (SP C) and a degenerating sacculiferous cell (SA C). x 5,500. **Fig. 5** - A choanocyte chamber. x 4,000. **Fig. 6** - Enlargment of a choanocyte from Fig. 5. Note the periflagellar sleeve (arrow). x 8,000.

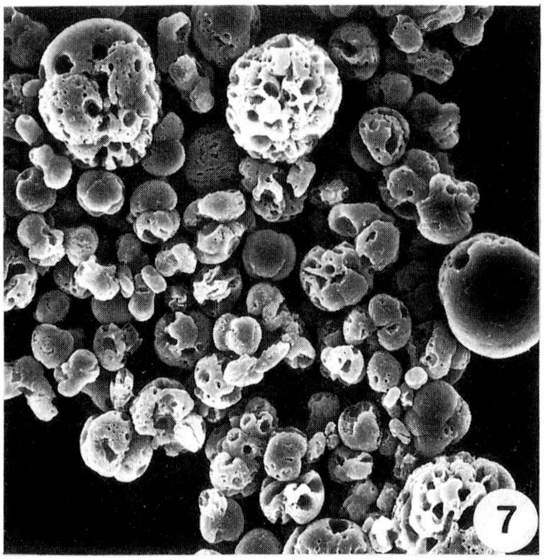
7

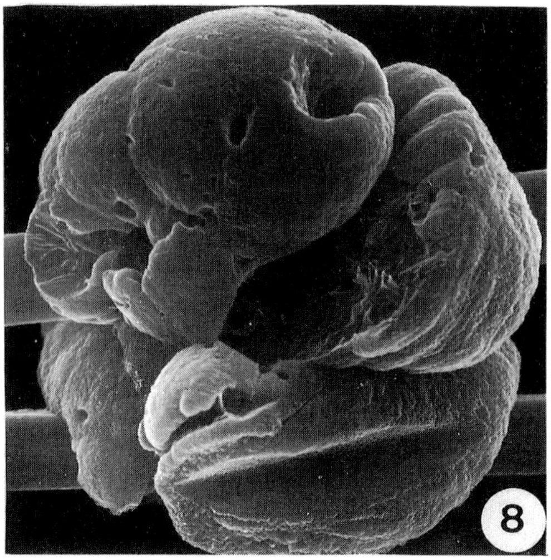
8

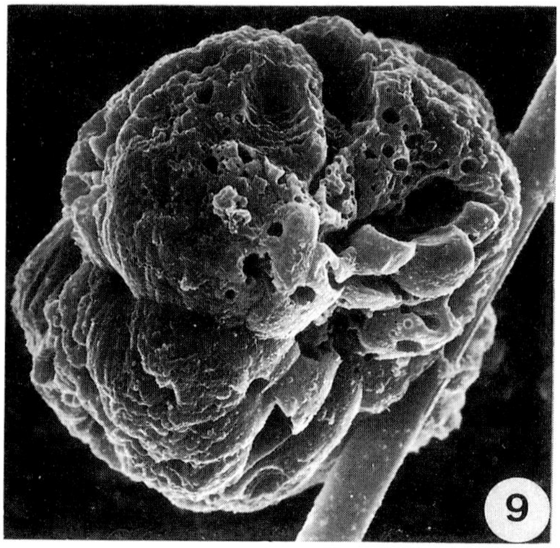
9

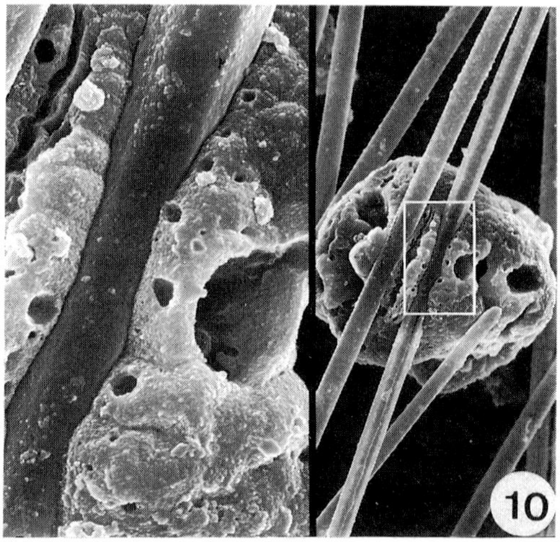
10

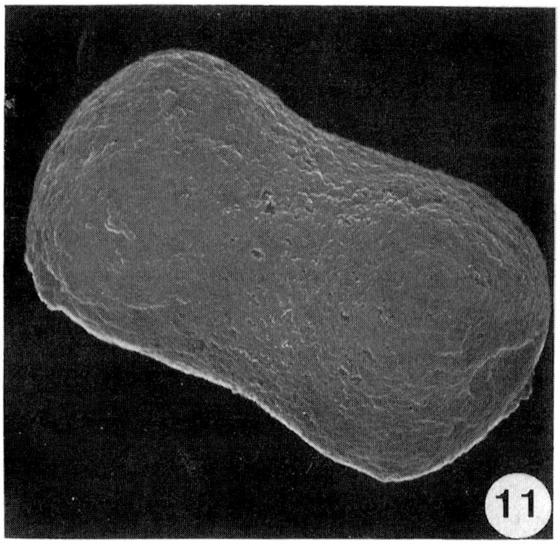
11

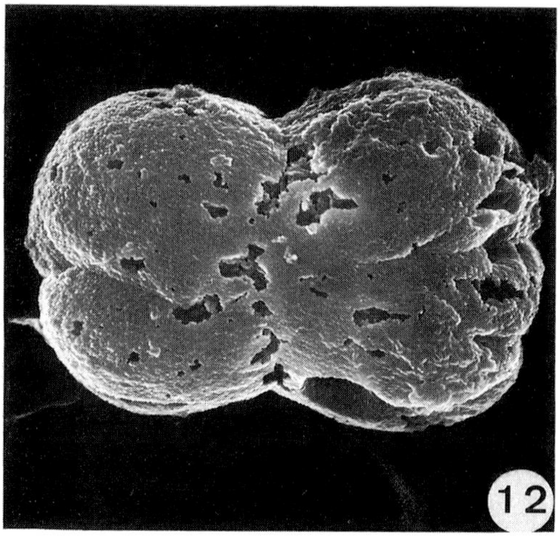
12

Calcium carbonate spherules.

The spherules exhibit different appearances according to the fixation procedure, and they are conspicuous structures only in specimens fixed in formalin or in glutaraldehyde. In these specimens, the spherules occur in high numbers both in the choanosome and in the superficial tissue, but their distribution is somewhat irregular and they are rare in places. In Mediterranean specimens, their diameter ranges from 2.4 to 55 µm, the most common size being between 15 to 30 µm (Fig. 7). In the specimen studied from The Channel, they are smaller and less numerous, although they are similar in general shape and characters. This size falls within the range succinctly indicated by Bowerbank (1874) for the "gemmules" (less than 21 µm and up to 80 µm) and by Forster (1955), who observed spherules smaller than those of Bowerbank. Topsent and other authors who have observed this sponge did not give any measurements.

In SEM, most of the smaller spherules have the shape of a short rod with rounded ends and a median depression (Figs 7, 11, 12). They are slightly lobate, with no or few small holes. Those which are larger than 15 µm are more irregular (Figs 7-10, 13-17). Most have large, deep holes or depressions which reach 10 µm in diameter and in which siliceous spicules may be entrapped (Figs 9, 10). These large concretions also bear smaller and more regular holes, 0.4 to 0.7 µm in diameter (Figs 10, 17). Many have two different faces : one which is rounded, regularly lobated with a smooth surface and with few or no holes ; the other is irregular, possessing many holes or deep depressions, and sometimes polyhedral spines (Figs 13, 15, 16). The surface of this irregular face may be finely granular (Fig. 14). Rhombohedric crystals characteristic of calcite are sometimes apparent (Fig. 17). These different appearances usually occur in the same specimen, but they are more or less common. However, the polyhedral spines and the rhombohedric crystals were observed only in a few specimens.

Fig. 7 - SEM image illustrating the variety in size and morphology of the spherules. x 425. **Figs 8, 9 & 10** - Large spherules entrapping spicules. 8 : x 2,800 ; 9 : x 1,500 ; 10 : x 650 and 3,250. **Figs 11 & 12** - Small spherules. 11 : x 5,000 ; 12 : x 6,500.

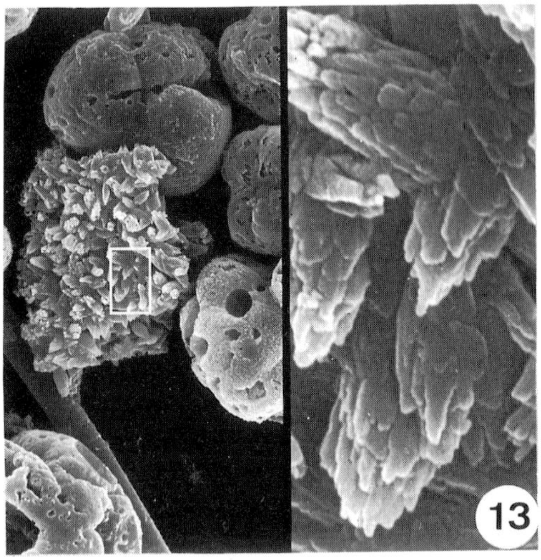
13

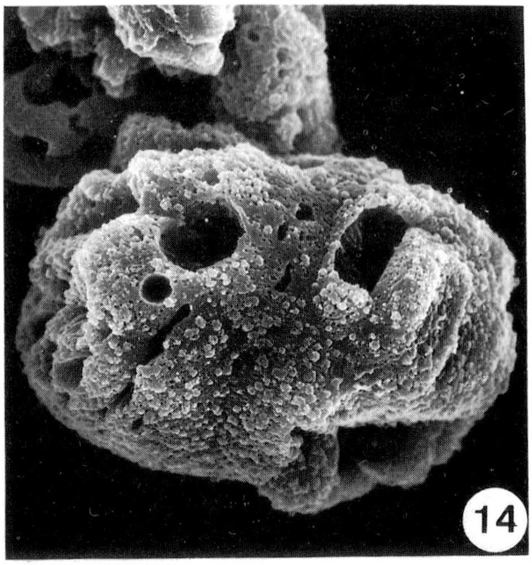
14

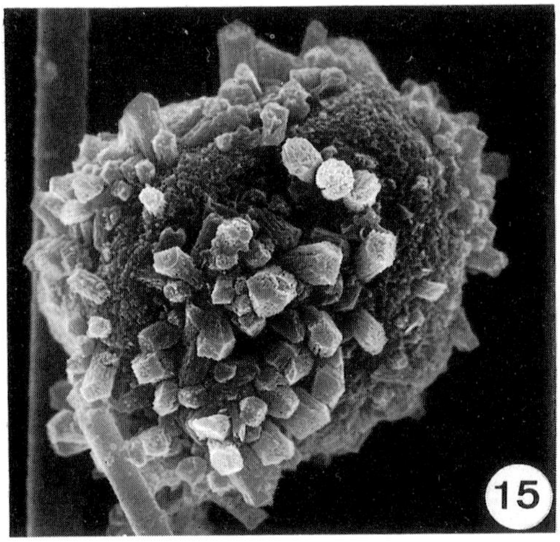
15

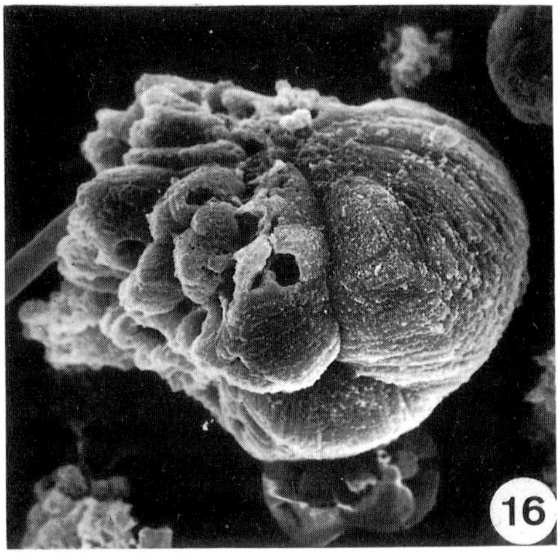
16

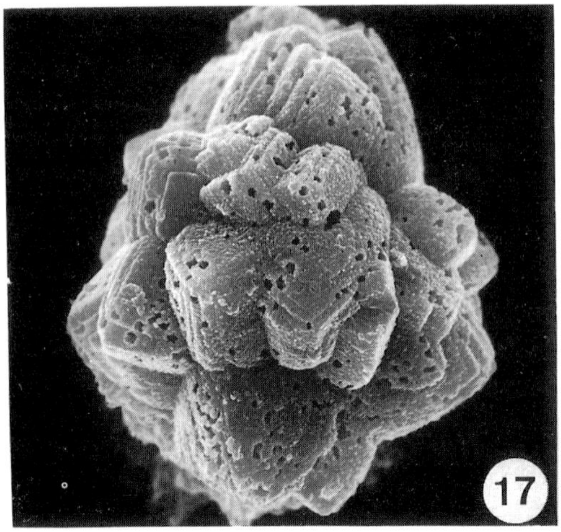
17

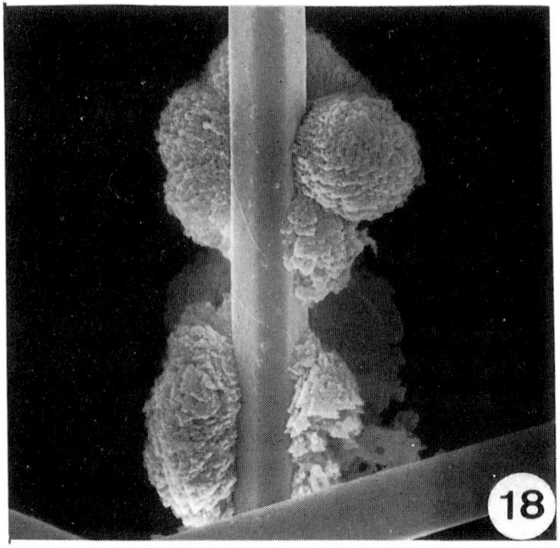
18

In light microscopy (Fig. 19), the spherules often display a darker center or internal dark lines. Most appear composed of several round granules, united by a common enveloppe. Radial fibrous texture or concentric layers are often apparent. With the polarizing microscope, the interference colour produced by the spherules when examined between cross nicols is a high-order white. Due to the radial fibrous structure, extinction images in black cross are apparent (Fig. 20).

X-ray analysis performed on spherules obtained after fixation in formalin and cleaning with sodium hypochlorite indicates that they are made of a pure calcite (major reflection at 3.035 Å), which, unlike most calcitic structure of biological origin, does not contain magnesium.

In specimens fixed in ethanol or boiled in a sodium hypochlorite solution without any fixation, there is no large spherules, but only small granules, 1.1 to 1.9 µm in diameter (Fig. 21). These are generally round, although they may sometimes be lobate or rarely polyhedral. In polarized light, these granules exhibit the same high-order white as the spherules. In specimens fixed in alcohol, the spherules only appear after a few hours in the fixative, first as small granules, which grow progressively up to a few micrometers in diameter.

In living specimens the spherules were not observed and it seems that they do not occur here, at least in the same form as in fixed specimens.

In histological sections and in TEM, subspherical bodies similar in size and general shape to the calcium carbonate spherules have been observed on undecalcified sections. In semi-thin sections (Fig. 22), these bodies are mineralized and very difficult to cut. They exhibit the same characteristics in polarized light as isolated spherules. They are easily demineralizated by dilute acetic acid, which leaves their organic matrix in evidence. In ultrathin sections (Figs 23-26), the bodies are devoid of electron opaque material and do not appear to be mineralized. Most probably, their calcium carbonate has been removed by distilled water and uranyl acetate during

---

Fig. 13 - Irregular crystals on a spherule. x 1,300 and 13,000. **Fig. 14** - Spherule exhibiting a granular surface. x 1,400. **Fig.** 15 - Crystals on a spherule face. x 2,400. **Fig. 16** - A spherule displaying its two faces (rounded, irregular). x 1,000. **Fig. 17** - Rhombohedric crystals of a spherule. x 2,800. **Fig. 18** - Concretions entrapping a spicule in an undescribed Red Sea sponge, presumably a new species in the genus *Hemimycale*. x 2,800.

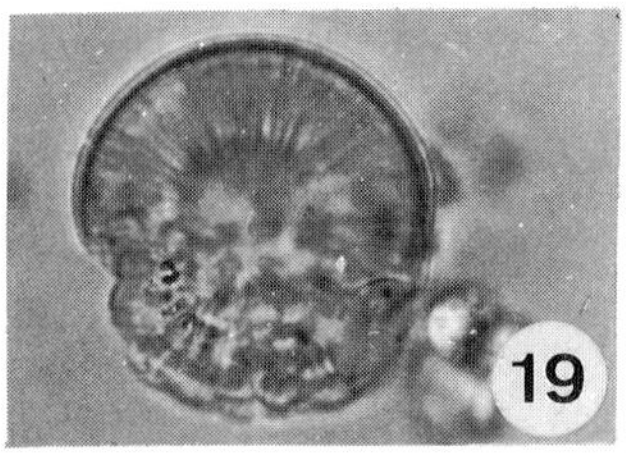
19

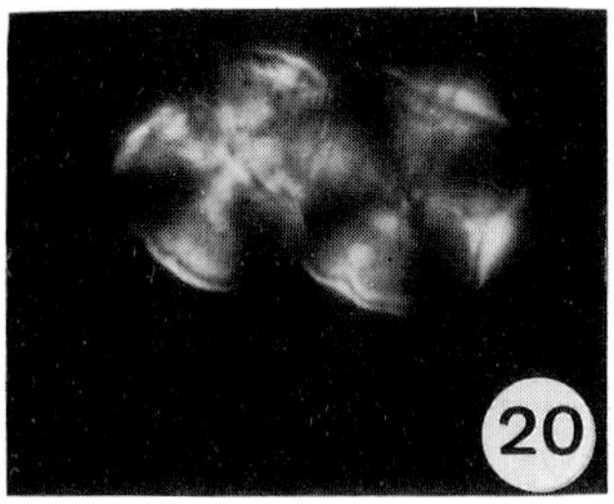
20

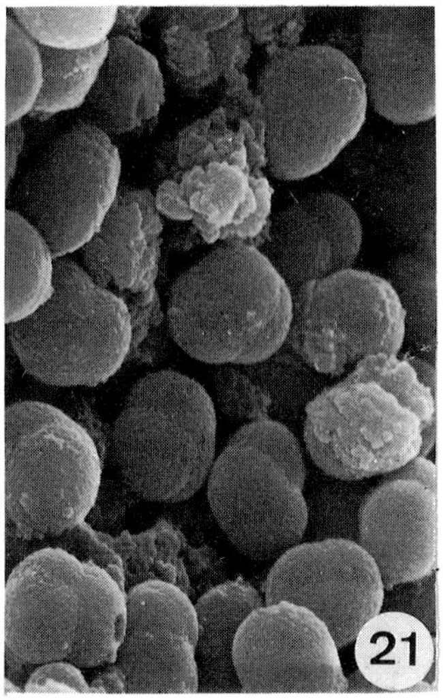
21

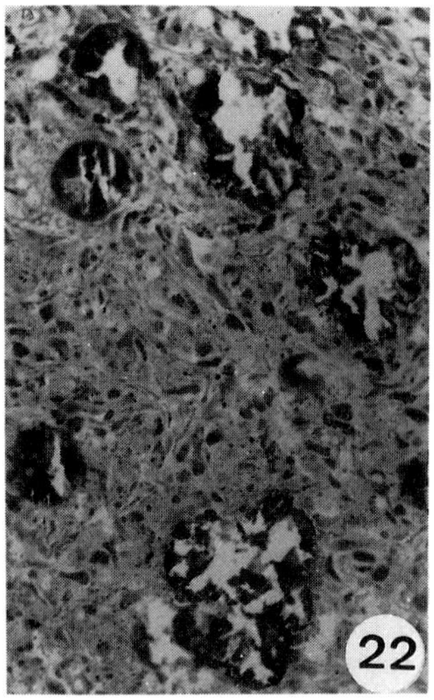
22

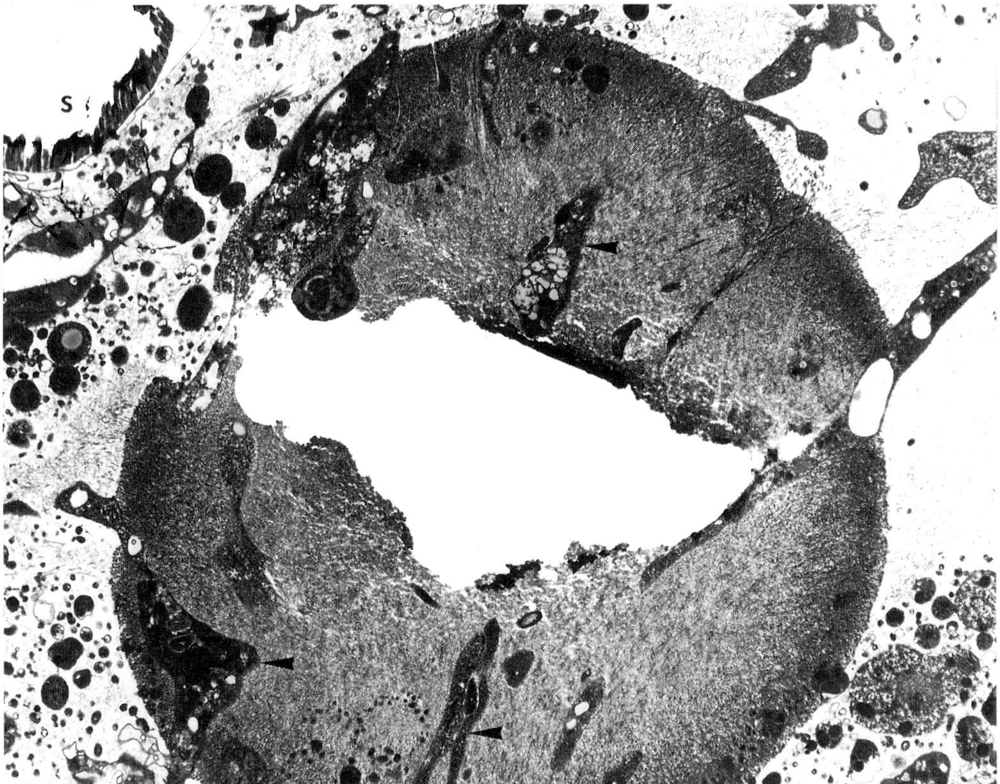
S
23

sectionning and staining.

The organic matrix is composed of a granulo-fibrillar material, in which collagen fibrils, cells and cellular debris are embedded. Most of the collagen fibrils are arranged in radial fascicles, about 300 nm in diameter, which could correspond to the smaller holes observed by SEM in the spherules. Some cells are lying on the surface of the bodies, or are partially embedded in large depressions, which again may correspond to the large holes of the spherules. These cells are mostly degenerative stages of cells with inclusions. They send long cellular processes inside the bodies, and their inclusions or parts of their cytoplasm also occur free within the matrix. Siliceous spicules are sometimes found lying on large depressions of the bodies, again as in the spherules which entrap spicules (Fig. 26). Two or three dense, concentric lines have been observed near the surface in a few of these bodies (Fig. 25).

DISCUSSION

This study was mainly limited to conventionnally preserved material, similar to that which is currently used in taxonomic studies. In these conditions, i.e. aldehyde fixed specimens, calcareous spherules are abundant in the tissue of this sponge and may be isolated, together with spicules, by treatment with sodium hypochlorite. The presently available data indicate that these spherules are considerably smaller or absent in alcohol preserved specimens or in living specimens. Thus they may be considered , at least in part, as artefacts. As such, these concretions cannot be assimilated to usual skeletal elements and given a comparable taxonomic value.

Fig. 19 -Light microscope image of a spherule. x 800. **Fig. 20** - Spherules as seen with the polarizing microscope,using crossed nicols. x 730. **Fig. 21** - Small calcium carbonate granules present after boiling the living sponge in sodium hypochlorite, without any fixation. x 7,500. **Fig. 22** - Semi-thin section of glutaraldehyde and osmium tetroxide fixed tissue. Most of the spherules have been broken during the cutting process. x 240. **Fig. 23** - TEM image of a spherical body, presumably corresponding to the organic matrix of a calcareous spherule. Note the cellular processes inside the matrix (arrows). S : siliceous spicule. x 5,500.

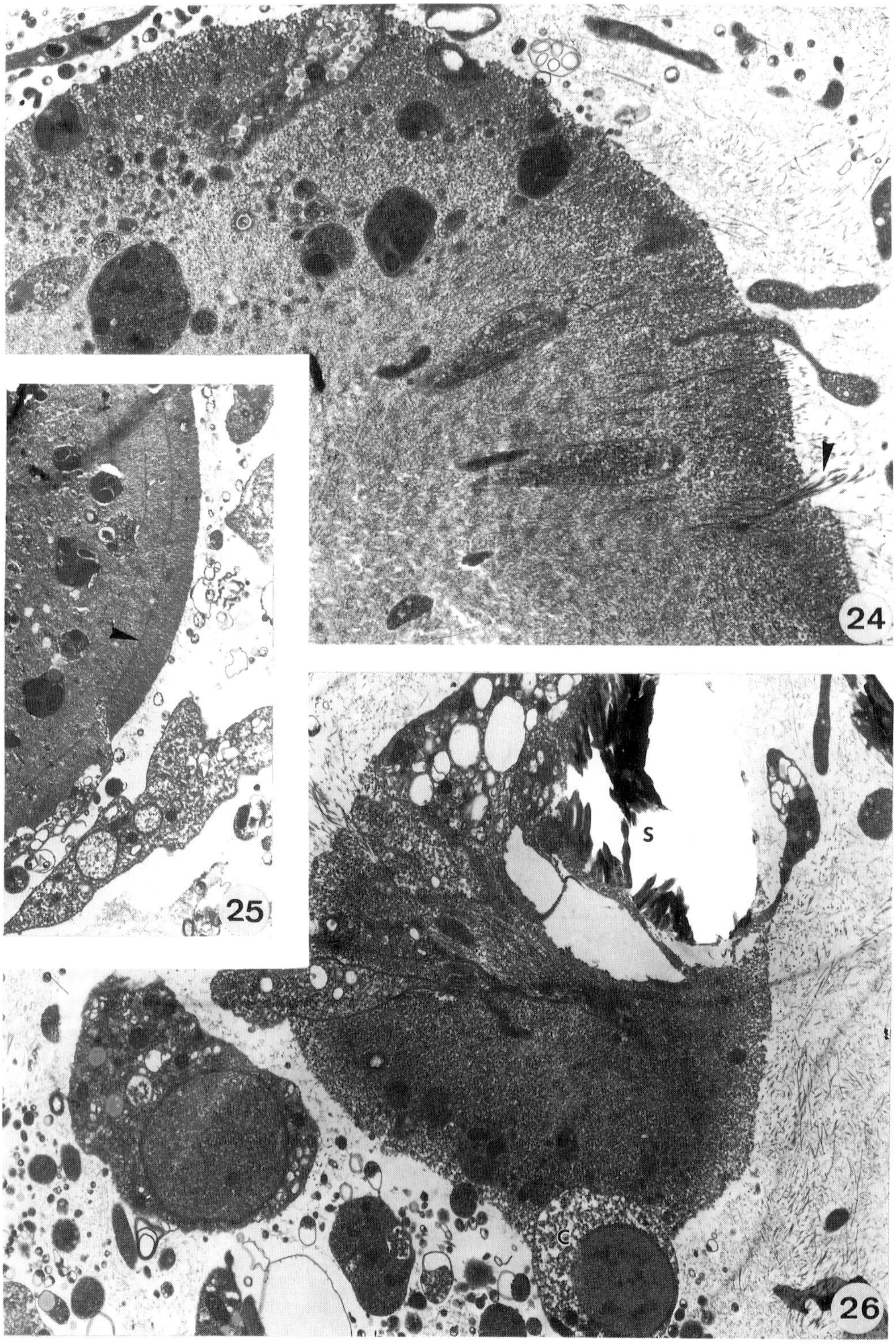
24
25
S
C
26

With this limitation in mind, it would nevertheless be interesting for taxonomic purposes to know if similar phenomena occur in other sponges. The problem of their biological significance will only be briefly discussed here, as it needs to be more thoroughly investigated in living material.

The only other case where calcareous spherules have been described in the tissue of a siliceous sponge is that of the calcified sponge *Astrosclera willeyana* Lister, 1900. Calcareous asters were mentionned in *Hymeniacidon calcifera* Row, 1911, from the Red Sea, but these probably belong to an ascidian ; we did not observe them on the type specimen of this species (British Museum). In *Astrosclera*, the spherules are aragonitic and are not artefacts. They are intracellular growth stages of the spherulitic elements of the solid skeleton (Lister, 1900 ; Gautret, 1986). In this sponge, it has been questionned (Weltner, 1910) whether the spherulitic elements of the skeleton are actually secreted by the sponge or if they have a foreign origin. It is now evident that this sponge and other calcified sponges are able to secrete both siliceous spicules and calcium carbonate material, but the question may be raised again for *H. columella*.

Although irregular in their distribution in the sponge tissue, the spherules as seen after formalin fixation are present in all the studied specimens. Their peculiar form, their somewhat unusual composition of pure calcite without magnesium, the fact that they entrap siliceous spicules of the sponge and that they only appear after fixation, show that they cannot be foreign materials selected by the sponge in its environment.

However, this preliminary study does not result in a clear picture of their development in the sponge tissue during fixation. The relationships between the fully developped spherules as seen in fixed material and the living material remain to be explained. As it may be inferred from TEM observations, the crystallization of the calcite occurs on organic and cellular bodies, which most probably are not artefacts, but which are highly unusual in sponges. The significance of these bodies remains unknown.

Fig. 24 - Spherical body (see Fig. 23). Note the radial fascicles of collagen fibrils (arrow) and the cellular debris inside the matrix. x 5,500. **Fig. 25** - Dense concentric lines observed in an organic body (arrow). x 6,500. **Fig. 26** - A spherical body entrapping a siliceous spicule (S) ; a degenerating cell (C) lying on the surface of the body. x 4,700.

It also remains to be investigated whether the mineralization always occurs on an organic matrix or if the smaller granules initiate growth without special organic structure. The relationships between the fully developped spherules and the inclusions of the sacculiferous cells would be worthwhile to investigate, as the size and the shape of the calcium carbonate subunits of the spherules are similar to those of the inclusions of these cells, which may correspond to their nucleating sites.

To our knowledge, no comparable mineralizations are known in other organisms. Diverse calcareous spherules have been described in many other phyla (Watabe *et al.*, 1976 ; Lowenstam, 1986). In Invertebrates, they usually are intracellular and are present in living material. However, the spherules of *H. columella* morphologically resemble the spherules of the gastropod, *Pomacea paludosa*, which were said to be composed of amorphous calcium carbonate (Watabe *et al.*, 1976).

The genus *Hemimycale*, genotype *H. columella*, was defined by Burton (1934, p. 556) as "Reduced Mycaleae with skeleton of loose fibres of styli, sometimes modified in anisostrongyla, running vertically to surface ; fibres tending to branch and anastomose ; no special dermal skeleton ; no microscleres". As stated above, this diagnosis makes no use of the calcareous spherules and this appears to be right if they are artefacts, as suggested by the present study. However a question which arises concerns whether these spherules actually do not exist in other species of the same genus or of the same family. Up to the present, it seems that there are no other described valid species in the genus *Hemimycale*. However, a sponge very common in the Red Sea, which will probably be described as a new species by Nobbe (personnal communication), may be related to the genus. We have observed that fixed specimens of this undescribed sponge also contain spherules (Fig. 18), which are dissolved by strong acids and are at least in part calcitic in nature. Their morphology, which seems to be variable according to the fixative, and their composition are quite different, however, and are presently under study. In any case, the occurence of unusual mineralizations in another sponge related to the genus *Hemimycale* strongly suggests the taxonomic interest of the spherules, even if they are artefacts.

The hypothesis of Burton (1934) that *H. columella* is a Poecilosclerid with a reduced skeleton is accepted here. A different opinion has been expressed by Pulitzer-Finali (1977), who classified the genus in the order Halichondrida, family Hymeniacidonidae. However, this treatment appears less consistent with the presence of cribs, which is a Poecilosclerid character.

It is worthwhile to note that such mineralizations could easily escape detection during routine taxonomic studies, not only because they may be in part artefacts, but also because of the conventionnal method of spicule cleaning for Demosponges, i.e. boiling in strong nitric acid. So they could actually be of more general occurence than previously believed. Routine treatment with sodium hypochlorite, or with a less drastic method by enzymatic digestion, as described by Fry and Gray (this book), could enlarge the number of known cases. It would be of special interest to check for the presence of these structures in some genera in which the spiculation is made only of strongyles, such as *Batzella* Topsent, 1893 or *Prianos* Gray, 1865 which are presently difficult to allocate to a family or even to an order, and which are probably relatives of *Hemimycale* as pointed out by Topsent (1934).

ACKNOWLEDGMENTS

We greatly acknowledge Dr. L. Cabioch's help for the collection and fixation of material from Roscoff. We also thank C. Bézac and M.R. Causi for their technical help, and S. Stone for kindly communicating type specimens from the British Museum.

LITERATURE CITED

Bowerbank J.S., 1874 - A Monograph of the British Spongiadae, 3, 367 p.

Boury-Esnault N., 1972 - Une structure inhalante remarquable des spongiaires : le crible. Etude morphologique et cytologique. *Archs Zool. exp. gén.*, 113 : 7-23.

Burton M., 1934 - Sponges. *Great Barrier Reef Exp. Sci. Rep.*, 14 : 513-614.

Forster G.R., 1955 - *Hemimycale columella* (Bowerbank) : a short description and history of the species. *J. mar. biol. Assoc.*, 34 : 553-557.

Fry P.D. & Gray D. (this book) - An enzymatic technique for the separation of spicules from alcohol-preserved sponge tissue.

Gautret P., 1986 - Utilisation taxonomique des caractères microstructuraux du squelette aspiculaire des Spongiaires. Etude du mode de formation des microstructures attribuées au type sphérolitique. *Ann. Paléont.*, 72: 75-110.

Hartman W.D. & Goreau T.F., 1970 - Jamaican coralline sponges : their morphology, ecology and fossil relatives : pp 205-243 in *The biology of the Porifera, Symp. zool. Soc. London, 25*. W.G. Fry(ed.), Academic Press, London.

Jones W.C., 1979 - The microstructure and genesis of sponge biominerals : pp 425-447 in *Biologie des Spongiaires, Colloq. int. C.N.R.S., 291*. C. Lévi & N. Boury-Esnault (eds), Paris.

Lévi C., 1956 - Etude des *Halisarca* de Roscoff. Embryologie et systématique des Démosponges. *Arch. Zool. exp. gén.*, 93 : 1-181.

Lister J.J., 1900 - *Astrosclera willeyana*, the type of a new family of Sponges. *Willey's zool. Results*, 4 : 459-482.

Lowenstam H.A., 1986 - Mineralization processes in monerans and protoctists: pp 1-17 in *Biomineralization in lower Plants and Animals, Systematic Assoc. special Vol. 30*. B.S.C. Leadbeater & R. Riding (eds), Clarendon Press, Oxford.

Pulitzer-Finali G., 1977 - Report on a collection of sponges from the Bay of Naples. III. Hadromerida, Axinellida, Poecilosclerida, Halichondrida, Haplosclerida. *Boll. Mus. Ist. Biol. Univ. Genova*, 45 : 7-89.

Topsent E., 1891 - Essai sur la faune de Spongiaires de Roscoff. *Archs Zool. exp. gén.*, 9 : 523-554.

Topsent E., 1925 - Etude des Spongiaires du golfe de Naples. *Archs Zool. exp. gén.*, 63 : 623-725.

Topsent E., 1934 - Eponges observées dans les parages de Monaco (Première partie). *Bull. Inst. océnogr. Monaco*, 659 : 1-42.

Vacelet J., 1985 - Coralline sponges and the evolution of Porifera : pp 1-13 in *The origins and relationships of lower invertebrates, Systematics Assoc. special Vol. 28*. S. Conway Morris, J.D. Georges, R. Gibson & H.M. Plattt (eds), Clarendon Press, Oxford.

Watabe N., Meenakshi V.R., Blackwelder P.L., Kurtz E.M. & Dunkelberger D.G., 1976 - Calcareous spherules in the gastropod, *Pomacea paludosa* : pp 283-308 in *The Mechanisms of Mineralization in the Invertebrates and Plants, The Belle W. Baruch Library in marine Science n° 5*. Watabe N. & Wilbur K.M. (eds), Univ. South Carolina Press.

Weltner W., 1910 - Ist *Astrosclera willeyana* Lister eine Spongie ? *Arch. Naturgesch.*, 76 : 128-134.

Willenz P., 1982 - Aspects cinétiques, quantitatifs et ultrastructuraux de l'endocytose, la digestion et l'exocytose chez les éponges. *Thèse, Zoologie, Université Libre de Bruxelles, Belgique*, pp 1-107, 180 figs.

SOME REMARKS ON THE MEDITERRANEAN SPECIES OF THE GENUS *APLYSINA* (DEMOSPONGIAE, VERONGIDA).

E. VOULTSIADOU-KOUKOURA

Department of Zoology, University of Thessaloniki
540 06 Thessaloniki - Greece

SYNOPSIS

The study of numerous specimens belonging to the genus *Aplysina* from the North Aegean Sea, illustrated that, although they have been found at different depths, they do not seem to belong to two separate species, as indicated by the bibliographical data . The existence of two separate species of this genus in the Mediterranean Sea should possibly be reexamined.

INTRODUCTION

Two species belonging to the genus *Aplysina* (= *Verongia*) have been described from the Mediterranean Sea : *A. aerophoba* Schmidt, 1862 and *A. cavernicola* (Vacelet, 1959). During a survey on the Demosponges of the North Aegean Sea, numerous specimens belonging to this genus were collected in various biotopes. An attempt was made to classify them into the two species mentioned above. The questions and the problems that arose from this attempt are presented in this paper.

MATERIALS AND METHODS

Thirty-nine specimens were collected from thirty-five sampling stations scattered all along the Greek coasts of the North Aegean Sea (Fig.1). The collection of the specimens was made in depths from 0.5 m to 150 m either by free and SCUBA diving or by dredging. Between depths of 0.5 to 30 m, 56 % of the specimens were collected and the remaining 44 % were taken from 30 to 150 m.

NATO ASI Series, Vol. G13
Taxonomy of Porifera
Edited by J. Vacelet and N. Boury-Esnault

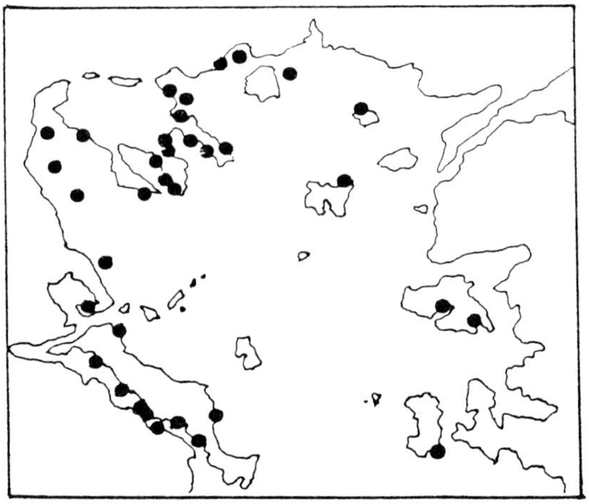

Figure 1 - Map showing the sampling stations in the North Aegean Sea.

RESULTS AND DISCUSSION

In order to identify the specimens Table 1 was constructed. It is based on the main characteristics used by Vacelet (1959) for the distinction of the two species. The ecological characteristics of the two species are based on Wilkinson & Vacelet (1979). It should be pointed out that no differences were found in the spongin skeleton by Vacelet (1959) for the two species.

Careful examination of the 39 specimens collected showed that the morphological characteristics which should distinguish the two species from one another, were neither related to the depth nor the illumination conditions of their habitat. Both in the specimens coming from depths above 30 m and in those collected from depths greater than 30 m we observed the whole range of characteristics for *A. aerophoba* and *A. cavernicola* (see Table 1). Therefore it was concluded that :

1/ The regularity of the general form and especially that of the digitations and the conules varies independently of the depth and the intensity of the light.

2/ The color of the sponges does not show a relationship with depth and illumination. Specimens from shallow depths showed various tones in alcohol between violet and black and the same was observed for the specimens from greater depths. The time needed for the transition of color from yellow to violet or black was not observed in some of the deeper water specimens. However, it was almost immediate in all the remaining specimens. It should be noted that the presence or absence of cyanobacteria in the specimens was not examined.

Table 1 - The main morphological and ecological characteristics distinguishing *A. aerophoba* from *A. cavernicola*.

		A.aerophoba	*A.cavernicola*
ECOLOGICAL CHARACTER.	Illumination	Fully exposed	slightly exposed
	Depth range	0.5 to 30 m	5 - 130 m
	Presence of cyanobacteria	always	very rarely
MORPHOLOGICAL CHARACTERISTICS	Surface	Conules small and irregularly distributed	conules more regularly distributed
	Digitations	irregular, often fused together, thinner towards the top, forming a depression in the center of the plane existing at the top of each digitation	more regular and thinner, exhibiting approximately the same diameter from their base to the top, forming a plane at the top but not a depression.
	Lateral projections	often present	never present
	Color	Yellow, turning dark violet or black quickly when exposed to air or placed in alcohol.	light yellow turning violet in alcohol. Remains yellow after a few hours in air.

3/ Specimens from greater depths (i.e. one collected from 150 m) exhibited an irregular form and irregular digitations which are characteristic of *A. aerophoba*. The usual form of the specimens of this genus in the Aegean Sea (independent of depth) is a form with several, regular digitations and with lateral projections rarely occuring. The irregular form with lateral projections was not observed often. These can be seen in the photographs given by Topsent (1927-29), and Pulitzer-Finali & Pronzato (1981).

4/ No difference in the skeleton of the various specimens has been observed.

5/ The depression on the top of the digitations existed in all the specimens from shallow depths. However, in specimens collected deeper than 30 m, some displayed this characteristic while others did not. Pulitzer-Finali & Pronzato (1977), (who have found *A. cavernicola* in depths between 40 and 50 m) state that : "we have not been able to form an opinion on a discriminating feature indicated by Vacelet : the apex of the digitations being a plane with or without a depression".

CONCLUSION

From the observations of this study it is difficult to accept that, at least in the Aegean Sea, there are two different species of *Aplysina*. It is perhaps easier to accept that the morphological differences observed by Vacelet (1959), discriminate between two separate ecological forms and not two separate species. This aspect is partially supported by the fact that such morphological differences appear in other sponge species. *Petrosia ficiformis* Poiret, for example, in the Aegean Sea (personal observations) and in the Western Mediterranean (according to Wilkinson & Vacelet, 1979) appears with three different morphological forms depending on the habitat conditions (i.e. light and substrate).

Further investigations need to be done with the application of different taxonomical criteria, besides morphology, in order to reach a conclusion on this subject.

LITERATURE CITED

Pulitzer-Finali G. & Pronzato R., 1977 - Report on a collection of sponges from the bay of Naples. II. Keratosa. *Pubbl. Stn. Zool. Napoli* 40(1) : 83-104.

Pulitzer-Finali G. & Pronzato R., 1981 - The Keratosa in a collection of Mediterranean sponges mainly from the Italian coasts. *Ann. Mus. Civ. st. nat. "G. Dovia"*, 83 : 127-158.

Topsent E., 1927-29 - Faune et flore de la Méditerranée. Porifera. *C.I.E.S.M.*, 11 pls.

Vacelet J., 1959 - Répartition générale des éponges et systématique des éponges cornées de la région de Marseille et de quelques stations méditerranéennes. *Recl Trav. Stn mar. Endoume*, 16(26) : 39-101.

Wilkinson C.R. & Vacelet J., 1979 - Transplantation of marine sponges to different conditions of light and current. *J. exp. mar. Biol. Ecol.*, 37 : 91-104.

SEXUAL REPRODUCTION, LARVAL MORPHOLOGY AND BEHAVIOUR IN DEMOSPONGES FROM THE SOUTHWEST OF THE NETHERLANDS.

M. WAPSTRA and R.W.M. van SOEST

Institute of Taxonomic Zoology, University of Amsterdam
P.O. Box 20125, 1000 HC Amsterdam, The Netherlands.

SYNOPSIS

The reproductive periods of sponge species from the SW of The Netherlands were determined. Morphological characteristics of the larvae of the recently described *Haliclona xena* and *Mycale micracanthoxea* are given. More detailed information is included on the previously described larvae of *Halichondria panicea, Halichondria bowerbanki, Hymeniacidon perlevis, Haliclona oculata, Haliclona rosea, Acervochalina loosanoffi* and *Dysidea fragilis*. Morphological changes in sponge larvae during their free life are described. Furthermore an overview of known larval descriptions is included along with a discussion of the taxonomic value of larval features.

INTRODUCTION

The mode of reproduction and larval morphology are important characteristics in the classification of sponges. At class- and subclass levels oviparity or viviparity (Lévi, 1956, 1973 ; Simpson, 1984) and the structure of the larvae (Lévi, 1956, 1973 ; Borojevic, 1970 ; Bergquist, 1978 ; Simpson, 1984) are characteristic. Many studies have been conducted on larvae of species of the Demospongiae. Results from these studies indicate that larval morphology may provide useful information in the classification of lower level categories (Topsent, 1911 ; Bergquist *et al.*, 1979). However it has been observed that the morphology of larvae may change during their free swimming life. One of the most obvious morphological characteristics, notably the ciliation pattern, is subject to change. Posterior and/or ante-

NATO ASI Series, Vol. G13
Taxonomy of Porifera
Edited by J. Vacelet and N. Boury-Esnault

rior cilia may be shed (Barrois, 1876 ; Marshall, 1882 ; Meewis, 1939 ; Lévi, 1956 ; Bergquist & Sinclair, 1968 ; Bergquist *et al.*, 1979 ; Uriz, 1982). The shape may change upon aging (Barrois, 1876 ; Topsent, 1888 ; Burton, 1931 ; Lévi, 1956 ; Uriz, 1982), but is also constantly changing during swimming (Topsent, 1911 ; Uriz, 1982). The colour may also vary (Lévi, 1956). Occasionally such larval changes during free life have not been noted, thus resulting in the larvae of the same species being described differently by different authors. One of the aims of this study is to get a better perspective of the reliability of features for taxonomic purposes. Therefore the morphology in different species of different ages has been observed. Groups of closely related species have been studied to determine infrageneric larval variation. Larvae from different orders were also compared.

METHODS

The study area

Near Yerseke in the Oosterschelde (narrow bay SW of the Netherlands), a study area was chosen in the upper sublittoral area of an oysterbank. In order to be able to sample from the same individuals each time, specimens of the following species were marked : *Halichondria panicea* (Pallas, 1766), *Halichondria bowerbanki* Burton, 1930, *Hymeniacidon perlevis* (Montagu, 1818), *Haliclona xena* De Weerdt, 1986, and *Cliona celata* Grant, 1835. *Hymeniacidon perlevis* grew on mud while the other species were attached to oysters. The sponges were never exposed to air. Sampling was carried out by wading during low tide.

Haliclona oculata (Pallas, 1766) was sampled elsewhere in the Oosterschelde. These specimens were taken from the floating decks outside the locks at Wemeldinge.

Prosuberites epiphytum (Lamarck, 1814) was sampled from the pillars of a bridge at Wilhelminadorp. This was located in a brackish channel closed off by locks from the Oosterschelde.

Mycale micracanthoxea Buizer & van Soest, 1977, was sampled at several localities in the Oosterschelde and also in Lake Grevelingen which is slightly brackish.

Larvae of *Dysidea fragilis* (Montagu, 1818) and *Haliclona rosea* (Bowerbank, 1866) were obtained from the NW of France. The former was sampled between Morgat and Camaret (Brittany) and the latter at Ambleteuse (Pas de Calais).

Due to its rare and ephemeral occurrence, *Acervochalina loosanoffi* (Hartman, 1958) could not be systematically sampled. However occasional observations of larvae were made from sponges sampled at Wilhelminadrop.

Haplosclerid generic assignments were based on recent revisions by De Weerdt (1985, 1986). *Haliclona rosea* and *Haliclona xena* would have been assigned to the genus *Reniera* (now considered a synonym of *Haliclona*) based on previous reviews of the order (e.g. van Soest, 1980)

Determination of the presence of reproductive elements

From April 1984 to April 1985, sponges were sampled at regular intervals : twice a month from April to December and once a month thereafter until the following April. In the field, sponge specimens (others than the ones marked) were examined for the presence of embryos by tearing fragments away. Six marked specimens per species were sampled each time for histological purposes. Samples were taken from the same individuals over a longer period from the sponges in the study area. If a specimen had disappeared, it was replaced by another of the same species. Small pieces of the basal tissues were collected and immediately fixed in Bouin's solution. After a fixation period of 24-48 hours, the samples were dehydrated in an alcohol series followed by Oleum Caryophylli and embedded in paraffin. They were then serially sectioned at a thickness of 10 µm. After deparaffining, the mounted sections were stained with Mallory-Heidenhain (Gurr, 1963), and examined for reproductive elements. For *Mycale micracanthoxea*, part of the material was collected and processed in October and November 1983 while other samples came from the collection of the Zoological Museum of the University of Amsterdam. This material collected in 1976 and 1977 was preserved in alcohol ; Por. 4269-4272, 4274, 4277, 4279, 4280, 4281, 4306 and 4326.

Larval morphology

Embryo containing sponges were collected and transported in local sea water to the laboratory aquaria. Within 12 hours they were moved to smaller trays to facilitate the sampling of larvae. If the larvae were not spontaneously released, the sponge was gently torn into smaller pieces to enhance liberation. To keep the larvae alive, they were maintained in small aquaria with the water changed regularly. Larvae were sampled at time intervals and were observed with a stereo microscope, a light microscope and a scanning electron microscope. For SEM-preparations larvae were fixed according to the following different techniques :

1/ Fixation in Bouin's solution for 24 hours (a fixative generally used for sponges).

2/ Fixation in a mixture of one part saturated solution of $HgCl_2$ and six parts 1% OsO_4 for 15 minutes (devised by Parducz, 1967 for ciliates, and used by Elvin, 1979 and Bergquist and Green, 1977 for sponge larvae).

3/ Fixation in 4% glutaraldehyde in 0.2 M cacodylate buffered solution for 1 hour, followed by 1 hour of postfixation in 1% OsO_4 (used by Bergquist and Green, 1977 for electron microscopical preparations of sponge larvae).

The different fixatives did not appear to yield markedly different results, so only the simplest method was applied, i.e. fixation in Bouin's solution.

The fixation was followed by an alcohol series dehydration and carbogen critical point drying. Then the larvae were mounted on SEM tables and coated with platinum. The preparations were scanned in a Cambridge scanning electron microscope.

RESULTS

Presence of reproductive elements

Table 1 illustrates the presence of reproductive elements from April 1984 till April 1985. Most sponges carried oocytes for a prolonged period of the year. It appeared that *Halichondria panicea* and *Haliclona oculata* retained oocytes year round. Spermatic cysts were more restricted in their period of development. Larvae were liberated at variable times, starting about one month after large embryos were observed in the sponges tissues.

Table 1 - The occurence of reproductive elements in sponges from the south-west of Holland. /// = no sponges found in our study area during these periods.

month:	A	M	J	J	A	S	O	N	D	J	F	M	A
Halichondria panicea											////	////	
oocytes	oooo	oooo	oooo	oooo	oooo	oooo	oooo	oo		oooo			oo
spermatic cysts			ssss	ssss	ssss	ssss	ssss	ss					ss
embryos		ee	eeee	eeee	eeee	eeee							
larvae			ll	llll	ll	llll							
H. bowerbanki											////	////	
oocytes	oooo		oooo	oooo	oo	oo	oo	oooo					
spermatic cysts			ssss	ssss	ssss	ss	ssss	ss					
embryos			eeee	eeee	eeee	eeee	eeee	ee					
larvae				ll	llll	llll	ll	ll					
Hymeniacidon perlevis												////	////
oocytes	oooo		oooo	oo	oo	oo	oooo	oo	oooo				
spermatic cysts			ss	ssss	ssss	ssss							
embryos			ee	eeee	eeee	ee							
larvae					llll	ll							
Haliclona oculata	//											////	
oocytes	oo	oooo	oooo	oooo	oooo	oooo	oooo	oooo		oooo			
spermatic cysts						ss							
embryos		ee	ee	eeee	ee	ee	eeee	eeee		eeee			
larvae			ll	ll			ll	ll					
Haliclona xena											////	////	
oocytes	oooo	oooo	oooo	oooo	oo								
spermatic cysts		ss	ss	ssss	ss								
embryos	ee	ee	eeee	eeee	eeee	eeee							
larvae			llll	llll		ll							
Mycale micracanthoxea	////	////	//	////							////	////	////
oocytes													
spermatic cysts													
embryos			ee		eeee	eeee	eeee	eeee					
larvae					llll		llll						
Cliona celata	////	////				//		//	////		////	////	////
oocytes			oooo		oo	oo	oooo	oo					
spermatic cysts						ss	ss						
embryos													
larvae													
Prosuberites epiphytum	////	////		//			//	//	////	////	////	////	////
oocytes					oo	oo							
spermatic cysts			ss		ss								
embryos													
larvae													
month:	A	M	J	J	A	S	O	N	D	J	F	M	A

Sexual differentiation

Generally, when spermatic cysts were observed in *Halichondria panicea*, oocytes and/or embryos were also present. One sample was found to contain only spermatic cysts. Often only oocytes and/or embryos were present. However, the five individuals that were examined, from April to November, all turned out to be hermaphrodite.

For *Halichondria bowerbanki* the same results were found. Five individuals were observed over an extended period and they all exhibited hermaphroditism, although during a certain period only spermatic cysts or oocytes and/or embryos were observed. Thus the possibility of protandrous or protogynous hermaphroditism existed.

Of six specimens of *Hymeniacidon perlevis* sampled over an extended period, two were hermaphrodites, only male elements were found in three, no reproductive elements were observed in the last. Six specimens of *Haliclona xena* were sampled from April to at least the beginning of June, some until August. Five were hermaphroditic. In one specimen only female reproductive elements were found. By September all the marked sponges had disappeared. In October many new specimens of *Haliclona xena* were observed (supposedly post larval sponges). In the specimens sampled after October, no reproductive elements were again observed.

Sampling of *Haliclona oculata* revealed some specimens with only spermatic cysts, while others had spermatic cysts and oocytes occurring simultaneously. Since the same individuals were not sampled each time, it can only be assumed that hermaphroditism is not uncommon in *Haliclona oculata*.

Cliona celata was not found from mid-June to the end of November. Four specimens were followed for a longer period and all four were hermaphroditic.

There were only occasional records for *Prosuberites epiphytum*. In some individuals oocytes and spermatic cysts were found to occur simultaneously.

Summarizing, it was observed that all the sponge species under investigation exhibited hermaphroditism. In the literature, several possibilities for the sex of sponge species are mentioned. They can be hermaphroditic with female- and male gametes found at the same time (simultaneous hermaphrodites), or female and male gamete production may be separated by time (protandrous, protogynous hermaphroditism). Others are said to be gonochoristic or a population can consist of a mixture of female, male, and hermaphroditic individuals. Often these conclusions are based on sampling of the general population, so that a sponge species may appear to be gonochoristic while in reality it is hermaphroditic with female and male gamete production separated by time. One method to obtain a more reliable picture of the sex of a sponge species is to check individuals for reproductive elements with an extended sampling regime (as in this study). However, following a sponge for a whole reproductive season was still not conclusive since it was found in *Spongilla lacustris* that specimens could change their sex after winter dormancy (Gilbert & Simpson, 1976).

According to Lévi (1951) the sexual stages of oviparous sponges are either well separated by time or they are gonochoristic. However, simultaneous hermaphroditism as observed in *Prosuberites epiphytum* and *Cliona celata* is not uncommon.

Reproductive periods

In table 2, the reproductive periods of the sponges in the southwest of Holland are listed along with records for other areas. Often the presence of embryos or the time that larvae were found to settle was referred to in discussions concerning reproductive seasonality in sponges. The actual reproductive time is assumed to be the time of larval release. Larval settlement corresponds highly with larval release since their free swimming life seldom exceeds three days (Bergquist & Sinclair, 1968 ; Fell, 1974). However, when larger embryos are present in sponges, it is not necessarily a cue for larval release. Larger embryos were found here in sponges between two weeks to one month before larvae were released. For *Halichondria panicea, Halichondria bowerbanki, Hymeniacidon perlevis, Haliclona oculata* and *Haliclona xena*, there existed periods when sponges contained larger embryos but larvae were not released. One possible explanation is that physical conditions (temperature being the most likely) changed in such a manner that larvae were retained during that time. But since these periods of non-release in different species do not overlap, this possibility seems less likely. Another possibility is that different generations of reproduction were studied, i.e. overwintered sponges and postlarval generations. The presence of larger embryos in the different generations overlapped, but larval release did not. The occurrence of extended periods of larval release in itself is an indication of postlarval reproduction, as Fell *et al.* (1984) found for New England populations of *Halichondria* sp. and *Haliclona* sp. The reproductive period of the same species from different localities appears to vary considerably. There is not only the influence of the difference in geographical locations, but also the varying weather conditions for different years. This is also illustrated by the reproductive periods observed in *Halichondria panicea* and *Halichondria bowerbanki*. These were determined for the same area based on fieldwork in 1978, 1979 and 1980 (Vethaak *et al.*, 1982). The periods that these sponges contained embryos were more restricted during 1978-1980 and their reproductive seasons were almost completely separated, whereas in 1984 there appeared to be an overlap of 3.5 months.

Table 2 - Reproductive periods of widespread sponges at different localities, based on literature data and own observations. 1 : European and North Atlantic species. 2 : *H. sanguinea* is described, but we assume it to be *H. perlevis*. embr = embryos, repr = reproduction, settl = settlement.

Locality + presence of	J	F	M	A	M	J	J	A	S	O	N	D	reference
Halichondria panicea													
Holland, embryos					****	****	****	****					Vethaak et al. 1982
Holland, embryos					****	****	****	****	****				Present study
Holland, larvae						****	****	****	****				Present study
Eur.& N.Atl.sp.[1], embr.				****	****	****							Burton 1932
France, larv. release					****	****							Topsent 1911
Roscoff, repr. per.						****							Levi 1956
New Zeal., settlement	****	****	****	****					****	****	****	****	Bergquist et al. '68
Halichondria bowerbanki													
Holland, embryos								****	****	****			Vethaak et al. 1982
Holland, embryos						****	****	****	****	****	****		Present study
Holland, larvae							****	****	****	****	****		Present study
Eur.& N.Atl.sp.[1],repr.				****	****	****	****	****					Burton 1932
idem, embryos							****	****					Burton 1932
France, larv. release								****	****				Topsent 1901, 1911
Roscoff, repr. per.							****	****					Levi 1956
New Engl., larv.release								****	****	****	****		Hartman 1958
Connecticut, repr.per.					****	****	****	****	****	****			Fell & Jacob 1979
Hymeniacidon perlevis													
Holland, embryos						****	****	****	****				Present study
Holland, larvae								****	****				Present study
Hampshire, embr.							****	****	****	****			Stone 1970
France, release								****	****				Topsent 1911
Roscoff, larvae								****					Levi 1956
New Zeal., repr.	****	****	****	****	****	****	****	****	****	****	****	****	Bergquist 1973
idem, larval settl.	****	****	****	****					****	****	****	****	Bergquist 1973
idem, larval settl.	****									****	****	****	Bergquist et al. '73
Costa Brava, release[2]								****					Uriz 1982
Alicante, release[2]						****	****	****					Uriz 1982
Haliclona oculata													
Holland, embryos	****				****	****	****	****	****	****	****		Present study
Holland, larvae						****	****			****	****		Present study
Eur.& N.Atl.sp.[1], repr.			****	****	****	****							Burton 1932
idem, embryos						****							Burton 1932
Calvados, larvae						****	****						Topsent 1887
Le Portel, embr. or larv.								****					Topsent 1887
Portsmouth, embryos								****	****	****	****	****	Rowe, unpubl. data
in 3 years	****	****	****	****	****	****		****	****	****	****	****	
	****	****	****	****									
New York, reprod.per.			****	****	****	****							Fell 1974
Cliona celata													
Holland, oocytes						****		****	****	****	****		Present study
Eur.& N.Atl.sp.[1],embr.				****	****	****	****	****	****				Burton 1932
Scotl. oocytes			****	****									Grant 1826
The Channel, oocyt									****	****			Topsent 1887
New England								****	****	****			Hartman 1958
New England, oocytes					****	****	****						Fell et al. 1984
New England, settl.						****	****						Fell et al. 1984

Larvae of *Acervochalina loosanoffi* were observed being released in the first half of October, 1983.

For *Mycale micracanthoxea* the reproductive period was based on samples taken from different years. Because the period of larval release may change over the years, the period of liberation may appear longer than in any given year.

On the basis of the presence of larger oocytes in the two tetractinomorph oviparous sponges, it seems reasonable to expect the larvae of *Prosuberites epiphytum* to occur in the second half of September, and larvae of *Cliona celata* in the beginning of November and possibly already as early as July.

Larval morphology

The anterior pole was assumed to be the forwardly directed pole, since it is difficult to establish it in the sense of embryogenesis and cell lineage (see also Simpson, 1984).

The orange-yellow larvae of *Halichondria panicea* (Figs 1-3) were oval as well as having a more oblong form. Size ranges measured from 180 µm by 340 µm up to 150 µm by 600 µm. When just released, the larvae were generally oval. As they got older, and changed from a swimming type movement to a gliding type, the larvae changed into a more oblong form. This may have happened (from 2 hours up to 3 days) after they had been swimming. After three days most larvae were settled. Sometimes the emerging larvae were observed as already oblong, as was seen in a sponge sampled on the 8th of August. These larvae did not swim but immediately started gliding along the bottom and sides of the tray. When oblong larvae were disturbed by turbulence they started swimming again and regained a more oval form. Soon after the turbulence stopped they returned to an oblong form. To a lesser degree form changes were also observed during swimming. Whether they were oval or oblong the larvae of *Halichondria panicea* were entirely ciliated with a posterior tuft of longer cilia. The density of cilia was close-set at the anterior pole and became thinner towards the posterior pole. This was more conspicuous in the oblong forms where a barely discernable ring was present around the tuft. The density of the tuft cilia was relatively meager. Samples of very young larvae (fixed immediately upon liberation) were found to be uniformly ciliated and no tuft could be distinguished (Fig. 1). The cilia were very close-set and the larvae were spherical to oval.

In table 3 *Halichondria panicea* larvae can be compared with those described by Topsent (1911) and Hartman (1958). Although the same species was studied there are some distinct differences between descriptions. Most obvious is that Topsent and Hartman did not observe a tuft on their larvae. The only larvae observed without the tuft were young ones and it was possible that they were premature.

The yellow *Halichondria bowerbanki* larvae illustrated the same caracteristics as *Halichondria panicea* larvae with respect to their form in relation to their age and type of movement. Thus, there were younger oval larvae that swam and older oblong larvae that glided. Larvae ranged from 140 x 240 um up to 140 x 320μm in size. They also have a tuft which consisted of rela-

---

Table 3 - Larval characteristics.
LARVA ; some sponges were described under a different name : 1 as *Halisarca*, 2 as *Euspongia*, 3 as *Verongia*, 4 as *Halichondria coalita*, 5 as *Esperia syrinx*, 6 as *Esperella sordida*, 7 as *Adocia*, 8 as *Reniera*, 9 as *Chalinula*, 10 as *Isodictya*, 11 as *Gellius*, 12 as *Haliclona*, 13 as *Chalina*.
CILIATION PATTERN ; uf = uniformly ciliated, bpp = bare posterior pole, bap = bare anterior pole, bpr = bare posterior ring, bar = bare anterior ring, r = ring of longer cilia around the posterior pole, ar = ring of longer cilia around the anterior pole, t = tuft of longer cilia at the posterior pole, cl→p = cilia become longer towards the posterior, cl→a = cilia become longer towards the anterior, cl→s = cilia become longer towards the sides, cd→a = cilia become denser towards anterior, short cilia, sc = scarcely ciliated, d = denser, () = sometimes, / = later during their free life.
SHAPE ; s = spherical, o = oval or ovoid, p = pear shape, co = conical, el = elongated, fl = flattened, e.pp = protruded posterior pole, bl = blastula, amb = amphiblastula, par = parenchymella.
COLOUR ; pi = pigmented, or = orange, yel = yellow, br = brown, vi = violet, whi = white, cr = creme, bl = black, tr = translucent, li = light, ap = anterior pole, pp = posterior pole, pr = posterior ring, a = anterior, p = posterior, r = ring, p-r = pole or ring.
SPICULATION ; + = spicula present, - = spicula not present.
LIGHT REACTION ; - = negative photo sensitive, + = positive photo sensitive, 0 = not photo sensitive.
TYPE OF MOVEMENT ; -ro = rotating in a clockwise direction, +ro = rotating in an anticlock direction, cs.sw = corckscrew swimming, bou = bouncing swimming, str = straight, di = directional swimming, cr = crawling, s = surface, b = bottom, / = later during their free life.
REFERENCE ; 14 = descriptions taken from Brien, 1973. 15 = d.o. from Bergquist *et al.*, 1970.

Larva	Cil. pat.	shape	colour	size	sp	lr	type of movement	reference
Subclass HOMOSCLEROMORPHA								
Oscarella sp.	uf, cl→a	bl, s	whi	100 - 200		+	+ro	Bergquist et al. 1979
Oscarella sp.	uf	hollow						Delage 1892
Oscarella lobularis[1]	uf, cl→a							Topsent 1888
O. lobularis	uf, cl→s	amb						Meewis 1939
Plakina sp.	uf, cl→a	bl, s	whi	100 - 200		+	-ro	Bergquist et al. 1979
Plakina monolopha	uf, cl→a	o, amb	pink, darker→pp			+		Schulze 1880[14]
Subclass TETRACTINOMORPHA								
Tetilla serica	no free swimming larval stage							Borojevic 1968[14]
Tetilla cranium	uf	bl, o						Burton 1931
T. cranium	no free swimming larval stage							Borojevic 1970
Tethya aurantium		par,s,el	yel				ro	Lévi 1956
Tethya sp.		bl/ par						Borojevic 1970
Polymastia robusta		bl, fl					cr	Borojevic 1968[14]
Polymastia granulosa						0	cr	Bergquist et al. 1970
Cliona celata						0	ro, bou→str at s /cr	Warburton 1966
Cliona viridis	uf	amb						Tuzet 1930, Warburton 1962[14]
Axinella crista-galli	bpp	o-el			+			Maas 1894[14]
Subclass CERACTINOMORPHA								
Dictyoceratida								
Spongia sp.	t, bar, bpr	el	whi, 2 dark r	500 - 1000		+	+ro, sw	Bergquist et al. 1979
Spongia reticulata		o-el						Bergquist et al. 1970
Spongia sp.[2]	t		whi, br.pr					Lévi 1956
Spongia officinalis[2]	t, bar, bpr		bl.pr	500 - 1000			+ro, sw	Bergquist et al. 1979
Hippospongia sp.	t		whi, br.pr					Maas 1894[14]
Hippospongia communis	t	o, e.pp	whi-yel, bl.pr					Lévi 1956
Phyllospongia foliascens	t	s-o	whi, br.pr					Lévi 1956
Ircinia sp.	t		whi, br.pr	420-480x600-650				Maas 1894[14]
Ircinia sp.	t, bar, bpr	el	whi, 2 dark r					Lévi 1956
Ircinia variabilis	t							Maas 1894[14]
Dysidea sp.	t		whi, br.pr					Lévi 1956
Dendroceratida								
Aplysilla sp.	r/bap		pi	500 - 600		+	ro, +di sw	Bergquist et al. 1979
Aplysilla sulfurea	t, sc-bpp	o	yel	500 - 880		-		Delage 1892
Aplysilla rosea[3]	r, bpp/bap	o	pi.pp					Barrois 1876
Halisarca sp.	uf/ bpp		whi	100 - 200		0	+ro	Bergquist et al. 1979
Halisarca dujardini	uf, cl→a							Topsent 1888
H. dujardini	uf, cl at pp/bpp	o/ el/ fl		90 - 95			ro, sw to bottom	Lévi 1956
Halisarca metschnikovi	uf/ bpp	o/ fl	whi	120 - 130			+ro, sw to surface	Lévi 1956
Halisarca nahatensis a	uf	bellsh.	whi	120-160x150-200				Chen 1976
b	uf	fl	whi	120-140x140-180				Chen 1976
Halichondrida								
Halichondria sp.						0	sw /cr	Bergquist et al. 1970
Halichondria sp. 1	uf	el	pi	200 - 1500		0	-ro, cr	Bergquist et al. 1979
Halichondria sp. 2	uf, (bpp)	o	pi	200 - 1500			+ro, sw	Bergquist et al. 1979
Larva	Cil. pat.	shape	colour	size	sp	lr	type of movement	reference

Table 3, part one.

Larva	Cil. pat.	shape	colour	size	sp	lr	type of movement	reference
Halichondria sp. 3	uf, (bpp)	o	pi	200 - 1500			+ro, sw	Bergquist et al. 1979
Halichondria sp.	t	o	yel					Lévi 1956
Halichondria bowerbanki[4]	t, cl+a	o	dark yel	175-250x330-440	+			Topsent 1911
H. bowerbanki[4]	sc.t	el	yel					Meewis 1941
H. bowerbanki[4]	t	o	yel					Lévi 1956
H. bowerbanki	cl+a, sh.c.pp	o		102-113x183-289	+			Hartman 1958
H. bowerbanki	t, cl+a	o-p-el	yel	150 x 275	+			Fell & Jacob 1979
H. bowerbanki	t, cl+a, (sc.pr)	o/el	ockre	180x340-150x600	+	0	+ro, cs.sw, bou /cr	Present study
Halichondria panicea	cl a	o-p,el	yel	165x484,187x517	(+)	+		Topsent 1911
H. panicea				99 x 357				Hartman 1958
H. panicea	t, (sc.pr), cd+a	o/el	yel	140x240-140x320	+	0	+ro, cs.sw, bou /cr	Present study
Halichondria moorei	uf						cs.sw /cr	Warburton 1966
H. moorei	uf	o-el	yel-or	518-740x1110-1926	-	0	cs.sw /cr	Bergquist et al. 1968, 1970
Hymeniacidon caruncula	bpp	o-p-el	dark red, or.pp	170x260,300x470	+			Topsent 1888, 1911
H. caruncula	uf		yel-or		-	+		Bergquist & Sinclair 1968
Hymeniacidon hauraki	uf	o	pi	200 - 1500		0	+ro, sw	Bergquist et al. 1979
Hymeniacidon perlevis	uf/ bpp	o	pi	200 - 1500		0	+ro, sw at surf /cr	Bergquist et al. 1973, 1979
H. perlevis	uf (sc.pp)	o	or	160 x 290	+	0	+ro, bou /cr	Present study
Hymeniacidon sanguinea a	uf/ bpp	co	or	350-450x550-600		0	+ro, cs.sw	Uriz 1982
H. sanguinea b	bpp	o-el	or	350-600x820-1300		0	+ro, cs.sw	Uriz 1982
H. sanguinea (2 sizes)	bpp	o-co	red	250-300,350-400				Lévi 1956
Ulosa	uf	el	pi			0	-ro, cr	Bergquist et al. 1979
Poecilosclerida								
Mycale sp.	bpp	s-o	pi, whi.pp	300 - 600	+	0	-ro	Bergquist et al. 1979
Mycale syrinx[5]						-/+	sw /cr	Wilson 1935[15]
M. syrinx	bpp	o, e.pp			+			Brien 1973
Mycale contarenii							cs.sw	Kinne 1970
Mycale macilenta	bpp	p	or-br, whi.pp	485-527x357-429	+	-	cs.sw, bou /cr	Bergquist & Sinclair 1968
M. macilenta	bpp	s-o		300 - 600	+	-	cr, cs, -ro	Bergquist et al. 1979
Mycale micracanthoxea	bpp	o	yel,wh.pp,(li ap)		+	-	cs.sw, ro on bottom	Present study
Mycale subclavata[6]	bpp	s-o/el	red, yel.pp	600 x 800	+	-		Delage 1892
Myxilla reses	bpp	e.pp	white		+		ro on bottom	Topsent 1911
Myxilla rosea	bpp	o, e.pp			+			Maas 1894[14]
Lissodendoryx sp.	bpp	s-o	pi, whi.pp	300 - 600	+	0	-ro	Bergquist et al. 1979
Lissodendoryx sp.						-	sw at s /cr /ro on b	Ali 1956[15]
Tedania sp.	bpp	s-o	pi, whi.pp	300 - 600		0	-ro	Bergquist et al. 1979
Tedania gurjanovae							cs.sw	Bakus 1964[15]
Ophlitaspongia	bpp	s-o	pi, whi.pp	300 - 600		0	-ro /cr	Bergquist et al. 1970 , 1979
Ophlitaspongia seriata	bpp	p-o	or, paler pp			+/0-	ro, cs.sw to s /cr	Fry 1971, Bergquist et al. 1970
O. seriata	bpp			214-384x272-429		0	ro, cs.sw /cr	Bergquist & Sinclair 1968
Microciona sp.	bpp	s-o	pi, whi.pp	300 - 600		0	-ro	Bergquist et al. 1979
Microciona prolifera			scarlet			0	ro, bou or str at s/cr	Warburton 1966
M. prolifera	(bare area)		red-or	270 - 300		0		Simpson 1968
Microciona coccinea	bpp	o	red, whi.pp	222-300x296-444	+	+,0	ro, cs.sw /cr	Bergquist et al. 1968, 1970
Microciona atrasanguinea	bpp							Lévi 1956
Desmacidon fructicosa	bpp, r/ bap	o	br-yel pp		+			Barrois 1876
D. fruticosa	bpp	o/el	yel, pp more so		+		ro on bottom	Lévi 1956
Larva	Cil. pat.	shape	colour	size	sp	lr	type of movement	reference

Table 3, part two.

Larva	Cil. pat.	shape	colour	size	sp	lr	type of movement	reference
Pronax plumosa	bpp	o/e.pp	or ap, red pp		+			Lévi 1956
Anchinoe sp.	bpp	s-o	pi, whi.pp	300 - 600	+	0	-ro	Bergquist et al. 1979
Anchinoe sp.			pale yel	146 x 179			cr	Ayling 1980
Phorbas sp.	bpp	s-o	pi, whi.pp	300 - 600		0	-ro	Bergquist et al. 1979
Coelosphaera sp.	bpp	s-o	pi, whi.pp	300 - 600	+	0	-ro	Bergquist et al. 1979
Stylopus sp.	bpp	o		210 x 240			cr	Ayling 1980
Stylopus sp.	bpp	s	or	392 x 407			cr	Ayling 1980
Stylopus coriaceus	uf	o	yel pp, tr ap					Lévi 1956
Paracornulum sp.	bpp	s-o	pi, whi.pp	300 - 600	+	0	-ro	Bergquist et al. 1979
Holoplocamia sp.	bpp	s-o	pi, whi.pp	300 - 600		0	-ro	Bergquist et al. 1979
Chondropsis sp.		fl.disk		1600			cr	Ayling 1980
<u>Haplosclerida</u>								
Haliclona sp.	r	o	pi.pp	100 - 1000	+	+	active sw.	Bergquist & Green 1977
Haliclona sp.						+/0	ro, di.sw /cr	Bergquist et al. '70, Warburton '66
Haliclona sp	r, bpp	o	whi, br.pp	143-271x314-500	+	+	ro, cs.sw /cr	Bergquist & Sinclair 1968
Haliclona sp. 1	r, bpp	o-el	whi, pi.pp	100 - 1000	(+)	+	+ro at surf	Bergquist et al. 1979
Haliclona sp. 2	r, bpp	o-el	whi, pi.pp	100 - 1000	(+)	+	+ro at surf	Bergquist et al. 1979
Haliclona sp. 3	r, bpp	o-el	whi. pi.pp	100 - 1000	(+)	+	+ro at surf	Bergquist et al. 1979
Haliclona sp.[7]	r, bpp, bap	o-el	whi, pi.pp	100 - 1000	(+)	+	+ro, sw	Bergquist et al. 1979
Haliclona sp. 1[8]	uf/ (bpp)	o	whi	100 - 1000	(+)	0	+ro at surf	Bergquist et al. 1979
Haliclona sp. 2[8]	uf/ (bpp)	o	whi	100 - 1000	(+)	0	+ro at surf	Bergquist et al. 1979
Haliclona sp. 3[8]	uf/ (bpp)	o	whi	100 - 1000	(+)	0	+ro at surf	Bergquist et al. 1979
Haliclona oculata[9]	r, bpp,bap, cl+p	o	white	260x420-220x470	+	0		Topsent 1911
H. oculata	r, sc.pp, sc.ap	s-o	white	165 x 250	+	0	+ro, cs.sw, bou	Present study
Haliclona indistincta	uf	o	pink-vi, whi.ap	250 x 500	–		sw irregular	Lévi 1956
Haliclona simulans[8]	r, bpp		whi, red-br.pp		+			Topsent 1888
H. simulans[8]	r, bpp	o	whi					Delage 1892
H. simulans[7]	bpp	o/el	whi, red-br.pp		+			Lévi 1956
Haliclona cinerea[7]	r,bpp, sc.+sh.c.ap	o-el	vi, darker pp		(+)			Meewis 1941
Haliclona rosea[10]	r, bpp/ (bap)	o	pink, red pp			+		Barrois 1876
H. rosea[8]	r, bpp		or.pp					Topsent 1888
H. rosea	r, bpp		pi, darker pr					Present study
Haliclona xena	uf/r, sc.pp, sc.ap	o,(e.pp)	whi, br pp-r	140 x 190	+	0	+ro at s, sw, ro on b	Present study
Haliclona angulata[11]	r, bpp	co-el	cr a, ockre p		+			Lévi 1956
Acervochalina sp.[9]	bpp, bap	s-o	whi	100 - 1000	(+)	0	+ro at surf	Bergquist et al. 1979
Acervochalina loosanoffi	r	o	li.br, dark pr					Present study
A. loosanoffi[12]	bpp	s	beige, br.pp or r	172 x 188	+			Fell 1976
Acervochalina limbata[13]	bpp/ bap	s-o	pink, br.pp	150 x 500				Meewis 1939, 1941
Callyspongia sp.	r, bpp	o-el	whi, pi.pp	100 - 1000	(+)	+	+ro	Bergquist et al. 1979
Callyspongia sp.							sw /cr, ro on b	Sivaramakrishnan 1951[15]
Callyspongia diffusa	uf, sh.c pp	o						Lévi 1956
"Reniera filigrana"	ar/ bpp	o/el	yel, br-vi ap-r	500				Marshall 1882
<u>Freshwater sponges</u>								
Spongilla lacustris	uf	s-o				–		Delage 1892
Ephydatia fluviatilis	uf	o						Topsent 1911
E. fluviatilis			white		+			Brien 1969[15]
Corvospongilla thyssi	uf	o-el			+			Brien & Meewis 1938[15]

Larva | Cil. pat. | shape | colour | size | sp | lr | type of movement | reference

Table 3, part three.

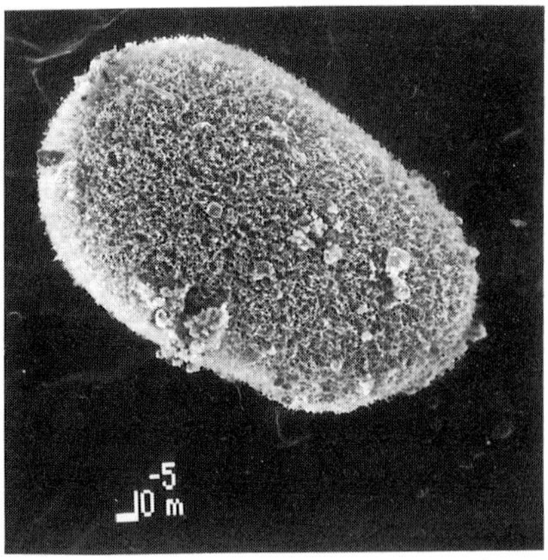

Fig. 1 - Premature *Halichondria panicea* larva.

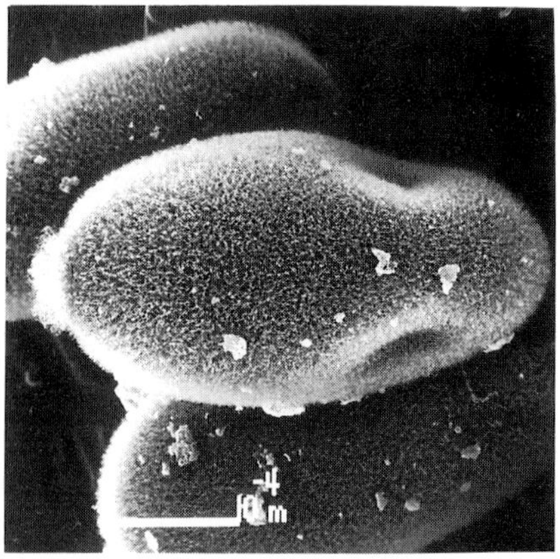

Fig. 2 - *Halichondria panicea* larva.

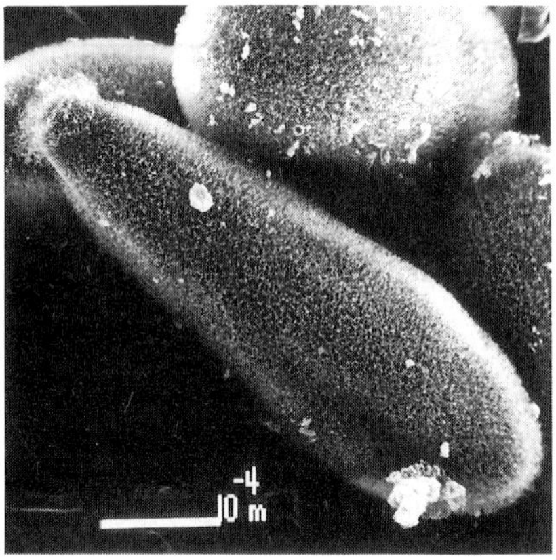

Fig. 3 - *Halichondria panicea* larva, crawling stage.

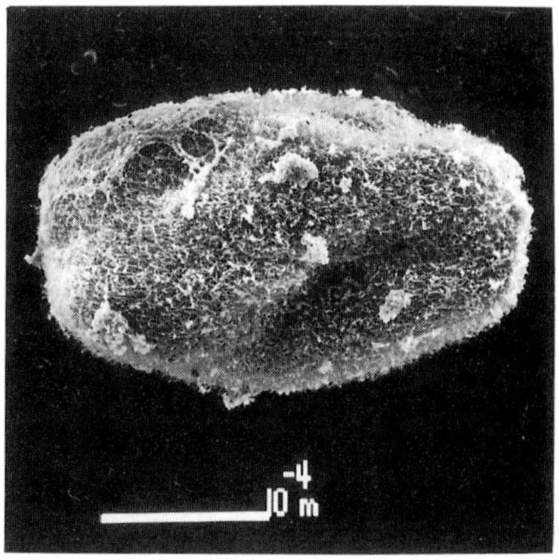

Fig. 4 - *Hymeniacidon perlevis* larva.

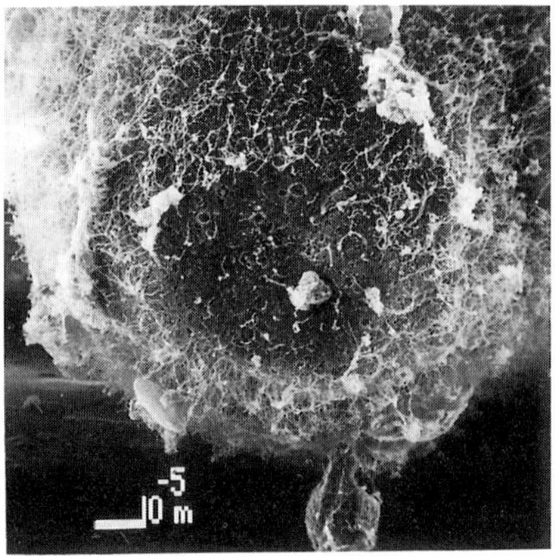

Fig. 5 - Posterior pole of *Hymeniacidon perlevis larva.*

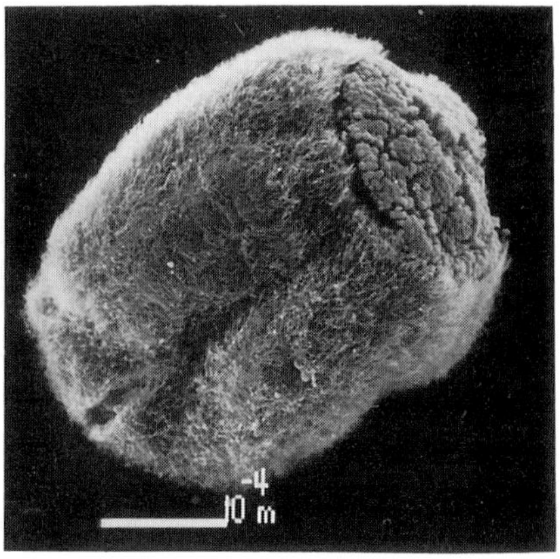

Fig. 6 - *Mycale micracanthoxea* larva.

tively few cilia. The cilia were not only caracteristically denser towards the anterior pole but also became longer towards the posterior half of the larvae. This ciliation pattern was observed in very young- as well as in older larvae. The only difference observed was that the thinly implanted cilia were more so in the oblong forms. On the whole, the cilia were not as close set as in *Halichondria panicea* larvae. Very young uniformly ciliated larvae were never found.

Table 3 illustrates the typical *Halichondria bowerbanki* larva observed compared to those described earlier either as *Halichondria bowerbanki* (Hartman, 1958) or as *Halichondria coalita* (Topsent, 1911 ; Meewis, 1941 ; Lévi, 1956). Marked differences were observed in size, but primarily in the absence of a tuft on the larvae observed by Hartman. He suggested that the larvae on opposite sides of the ocean could belong to different subspecies. However, Fell & Jacob (1979) described a *Halichondria* sp. from New England (formerly designated by them as *Halichondria bowerbanki*), that possessed a posterior tuft of longer cilia. Based upon this, they assumed that the sponge was probably a different species or subspecies compared to the *Halichondria bowerbanki* that Hartman found. They observed the larvae to be similar to the *Halichondria coalita* larvae described by Meewis (1941), but they hesitated to refer to the sponge as such, because the embryonic development appeared to be markedly different. However, *Halichondria bowerbanki* and *Halichondria coalita* are synonyms of the same species.

Hymeniacidon perlevis larvae (Fig. 4) were orange and measured 290 µm by 160 µm with little variation. All the larvae observed were oval and uniformly ciliated over the entire surface except for the posterior pole which was very sparsely ciliated, almost bare (Fig. 5). Also very young (not fully developed larvae) which came out of the sponge after it was torn apart, possessed a sparsely ciliated posterior pole, and would not start swimming immediately. No differences were found in the morphology of younger and older larvae.

Hymeniacidon perlevis larvae were already described by Bergquist *et al.* (1979). According to their results the scarcely ciliated posterior pole developed by the tendency of these larvae to shed their posterior cilia. This was also mentioned for other *Hymeniacidon* larvae (Topsent, 1911 ; Uriz, 1982). If this was the case with our *Hymeniacidon perlevis*, then the posterior cilia would have been shed while the larvae were still in the mother sponge. On the basis of other characteristics it was assumed that the *Hymeniacidon sanguinea* of Uriz (1982) was the same as the *Hymeniacidon perlevis*

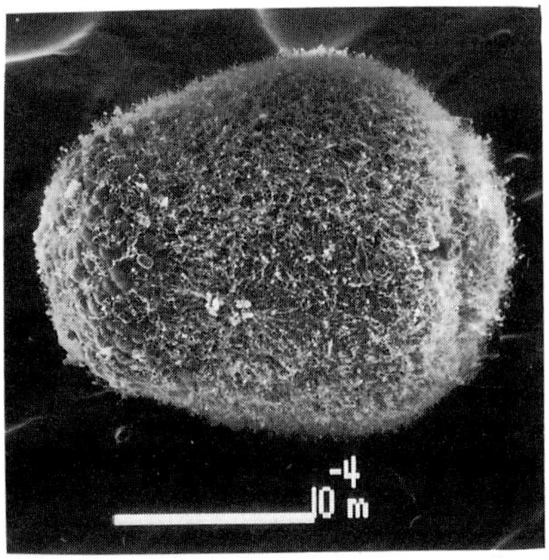

Fig. 7 - Premature *Haliclona oculata* larva.

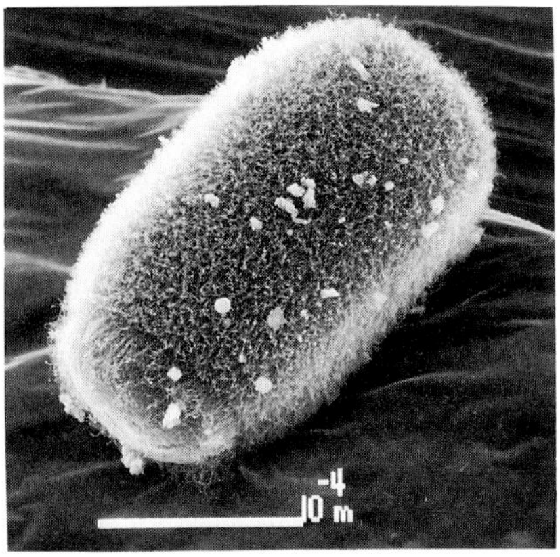

Fig. 8 - *Haliclona oculata* larva.

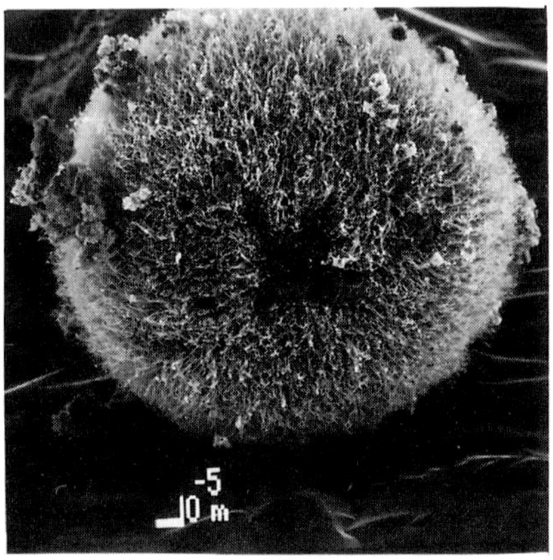

Fig. 9 - Anterior pole of *Haliclona oculata* larva.

Fig. 10 - Premature *Haliclona xena* larva.

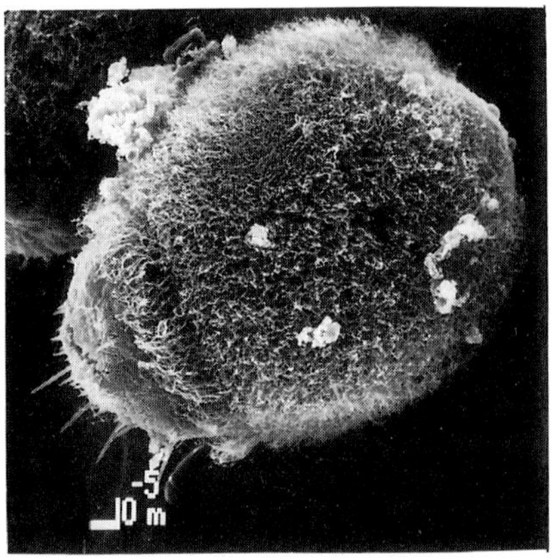

Fig. 11 - *Haliclona xena* larva.

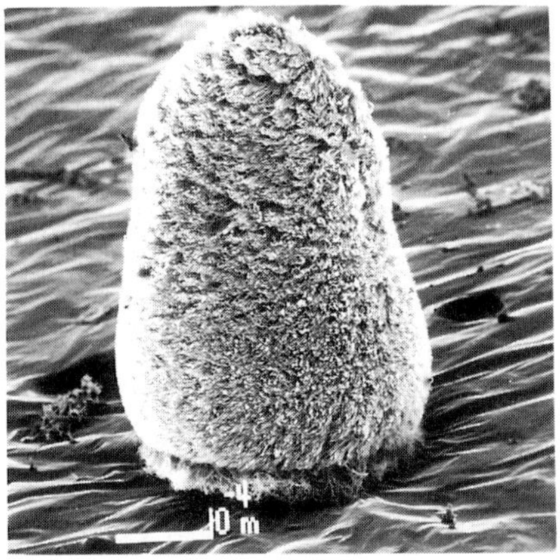

Fig. 12 - *Haliclona rosea* larva.

of this study. However, her larvae were enormous (up to ten times as big) compared to the present observations. She described two larval types : 1/ small round or conical ones that were uniformly ciliated upon liberation but lost their posterior cilia within five minutes ; 2/ larger more elongated larvae that exhibited a bare posterior pole when released. The former were the more vigorous swimmers. It is suggested that the larger larvae, once mature, stayed longer in the sponge before release, and were thus "older" at the beginning of their free life.

Mycale micracanthoxea larvae (Fig. 6) were yellow, with a white posterior pole. The anterior pole was sometimes of a brighter yellow. They were round to oval shaped, approximately about 360 µm in size, and uniformly ciliated except for a completely bare posterior pole. The posterior pole was bare both in oval and round, just released larvae as well as in larvae that were still swimming after three days. No morphological changes were observed during their free life.

The larvae of *Haliclona oculata* (Figs 7-9) measured approximately 250 um by 165 µm. Whether freshly released or a few hours old, they were all oval, totally white and possessed a ring of longer cilia around the posterior pole. The posterior pole was sparsely ciliated. The anterior pole also had a sparsely ciliated area which was larger in the younger specimens (comp. Figs 6 & 8). Larvae that were worked loose from the sponge (premature), lacked the ring of longer cilia, or if a ring of longer cilia was observed, it was markedly shorter than that of the older larvae.

The *Haliclona oculata* larvae described by Topsent (1911) as *Chalinula oculata* differed in that they appeared to possess a totally bare posterior pole. The reason for this could be that scanning electron microscopy revealed the few cilia on the otherwise bare looking posterior pole. An other observed difference was the size of the larvae. The larvae studied measured approximately 250 µm by 165 µm. Topsent's measurements of 240 µm by 260 µm and 470 µm by 220 µm showed that his larvae were considerably larger.

The larvae of *Haliclona xena* were oval, they measured 190 µm by 140 µm and they did not vary much in shape (Fig. 11). They were white with a brown cap at the posterior pole. Their ciliation pattern resembled that of *Haliclona oculata*, with a ring of longer cilia around the sparsely ciliated posterior pole, a sparsely ciliated area on the anterior pole, and the remaining surface uniformly ciliated. The posterior pole was more or less protruding and spicules could stick out. This phenomenon was not correlated with the age of the larvae however. Some very young specimens (less than 30 minutes old) were found to be uniformly ciliated except for a sparsely ciliated area at the anterior pole (Fig. 10). The larvae of *Haliclona xena* settled rather quickly. Two hours after release more than half of the larvae had already settled.

Haliclona rosea larvae were described by Topsent (1888) as having an orange posterior pole surrounded by a ring of longer cilia. The larvae observed during this study were whitish pink and had a darker ring of longer cilia around the posterior pole (Fig. 12). These cilia were shorter than those of *Haliclona oculata*. The posterior pole was completely bare in contrast to the other Haplosclerid larvae investigated. Unfortunately morphological changes during their larval free life were not observed.

Acervochalina loosanoffi larvae were not studied in detail, but examinations of the free swimming light brown larvae revealed a somewhat darker ring of long cilia. This contrasted with the findings of Fell & Jacob (1976) who did not mention the existence of a ring of longer cilia in their larvae.

From *Dysidea fragilis* sampled in August, the numerous larvae were very dificult to extract from the mother sponge. When the sponge was torn apart and water turbulence was created, only 5 larvae left the sponge. These were rather large (490 µm by 240 µm) and uniformly ciliated all over. Another *Dysidea* sp. larva described in literature (Lévi, 1956) was said to carry a tuft of longer cilia at the posterior pole. It is very possible that the *Dysidea fragilis* larvae of this study were premature since they were difficult to liberate. Therefore, a tuft might not have been fully developed at that time.

Larval behavioural characteristics

Both *Halichondria* and *Hymeniacidon* larvae exhibited what was called "corkscrew swimming" and while making this spiral path through the water they were constantly rotating in a clockwise direction. "Bouncing" swimming while rotating was also frequently observed, especially for *Halichondria panicea* larvae. In that case they swam upwards but when contact with the surface was made they immediately swam away from it and soon started over again, thus bouncing at the surface. In a later stage these larvae became less active and turned to a crawling stage. They were seen gliding or just hanging along the sides or bottom of the aquarium. They moved about still rotating, with the anterior pole pointed forwards or slantingly upwards and the posterior pole in contact with the substrate. When they were disturbed by water turbulence, they would resume swimming for a while. Of these larvae, *Halichondria panicea* and *Hymeniacidon perlevis* were more active swimmers and crawlers than *Halichondria bowerbanki*. During the crawling stage of the latter they were more frequently observed rotating in one place than moving about. Both the swimming stage and the crawling stage of these species lasted up to four hours. After three days most larvae were settled. There were no reactions to changes in illumination.

Haliclona oculata larvae were also observed swimming, using bouncing and corkscrew fashion while rotating in a clockwise direction. Most larvae swam near the surface, others swam over the bottom. One day after release most larvae had metamorphosed. These larvae did not have a light reaction.

Many of *Haliclona xena* larvae were observed settling immediately after release. They would spin around in a clockwise direction with their anterior pole in contact with the water surface. When larvae came into contact with each other they would circle and fuse while settling against the water surface. Groups consisting of as many as 39 individuals were counted. Other larvae swam through the water or over the bottom, either directionally or on a slightly corkscrew path. They were not very active swimmers. Often they would sink to the bottom with their posterior pole pointed downwards and rotated for a while before they resumed a short swimming period. These larvae were the fastest settlers observed. All were settled within a few hours after release. They exhibited no light reaction.

The larvae of *Mycale micracanthoxea* were not such active swimmers. They rotated while swimming in a corkscrew path but often they hardly moved from their position, so that only the anterior pole was making circles while the larvae still rotated around their axes. If they did not swim fast enough they sank to the bottom and rotated there with the anterior pole in contact with the substrate. Changes in illumination produced obvious changes in behaviour. If a light was turned on above the aquarium, the larvae stopped swimming immediately and sank to the bottom. When the light was turned off, they became more active and started swimming upwards again. These larvae could have a very long free life. After one day some larvae had settled. Others however, were observed still swimming or crawling after four days.

DISCUSSION

Changes during larval life

1/ Shape — A morphological change often mentioned in literature is a change of shape. It appears to be quite common for many larvae to change from an oval to an oblong form as they matured. This phenomenon was reported for *Mycale subclavata* (Delage, 1882 as *Esperella sordida*), "*Reniera filigrana*" (Marshall, 1882), *Halichondria panicea* (Topsent, 1888, 1911), *Halisarca dujardini, Haliclona simulans, Haliclona angulata*, and *Desmacidon fruticosum* (Lévi, 1956). This rather pronounced change in *Halichondria panicea* and *Halichondria bowerbanki* was observed, but data from this study show that it was only indirectly related to their age. It was related to their type of movement, which changed during free life from a swimming to a gliding stage. Generally, freshly released larvae started out by swimming for a couple of hours or even days, followed by a later stage of gliding, presumably when they were ready to settle. When the water was disturbed at that moment, they changed back to an oval form and started swimming towards the surface again. Even during swimming they were capable of changing their appearance by forming sudden folds, a feature also mentioned by Uriz (1982) for *Hymeniacidon sanguinea*.

Another change of form that took place was the protrusion of the posterior or anterior poles. This was mentioned respectively for *Haliclona rosea* (Topsent, 1888) and *Aplysilla rosea* (Barrois, 1876). A protruding posterior pole was observed in *Haliclona xena*, but there was no correlation with the age of the larvae.

In contrast it is also possible that an initially protruded pole may become depressed as was seen at the anterior pole of an *Aplysilla* sp. (Bergquist *et al.*, 1980). Lévi (1956) mentioned a flattening of the larvae of *Halisarca dujardini* and *Halisarca metschnikovi* during settlement. *Haliclona xena* exhibited this characteristic very pronouncedly. When these flattened, just settling larvae were disturbed, they resumed swimming and their form rounded off again. However this did not last long. Often within a minute after the disturbance they were settling again.

2/ Ciliation pattern — The second feature that exhibited change was the ciliation pattern. Many larvae were found to become less ciliated at their posterior pole. It was observed for *Desmacidon fruticosum, Aplysilla rosea*

(Barrois, 1876), "*Reniera filigrana*" (Marshall, 1882), *Haliclona angulata*, *Halisarca dujardini* (Lévi, 1956), *Hymeniacidon hauraki*, *Hymeniacidon perlevis* (Bergquist & Sinclair, 1968), *Aplysilla* sp., *Halisarca* sp, *Halichondria* sp., three *Haliclona* sp. (Bergquist *et al.*, 1980, as *Reniera* sp.) and *Hymeniacidon sanguinea* (Uriz, 1982). Other larvae lost their cilia at the anterior pole. This was characteristic of *Desmacidon fruticosum*, *Haliclona rosea*, *Aplysilla rosea* (Barrois, 1876) and *Acervochalina limbata* (Meewis, 1939). Most authors agreed to a loss of cilia. Lévi (1956) suggested that a lesser ciliated area developed by the spreading of the cilia as a result of larval elongation. Results from this investigation revealed that the scarcely ciliated areas at the posterior poles of *Halichondria panicea*, *Halichondria bowerbanki*, *Haliclona oculata* and *Haliclona xena* developed later. For *Halichondria panicea* and *Halichondria bowerbanki* it surely had to do with spreading of cilia due to elongation of the larvae, but possibly cilia were shed too. For *Haliclona oculata* and *Haliclona xena* it took place very early in their free swimming life and most of it had already occurred before release. Scarcely ciliated areas were however not an indication that the area was fully ciliated before. The scarcely ciliated posterior pole in *Hymeniacidon perlevis* was a constant feature. A scarcely ciliated anterior pole was characteristic of *Haliclona oculata* and *Haliclona xena*. The area was quite large in the youngest larvae and smaller in older ones which suggested that the development of the cilia started at the posterior pole and moved on to the anterior pole, but it was never completed.

Apart from areas which became less ciliated, it also occurred that larvae started out being uniformly ciliated with a ring or tuft of longer cilia developing later (as observed in some larvae of *Haliclona oculata*, *Haliclona xena* and *Halichondria panicea*).

It is suspected that these larvae left the sponge prematurely. Generally, their final ciliation pattern was exhibited by the time they were released.

3/ Colour — The colour of *Haliclona indistincta* larvae fades when they come out of the sponge, and the posterior poles of *Desmacidon fruticosum* larvae deepen in colour (Lévi, 1956). No colour changes were observed for any of the species studied. The larvae of *Mycale micracanthoxea* sometimes exhibited a cap at the anterior pole brighter yellow than the rest of the larve, but this appears to be a mere variation.

4/ Behaviour — In the orders of Halichondrida, Poecilosclerida and Haplosclerida, several larvae were reported to possess a swimming stage first, followed by a crawling stage (Wilson, 1935 ; Warburton, 1966 ; Bergquist & Sinclair, 1968 ; Bergquist *et al.*, 1970). This was consistent for all the study observations made on Halichondrida larvae.

A reverse in light sensitivity was mentioned for two species ; *Mycale syrinx* (Wilson, 1935) and *Ophlitaspongia reticulata* (Bergquist *et al.*, 1970). The larvae of the first were negatively phototrophic in the beginning and later became less distinctly phototrophic. The larvae of the latter were positively phototactic during their swimming period and later became negatively phototactic during the crawling phase.

Summarizing, many changes were observed during the free life of the sponge larvae. If larval features are to be used as a taxonomic character, one has to be aware that larvae may not be fully mature upon release, and that they exhibit different characteristics during their presettlement stages.

Larval features in sponge classification

Table 3 is an overview of the morphological characteristics of sponge larvae found in the literature.

1/ Ordinal characteristics — At the ordinal level it can be seen that Halichondrida are characterized by being uniformly ciliated sometimes with a bare posterior pole or a posterior tuft of longer cilia. The only other order in which a tuft may occur is Dictyoceratida. Furthermore in this order larvae tend to be uniformly pigmented. The differences in larval morphology of *Halichondria* and *Hymeniacidon* may be an indication of the polyphyletic nature of this order (see below).

Most Haplosclerida possess a ring of longer cilia around the posterior pole. It is the only order in which such a ring may be present, but not all species described appeared to possess one (although there is doubt about the ripeness of some larvae). The anterior pole may have a bare spot and the posterior pole is either completely bare or scarcely ciliated.

Poecilosclerida larvae are characterized by the possession of a completely bare posterior pole.

Dictyoceratida larvae have a tuft of longer cilia at the posterior pole and the New Zealand larvae have a bare anterior and posterior ring. They are all white with one or two dark rings. This order produces the biggest larvae.

The ciliation pattern and sometimes the colour are the only characteristics that can be used on at the ordinal level. Other characteristics such as shape, size, light reaction, or type of movement are highly diverse. Of the different types of movement, rotation against the surface occurs only within the Haplosclerida but not all larvae do so.

To conclude, evidence for an ordinal type of morphology is certainly present.

2/ Family differences — Of the Halichondrida larvae investigated, the two *Halichondria* and one belonging to the Hymeniacidonidae were clearly distinguishable. The *Halichondria* had a tuft whereas *Hymeniacidon* was without a tuft and had a scarcely ciliated area at the posterior pole. From the literature it is seen that other Hymeniacidonidae are alike, but not all *Halichondria* species are described with a tuft. It is of course possible that these differences reflect the uncertain status of the Halichondrida. Only within the Dendroceratida is a clear division of families found. The Halisarcidae have small larvae that are uniformly ciliated with a possible bare posterior pole and they are totally unpigmented. The Aplysillidae vary in all these features and their larvae are much larger. Bergquist *et al.* (1979) suggested to withdraw the Halisarcidae from the Dendroceratida. Apart from these examples it can be concluded that family differences are not very clear.

3/ Generic differences — From the literature *Myxilla* larvae are known to be totally white, while the larvae of other genera in the Myxillidae are pigmented with a white posterior pole. Within the Haplosclerida, four larvae belonging to two genera were investigated. There was no apparent generic character. Bergquist *et al.* (1979) reported differences between "*Reniera*", "*Chalinula*", "*Adocia*" and *Callyspongia* + *Haliclona*. They distinguished four larval types which in their view would be the basis for two groupings : one with a ring of longer cilia and a bare posterior pole which would include "*Adocia*", *Callyspongia* and *Haliclona* ; another group with the absence of such a ring containing "*Chalinula*" and "*Reniera*". De Weerdt (1985, 1986) has shown that *Adocia* and *Reniera* are synonyms of *Haliclona* and that *Chalinula* (sensu Topsent) is a synonym of *Acervochalina* ; studied representatives of these genera show a uniform larval type. It is here predicted that Bergquist *et al.*'s "*Reniera*" will prove to belong to the Halichondrida, and "*Chalinula*" to the Poecilosclerida. Generic differences have not been clearly established at present.

4/ Specific differences — Specific differences in larvae investigated were either very obvious or more obscure. *Haliclona rosea* was pink with a dark ring and a scarcely ciliated posterior pole whereas *Haliclona xena* was white with a totally bare and dark brown posterior pole.

Halichondria panicea and *Halichondria bowerbanki* were distinguished by the smaller size in *Halichondria bowerbanki*. *Halichondria bowerbanki* also possessed a lighter yellow colour, a thinner tuft, and the cilia towards the anterior pole were longer whereas in *Halichondria panicea* they were the same size.

Also in literature specific characters are clear. Most of the time the major features concerned were the ciliation pattern, the colour, the size, to a lesser degree the shape, and sometimes the behaviour.

LITERATURE CITED

Ayling A.L., 1980 - Patterns of sexuality, asexual reproduction and recruitment in some subtidal marine Demospongiae. *Biol. Bull.*, 158 : 271-282.

Barrois C., 1876 - Embryologie de quelques éponges de la Manche. *2e Thèse, Ann. Sc. nat.* (6) III, Art. 11 : 60-84.

Bergquist P.R., 1978 - Sponges. Hutchinson Univ. Libr., London. 268p.

Bergquist P.R. & Green C.R., 1977 - An ultrastructural study of settlement and metamorphosis in sponge larvae. *Cah. Biol. mar.*, 18 : 289-302.

Bergquist P.R. & Sinclair M.E., 1968 - The morphology and the behavior of larvae of some intertidal sponges. *N. Z. J. mar. Freshw. Res.*, 2 : 426-437.

Bergquist P.R., Sinclair M.E. & Hogg J.J., 1970 - Adaptation to intertidal existence : reproductive cycles and larval behaviour in Demospongiae : pp 247-271 in *The biology of the Porifera, Symp. zool. Soc. Lond., 25.* W.G. Fry (ed.), Academic Press, London.

Bergquist P.R. & Sinclair M.E., 1973 - Seasonal variation in settlement and spiculation of sponge larvae. *Mar. Biol.*, 20 : 35-44.

Bergquist P.R., Sinclair M.E., Green C.R. & Silyn-Roberts H., 1979 - Comparative morphology and behaviour of larvae in Demospongiae : pp 103-111 in *Biologie des Spongiaires, Colloq. int. C.N.R.S., 291*. C. Lévi & N. Boury-Esnault (eds), Paris.

Borojevic R., 1970 - Différentiation cellulaire dans l'embryogénèse et la morphogénèse chez les spongiaires : pp 467-490 in *The biology of the Porifera, Symp. zool. Soc. Lond., 25*. W.G. Fry (ed.), Academic Press, London.

Brien P., 1973 - Les démosponges, morphologie et reproduction : pp 133-461 in *Traité de Zoologie*, 3(1). P.P. Grassé (ed.), Masson et Cie, Paris.

Burton M., 1931 - The interpretation of the embryonic and post-larval characters of certain Tetraxonid sponges, with observations on post-larval growthstages in some species. *Proc. Zool. Soc. Lond.* : 511-525.

Burton M., 1932 - Sponges. *Disc. Rep.*, 7 : 237-392.

Chen W.T., 1976 - Reproduction and speciation in *Halisarca* : pp 113-140 in *Aspects of sponge biology*. F.W. Harrison & R.R.Cowden (eds), Academic Press, New York.

Delage Y., 1892 - Embryogénie des éponges. Développement post-larvaire des éponges siliceuses et fibreuses marines et d'eau douce. *Archs Zool. exp. gén.*, 10 : 345-498.

Fell P.E., 1974 - Diapause in the gemmules of the marine sponge *Haliclona loosanoffi* with a note on the gemmules of *Haliclona oculata*. *Biol. Bull.*, 147 : 333-351.

Fell P.E., 1974b - Porifera : pp 51-132 in *Reproduction of marine invertebrates, 1*. A.C. Giese & J.S. Pearse (eds), Academic Press, New York.

Fell P.E., 1976 - The reproduction of *Haliclona loosanoffi* and its apparent relationship to water temperature. *Biol. Bull.*, 150 : 200-210.

Fell P.E. & Jacob W.F., 1979 - Reproduction and development of *Halichondria* sp. in the Mystic estuary, Connecticut. *Biol. Bull.*, 155 : 62-75.

Fell P.E., Lewandrowski K. & Lovice M., 1979 - Post-larval reproduction and reproductive strategy in *Haliclona loosanoffi* and *Halichondria* sp. : pp 113-119 in *Biologie des Spongiaires, Colloq. int. C.N.R.S., 291*. C. Lévi & N. Boury-Esnault (eds), Paris.

Fell P.E., Parry E.H. & Balsamo A.M., 1984 - The life histories of sponges in the Mystic and Thames estuaries (Connecticut), with emphasis on larval settlement and post-larval reproduction. *J. exp. mar. Biol. Ecol.*, 78 : 127-141.

Fry W.G., 1971 - The biology of larvae of *Ophlitaspongia seriata* from two North Wales populations : pp 155-178 in *Fourth European marine Biology Symposium*. D.J. Crisp (ed.), Cambridge Univ. Press.

Gilbert J.J. & Simpson T.L., 1976 - Sex reversal in a freshwater sponge. *J. exp. Zool.*, 195 : 145-151.

Gurr G.T., 1963 - Biological staining methods. 7th edition, G.T. Gurr, London.

Hartman W.D., 1958 - Natural history of the marine sponges of southern New England. *Bull. Peabody Mus. Hist.*, 12 : 1-150.

Lévi C., 1951 - L'oviparité chez les spongiaires. *C.R. Acad. Sc. Paris*, 233 : 272-274.

Lévi C., 1956 - Etude des *Halisarca* de Roscoff. Embryologie et systématique des Démosponges. *Archs Zool. exp. gén.*, 93 : 1-181.

Lévi C., 1973 - Systématique de la classe des Demospongiaria (Démosponges) : pp 577-631 in *Traité de Zoologie 3(1)*. P-P. Grassé (ed.), Masson et Cie, Paris.

Kinne O., 1970 - Temperature-invertebrates : pp 407-517 in *Marine ecology*, I. O. Kinne (ed.), J. Wiley & sons, Chichester, New York.

Marshall W., 1882 - Die ontogenie von *Reniera filigrana*, Schmidt. *Z. wiss. Zool.*, 37 : 221-246.

Meewis H., 1939 - Contribution à l'étude de l'embryogénèse des Chalinidae : *Haliclona limbata* (Montagu). *Ann. Soc. r. zool. Belg.*, 70 : 201-247.

Meewis H., 1941 - Contribution à l'étude de l'embryogénèse des éponges siliceuses. Développement de l'oeuf chez *Adocia cinerea* (Grant) et *Halichondria coalita* (Bowerbank). *Ann. Soc. r. Zool. Belg.*, 72 : 126-147.

Parduzc B., 1967 - Ciliary movement and coordination in ciliates. *Int. Rev. Cytol.*, 21 : 91-128.

Simpson T.L., 1968 - The biology of the marine sponge *Microciona prolifera* (Ellis & Sollander). II. Temperature related, annual changes in fonctional and reproductive elements with a description of larval metamorphosis. *J. exp. mar. Ecol.*, 2 : 252-277.

Simpson T.L., 1984 - The cell biology of sponges. Springer-Verlag, New york. 662 p.

Stone A.R., 1970 - Growth and reproduction of *Hymeniacidon perleve* (Montagu) (Porifera) in Langstone Harbour, Hampshire. *J. Zool.*, 161 : 443-459.

Soest R.W.M. van, 1980 - Marine sponges from Curaçao and other Caribbean localities. Part II. Haplosclerida. *Stud. Fauna Curaçao Caribb. Islands*, 62(104) : 1-174.

Topsent E., 1888 - Contribution à l'étude des Clionides. *Archs Zool. exp. gén.*, 5(2) : 1-166.

Topsent E., 1911 - Sur les affinités des *Halichondria* et la classification des Halichondrines d'après leurs formes larvaires. *Archs Zool. exp. gén.*, 7(5) : 1-15.

Uriz M.J., 1982 - Reproducción en *Hymeniacidon sanguinea* (Grant, 1926) : biologia de la larva y primeros estadios postlarvarios. *Inv. Pesq.*, 46(1) : 29-36.

Vethaak A.D., Cronie R.J.A. & Soest R.W.M. van, 1982 - Ecology and distribution of two sympatric, closely related sponge species, *Halichondria panicea* (Pallas, 1766) and *H. bowerbanki* Burton, 1930 (Porifera, Demospongiae), with remarks on their speciation. *Bijdr. Dierk.*, 52(2) : 82-102.

Warburton F.E., 1958 - Reproduction of fused larvae in the boring sponge *Cliona celata* Grant. *Nature*, 181 : 493-494.

Warburton F.E., 1966 - The behaviour of sponge larvae. *Ecology*, 47 : 672-674.

Weerdt W.H. de, 1985 - A systematic revision of the north eastern Atlantic shallow-water Haplosclerida (Porifera, Demospongiae), Part I : Introduction, Oceanapidae and Petrosiidae. *Beaufortia*, 35(5) : 61-91.

Weerdt W.H. de, 1986 - A systematic revision of the north eastern Atlantic shallow-water Haplosclerida (Porifera, Demospongiae), Part II : Chalinidae. *Beaufortia*, 36 : 81-165.

Wilson H.V., 1935 - Some critical points in the metamorphosis of the Halichondrine sponge larva. *J. Morphol.*, 58 : 285-345.

A REVIEW OF NORTH-EASTERN ATLANTIC *HEMIGELLIUS* (NIPHATIDAE, HAPLOSCLERIDA)

W.H. de WEERDT & R.W.M. van SOEST

Institute of Taxonomic Zoology, University of Amsterdam
P.O. Box 20125, 1000 HC Amsterdam, The Netherlands.

SYNOPSIS

Eight nominal species referable to *Hemigellius* have been described from the north-eastern Atlantic area, but only two of them are well-established, viz. *Hemigellius arcoferus* (Vosmaer) and *H. pumiceus* (Fristedt). The remaining species are either synonyms of these two, or insufficiently documented.

INTRODUCTION

Among the Haplosclerids occurring in the north-eastern Atlantic Ocean, the Niphatidae van Soest, 1980, are the least documented. This is primarily due to the deep-water distribution of the family for this area. At present, no shallow-water records exist. Furthermore, the family is poorly represented. Only four species can be recognized, two of which are of doubtful identity. They are all referable to the genus *Hemigellius*, Burton, 1932. This genus is characterized by a skeleton of coarse multispicular primary tracts which are connected by uni-multispicular secondaries. The dermal skeleton is "formed by ends of primary fibres spreading out beneath the ectosome in a paniculate manner" (Burton, 1932 : 272).

The status of this genus is somewhat clouded by the fact that Burton indicated Topsent's (1901) *Gellius rudis* as the type, but made it clear that he however based his findings on specimens indentified as *G. rudis* by Kirkpatrick (1908). Yet, on the same page Burton also synonymized *Gellius fimbriatus* Kirkpatrick (1908) with Topsent's species. Vacelet & Arnaud (1972)

NATO ASI Series, Vol. G13
Taxonomy of Porifera
Edited by J. Vacelet and N. Boury-Esnault

(later confirmed by Boury-Esnault & van Beveren, 1982) discovered that Topsent's species did not possess the characters ascribed to it by Burton in his generic diagnosis. They concluded that Kirkpatrick's *Gellius rudis* was not conspecific with Topsent's type and that Burton's generic diagnosis referred to Kirkpatrick's rather than to Topsent's specimen. Fortunately, Kirkpatrick's *Gellius fimbriatus* also turned out to be clearly conspecific with his alleged *G. rudis*, which provided the opportunity to accept *fimbriatus* as the type species of *Hemigellius* with the unchanged diagnosis.

Both *Gellius arcoferus* Vosmaer (1885) and *G. pumiceus* Fristedt (1885) exhibit the same skeleton type as *H. fimbriatus*, and are therefore transferred to that genus. The generic definition may be slightly widened with respect to the presence of microscleres. Burton (1932) allowed for the presence of sigmata only, but *H. arcoferus* has toxa in addition to the sigmata and *H. pumiceus* has only toxa. *Gellius hartlaubi* Hentschel (1929) fits in *Hemigellius*, but only displays raphides. All the typical haplosclerid microscleres within the genus *Hemigellius* will be accepted in the description.

Despite the fact that the available material is very scarce, it is worthwhile to give a review of the *Hemigellius* species described from the north-eastern Atlantic. The study also fits in a taxonomic revision of the shallow-water Haplosclerida of that region (de Weerdt, 1985, 1986, in press).

MATERIAL AND METHODS

The material consists of original specimens from the collections in the Zoologisk Museum, København (ZMK) and the Zoölogisch Museum, Amsterdam (ZMA). In addition, a few specimens have been collected by dredging near Bergen, Norway.

Two microscopical preparations were made of the preserved specimens : one tangential and one cross-section. The sections were dried and mounted on a microscope slide with Canada balsam. Ten measurements were made of each spicule category. Only mature spicules were measured.

SYSTEMATICS

Order HAPLOSCLERIDA Topsent, 1928
Family Niphatidae van Soest, 1980
Genus *Hemigellius* Burton, 1932

Hemigellius arcoferus (Vosmaer, 1885) (n.comb.)
(Pl.I, Figs 1, 3, 4).

- *Gellius arcoferus* Vosmaer, 1885 : 29, Pl.IV, Figs 18, 19 ; Pl.V, Figs 87-90 - Fristedt, 1887 : 438, Pl.24, Figs 29-31 ; Pl.28, Fig.16 - Lambe, 1896 : 184, Pl.I, Figs 3, 3 a-b - Lundbeck, 1902 : 62, Pl.XII, Figs 11a-c, 1909 : 433 - Hentschel, 1929 : 898 - Burton, 1930 : 498.
- *Haliclona arcofera* ; Burton (pars), 1948 : 282 ; 1959 : 18.
- *Gelliodes plexa* Lundbeck, 1902 : 75, Pl.V, Figs 3, 4 ; Pl.XIV, Figs 3-5 ; 1909 : 434.
- *Gelliodes consimilis* Lundbeck, 1902 : 77, Pl.XIV, Figs 6a-c.
- *Orina consimilis* ; Laubenfels, 1942 : 263.
- *Gellius esperi* Arnesen, 1903 : 7, Pl.I, Fig.2.
- *Gellius jugosus* ; Koltun, 1959 : 211, Fig. 167 [Not : *Isodictya jugosa* Bowerbank, 1866 = *Haliclona fibulata* (Schmidt, 1862)].

Material examined.

— **Barents Sea** : ZMA POR. 1092, Stat.8, W. Barents Exp., 72°36'5 N, 24°57'5 E, 140 fms (vide Vosmaer, 1885 : 29), POR 1093, POR 6075, Stat. 20, W. Barents Exp., 77°7' N, 49°37'5 E, 170 fms (vide Vosmaer, 1885 : 29) ; POR. 1094, stat. 2, W. Barents Exp., 74°25'41'' N, 29°21' E, 215 fms, det. E. Arnesen ; Por 1095, Bodö, Leg. M. Weber, det. E. Arnesen.

— **Iceland** : ZMK, *Gelliodes plexa* Lundbeck, Ingolf Stat.3, 63°35' N, 10°24' W, 512 m, (vide Lundbeck, 1902 : 77).

— **Jan Mayen** : ZMK, *Gelliodes consimilis* Lundbeck, Ingolf Stat.113, 69°31' N, 7°06' W, 2465 m, 21-22-I-1896 (vide Lundbeck, 1902 : 78).

— **Canada** : ZMA, microscope slide made from Laubenfels' material of *Orina consimilis* (vide Laubenfels, 1942 : 263), USNM 23269, 66°46' N, 79°15' W, Fox Basin, Baffin Island.

Description.

Size and shape : ZMA POR. 6075 (Pl.I, Fig.1), designated here as the lectotype, is one of the specimens described by Vosmaer (1885 : 29). It is the specimen from Sta.20 of the Willem Barents Exp., 77°7' N 49°37'5 E, 170 fms. It is fan-shaped and measures 12 x 7 x 1.5 cm. Equal thickness is constant throughout. The basal part is damaged, so that the manner of attachment to the substratum cannot be observed. ZMA POR. 1092 is the other specimen described by Vosmaer (1885) ; this sponge is also fan-shaped and firmly attached to a small stone by a massive base. The height is 9.5 cm, the largest breadth 9 cm ; the thickness is 3 cm at the base and 0.5 cm at the top.

POR. 1095 is the largest specimen in the ZMA collection ; the fan, like the preceding specimen, is firmly attached to a stone with a broad and thick base. The total height of the sponge is 15 cm ; the largest breadth 8.5 cm ; the base of the specimen is of equal thickness, about 7 cm ; 2-2.5 cm up from the base, the sponge suddenly decreases in thickness to approximately 2 cm ; towards the top the thickness gradually tapers to 1 cm. The specimen is strongly fibrous ; at the surface of the fan, the spicule fibers are more or less tangentially orientated, but with numerous fibres projecting through the surface ; numerous circular and oval canal openings of 1-4 mm are situated at both sides of the fan, giving the surface a very open and loose appearance ; part of these openings is covered by a thin, transparent dermal membrane.

Colour : greyish-yellow (in alcohol).

Consistency : soft, fragile.

Surface : very open and loose with numerous canal openings ; strongly hispid by projecting fibres.

Ectosome : part of the choanosomal spicule fibres becomes tangentially oriented at the surface. They form an irregular reticulation around the canal openings.

Choanosome : the choanosomal skeleton is a rather close-meshed reticulation of strong multispicular primary lines, 125-400 µm thick. These are irregularly connected by unispicular secondar lines.

Spongin : very scarce, confined to the nodes of the spicules.

Megascleres : oxea : robust, slightly curved, with long, sharp points, 450-650 x 15-22 µm.

Microscleres : sigmata : slightly and unevenly curved, but with strongly curved points ; abundant, 12-28 x 1.0-1.9 µm. Toxa : rather thick, robust, evenly bent, with recurved apices, numerous, 95-180 x 3.0-6.8 µm.

Distribution.

Barents Sea (Vosmaer, 1885), Kara Sea (Koltun, 1959), NE of Taimur, Russia (Fristedt, 1887), Greenland (Fristedt, 1887 ; Lundbeck, 1902, 1909), Gulf of St. Lawrence (Lambe, 1896), Trondheim, Norway (Burton, 1930), Spitsbergen (Hentschel, 1929), between Iceland and Faroe (Lundbeck, 1902), Baffinland (Laubenfels, 1942).

Ecology.

Deeper than 160 m, on stones.

Remarks.

Hemigellius arcoferus is characterized by its robust toxa of variable size. In habit it is quite similar to *Hemigellius pumiceus* (see below), but this species has much thinner toxa, which lack the recurved apices. Apart from this *H. pumiceus* has no sigmata, and much larger oxea.

The conspecificity of *Gelliodes plexa* and *Gelliodes consimilis* with *H. arcoferus* has already been suggested by Burton (1948), and is confirmed by the present study. The spicules, both megascleres and microscleres, of the ZMK specimen of *G. consimilis* are somewhat larger than those of *G. plexa* (Table 1). The similarity in habit, skeletal architecture and form of the spicula of three species is, however, evident.

Table 1 - Spicule sizes of *Hemigellius arcoferus*.

SPECIMEN	OXEA	SIGMATA	TOXA
ZMA POR. 6075 *Gellius arcoferus* Lectotype	408.5-431.3(13.4)-446.5 / 15.2-20.9(2.3)-22.8	12.0-16.4(2.9)-21.6 / 1.0-1.4(0.3)-1.9	98.8-120.1(15.8)-152.0 / 3.4-4.5(0.7)-5.7
ZMK *Gelliodes plexa*	427.5-467.9(20.2)-503.5 / 20.0-21.1(1.1)-22.8	14.4-16.5(2.0)-19.2 / 0.8-1.3(0.4)-1.9	95.0-129.6(17.0)-152.0 / 3.0-4.7(1.3)-6.8
ZMK *Gelliodes consimilis*	593.8-624.5(17.7)-641.3 / 11.4-15.8(2.3)-18.1	15.6-22.0(5.3)-28.5 / 1.0-1.2(0.3)-1.9	110.2-146.5(23.1)-178.6 / 3.8-4.5(0.4)-4.9

Burton (1932) considered *Gellius esperi* Arnesen also a synonym of *H. arcoferus*. Arnesen's description of the species strongly resembles that of *H. arcoferus*. The species was described as fan-shaped, 20 x 15 cm, with a choanosomal skeleton consisting of coarse, multispicular primary lines. Megascleres evenly bent oxea, 450-560 x 16 µm and smaller oxea of 300 µm. Microscleres sigmata of 8 um and toxa of 160 µm. Arnesen described the species from a specimen in the collection of the Trondheim Museum but of which no further data were available.

Hemigellius pumiceus (Fristedt, 1885) (n.comb.)
(Pl.I, Figs 2,5,6)

- *Desmacella pumicea* Fristedt, 1885 : 29, Pl.II, Figs 9a-d.
- *Gellius pumiceus* ; Arndt, 1935 : 92, Fig. 198.
- *Toxadocia pumicea* ; Alander, 1942 :27.
- *Gellius massa* Arnesen, 1903 : 7, Pl.I, Fig.2 [=*Gellius arnesenae* Arndt, 1927 : 151].

<u>**Material examined**</u>.

Norway : ZMA POR. 6074, Vaagegrund, Bergen, 120-80 m, dredge, coll. W.H. de Weerdt & R.W.M. van Soest.

<u>**Description**</u>.

Size and shape : The ZMA material consists of two larger and one smaller, fan-shaped specimens, the largest of which measures 6.5 x 3 x 1.2 cm ; two of the specimens have a somewhat elongated form, the third is broader and shorter ; they are all of equal thickness. The structural features of the place of attachment are not observable, but the base of the three specimens is very small, which gives the impression that they were only weakly attached to the substratum. In each of the specimens there is one osculum ; the smallest measures 3 mm wide and the largest 5 mm. The interior of the oscular canal is pierced by numerous openings of the aquiferous system.

Colour : light brown (alive), off white (in alcohol).

Consistency : brittle, but not fragile.

Surface : rough and strongly hispid, very irregular with numerous canal openings, which vary in size from 0.1 to 2.0 mm.

Ectosome : the ectosomal skeleton consists of short spicule tracts which construct an irregular reticulum.

Choanosome : a strong, somewhat confused reticulation of multispicular primary lines, which are irregularly connected by unispicular secondary lines.

Megascleres : oxea : long and slender, slightly curved, ending in a long and sharp point, 910-985 x 21-28.5 µm.

Microscleres : toxa : very slender, sharply bent, with straight legs or with slightly recurved apices, in two size categories : 48-100.8 x 1.9-2.4 µm and 26.4-40.8 x 0.2-1.9 µm ; microtoxa : 6-19.2 x 0.2-0.7 µm.

Distribution.

Gullmarsfjord, Swedish West Coast (Fristedt, 1885), Hardangerfjord (Alander, 1942) and Bergen, Norwegian West Coast (ZMA material).

Ecology. 65-120 m.

Remarks.

The ZMA material is the third record of *Hemigellius pumiceus* since Fristedt's description of the species. The species is well characterized by its huge oxea, and by the presence of microtoxa. In skeletal architecture it is rather similar to *Gelliodes* (= *Hemigellius*) *bifacialis* Topsent, 1904, a species described from the Azores. Apart from their geographic distance, the species are also separated by their habit, by the smaller oxea, and absence of toxa in *G. bifacialis*.

Gellius massa Arnesen was considered as a synonym of *H. arcoferus* by Burton (1948), but we do not agree with this. Arnesen's description of the species is very meagre, but she describes the oxea as 900 x 28 µm, which is exactly the size in *H. pumiceus*. The size of the toxa was described as 160 µm. Arnesen did not mention the presence of sigmata, but she figured them in Pl.I, Fig. 3c. Possibly, she has given a wrong number to this figure. *Gellius massa* was described from Hjeltefjord and Bergen in Norway, at 130 m, thus very near our collecting site. Tentatively *G. massa* is considered conspecific with *H. pumiceus* in the opinion of the authors.

Hemigellius spec. aff. *flagellifer* Ridley & Dendy, 1886.

- *Gellius flagellifer* Lundbeck, 1902 : 71, Pl.II, Fig. 9, Pl.XIV, Figs 1a-d - Lambe, 1896 : 185, Pl.I, Figs 4a-d - Stephens, 1916 : 233 ; 1917 : 5 ; 1921 : 5 - Hentschel, 1929 : 978 - Koltun, 1959 : 212, Fig. 170.

- *Desmacella porosa* Fristedt, 1887 : 440, Pl.24, Figs 36, 37 ; Pl.28, Fig. 15.
- *Gellius porosus* ; Lundbeck, 1902 : 73, Pl.XIV, figs 2a-c - Koltun, 1959 : 213, Fig. 171.

Gellius flagellifer was originally described by Ridley & Dendy (1886) from Marion Island, southern Indian Ocean, but the species has been regularly recorded from north-eastern Atlantic localities. The species has large flagellate sigmata in addition to the small, normal ones. The skeleton is described as an irregular reticulation of very loose spicule fibers, with oxea of 420 x 18 µm. Most of the North Atlantic records of *G. flagellifer* agree with this description, but yet, it is doubtful whether one single species is concerned. However, it is a fact that a *Hemigellius* species occurs with flagellate sigmata in the North Atlantic Region. This species may turn out to be the species described by Fristedt (1887) as *Desmacella porosa*. It has been considered conspecific with *G. flagellifer* by many authors (Lambe, 1896 ; Topsent, 1904, 1928 ; Dendy, 1922). *Desmacella porosa* was originally described from Davis Strait as a cushion-shaped sponge with large, irregularly C- or S-shaped sigmata of 120 µm, and with oxea of 350 µm.

Finally, there is a third species which has been associated with *G.flagellifer*, i.e. *Desmacella vagabunda* Schmidt, 1870, originally described from Florida. Vosmaer (1885), who transferred this species to *Gellius*, described several specimens from the Barents Sea. Vosmaer's material is present in the ZMA (ZMA POR. 1117-1121), and it belongs to the Desmacellidae (probably a *Biemna* species). Topsent (1928) synonymized both *G. flagellifer* and *D. porosa* with *D. vagabunda*, but this does not seem justified.

Hemigellius hartlaubi (Hentschel, 1929) (n.comb.)

- *Gellius hartlaubi* Hentschel, 1929 : 897.

The species was described as follows :

Size and shape : massive, not exceeding 2 cm.

Colour : whitish, dull yellow.

Consistency : rather soft.

Surface : uneven.

Ectosome : a loose reticulation with 3-6 sided meshes, also with roundish

meshes formed by short, loose spicule tracts which consist of 1-4, tangentially orientated spicules.
Choanosome : irregular, consisting of loosely organized spicule tracts which cross irregularly.
Spongin : small amount, at nodes of the spicules.
Megascleres : oxea : with short points, strongly curved, 336-406 x 9-13 µm.
Microscleres : raphids : not organized in dragmata, 72-85 µm.
Locality : North Norway, 71°55' N, 20°54' E.
Depth : 192 m.

Remarks.

Hemigellius hartlaubi is the only *Hemigellius* known at present with raphides as microscleres. The species has not been recorded again since Hentschel's description.

DISCUSSION OF THE GENUS *HEMIGELLIUS*

Summarizing, there are two well recognizable *Hemigellius* species in the north-eastern Atlantic Region, (i.e. *Hemigellius arcoferus* and *H. pumiceus*), and two species of uncertain identity. The marked presence of different types of microscleres is conspicuous in these species. Another interesting fact is the predominantly deep-water distribution of the genus in this area. When comparing the area with the neighbouring south-eastern part of the North Atlantic, it appears that the same holds true for that area. There are two niphatids in the region, i.e. *Gelliodes fayalensis* Topsent, 1892, and *Gelliodes bifacialis* Topsent, 1904, both described from the Azores. Through the kindness of Dr. N. Boury-Esnault we could borrow Topsent's slides which are in the Paris Museum (D.T. 1058, *G. fayalensis*, and D.T. 1264, *G. bifacialis*). *Gellius bifacialis* has here been transferred to *Hemigellius* since its skeletal architecture agrees completely with *Hemigellius* and the primary fibers are far more loosely organized than in *Gelliodes*. *Hemigellius bifacialis* was recorded at 523 m (between Pico and São Jorge), 208 m (Banc de la Princesse Alice), and 1229 m (Florès) respectively (Topsent, 1892, 1904, 1928), thus also from considerable depth. Therefore it may be concluded that *Hemigellius* has a deep-water distribution in the entire eastern part of the North Atlantic Ocean. Outside this area the genus is represented in the Antarctic Region.

ACKNOWLEDGEMENTS

We wish to thank Dr T. Brattegard (Institutt for Marinbiologi, Bergen, Norway) for his hospitality during our stay at the Institute and for providing laboratory and dredging facilities. For the loan of specimens and microscopice slides we thank Dr O. Tendal (Zoologisk Museum, København) and Dr N. Boury-Esnault (Muséum National d'Histoire Naturelle, Paris).

The senior author was supported by a grant of the Foundation for Fundamental Biological Research (BION), which is subsidized by the Netherlands Organisation for the Advancement for Pure Research (ZWO).

LITERATURE CITED

Alander H., 1942 - Sponges from the swedish west-coast and adjacent waters. Thesis. (Lund). 95 p.

Arndt W.A., 1927 - Kalk- und Kieselschwämme von Curaçao. *Bijdr. Dierk.*, 25 : 133-158.

Arndt W.A., 1935 - Porifera (systematischer Teil). IIIa : pp 1-140 in *Die Tierwelt der Nord- und Ostsee*. G. Grimpe & E. Wagler (eds).

Arnesen E., 1903 - Spongien von der norwegischen Küste. *Bergens Mus. Arbog.*, 1 : 1-39.

Boury-Esnault N. & van Beveren M., 1982 - Les Démosponges du plateau continental de Kerguelen-Heard. *CNFRA*, 52 : 1-175.

Bowerbank J.S., 1866 - A monograph of the British Spongiadae II, I-XX, 1-388. (Ray Society, London).

Burton M., 1930 - Norwegian sponges from the Norman collection. *Proc. zool. Soc. London*, 2 : 487-546.

Burton M., 1932 - Sponges. *Disc. Reps.*, 6 : 237-392.

Burton M, 1948 - The synonymies of *Haliclona angulata* (Bowerbank) and *H. arcoferus* (Vosmaer). *Ann. Mag. nat. Hist.*, 12(1) : 273-284.

Burton M., 1959 - Spongia. *Zoology Iceland*, 2(304) : 1-71.

Dendy A., 1922 - Report on the Sigmatotetraxonida collected by H.M.S. "Sealark" in the Indian Ocean. *Trans. Linn. Soc. London, Zoology*, 18(1) : 1-164.

Fristedt K., 1885 - Bidrag till kännedomen om de vid sveriges vestra kust lefvande spongiae. *K. Svensk. Vet. -Akad. Handl.*, 21(6) : 1-56.

Fristedt K., 1887 - Sponges from the Atlantic and Arctic Oceans and the Behring Sea. *VEGA-Exped., vetensk. Arbeten*, 4 : 403-471.

Hentschel E., 1929 - Die Kiesel- und Hornschwämme des Nördlichen Eismeers. *Fauna Arctica*, 5 : 859-1042.

Koltun V.M., 1959 - [The Cornacuspongida of the northern and fareastern seas of the USSR. Keys to the fauna of the USSR], SSSR (Moscow, Leningrad), 1-236.

Kirkpatrick R., 1908 - Porifera 2. Tetraxonida. *Nation. antarct. Exped. 1901-1903, London, Zool. Botan.*, 4 : 1-56.

Lambe L.M., 1896 - Sponges from the Atlantic Coast of Canada. *Trans. r. Soc. Canada*, 4 : 181-211.

Laubenfels, M.W. de, 1942 - Porifera from Greenland and Baffinland collected by Capt. Robert A. Bartlett. *Journ. Washington Acad. Sci.* 32(9) : 263-269.

Lundbeck W., 1902 - Porifera. Part I. Homorrhaphidae and Heterorrhaphidae. *Danish Ingolf Exped.*, 6 : 1-108.

Lundbeck W., 1909 - The Porifera of East-Greenland. *Meddr. Grønl.*, 29 : 423-464.

Ridley S.O. & Dendy A., 1886 - Preliminary report on the Monaxonida collected by the H.M.S. "Challenger". *Ann. Mag. nat. Hist.*, (5)18 : 325-351.

Schmidt O., 1862 - Die Spongien des Adriatischen Meeres, Leipzig, I-VIII, 1-88.

Soest R.M.W. van, 1980 - Marine sponges from Curaçao and other Caribbean localities. Part II. Haplosclerida. *Stud. Fauna Curaçao Caribb. Islands*, 62(104) : 1-174.

Stephens J., 1916 - Preliminary notice of some Irish sponges. The Monaxellida (suborder Sigmatomonaxellida) obtained by the fisheries branch of the department of agriculture and technical instruction, Ireland. *Ann. Mag. nat. Hist.*, (8)20 : 232-242.

Stephens J., 1917 - Sponges collected by the dredging expeditions of the Royal Irish Academy and Royal Dublin Society. *Proc. R. Irish Acad.*, 34(1) : 1-16.

Stephens J., 1921 - Sponges of the coast of Ireland. II. The Tetraxonida (concluded). *Dept. Agr. Ireland, Fisheres sci. Invest.*, 2 : 1-75.

Topsent E., 1892 - Contribution à l'étude des Spongiaires de l'Atlantique Nord. *Rés. Camp. sci. Prince Monaco*, 2 : 1-165.

Topsent E., 1901 - Spongiaires. *Rés. Voy. S.Y. Belgica (1897-1898)*, 1-54.

Topsent E., 1904 - Spongiaires des Açores. *Rés. Camp. sci. Prince Monaco*, 25 : 1-280.

Topsent E., 1928 - Spongiaires de l'Atlantique et de la Méditerranée provenant des croisières du Prince Albert Ier de Monaco. *Rés. Camp. sci. Prince Monaco*, 74 : 1-376.

Vacelet J. & Arnaud F., 1972 - Invertébrés marins des XIIème et XV ème Expéditions antarctiques françaises en Terre Adélie. 2. Démosponges. *Téthys*, suppl.4 : 9-24.

Vosmaer G.C.J., 1885 - Sponges of the "Willem Barents" Expedition 1880 and 1881. *Bijdr. Dierk.* 12 : 1-47.

Weerdt W.H. de, 1985 - A systematic revision of the north-eastern Atlantic shallow-water Haplosclerida (Porifera, Demospongiae), Part I : Introduction, Oceanapiidae and Petrosiidae. *Beaufortia*, 35(5) : 61-91.

Weerdt W.H. de, 1986 - Ibid, Part II : Chalinidae. *Beaufortia*, 36(6) : 81-165.

Weerdt W.H. de, in press - The marine shallow-water Chalinidae (Haplosclerida, Porifera) of the British Isles : in *European contributions to the taxonomy of sponges*. W.C. Jones (ed.).

Plate 1 - **Fig.1** : *Hemigellius arcoferus* (Vosmaer), lectotype, ZMA POR 6075. **Fig.2** : *Hemigellius pumiceus* (Fristedt), ZMA POR 6074. **Fig.3** : *Hemigellius arcoferus* (Vosmaer), cross-section of choanosome. **Fig.4** : *Hemigellius arcoferus* (Vosmaer), microscleres. **Fig.5** : *Hemigellius pumiceus* (Fristedt), cross-section of choanosome. **Fig.6** : *Hemigellius pumiceus* (Fristedt), microscleres.

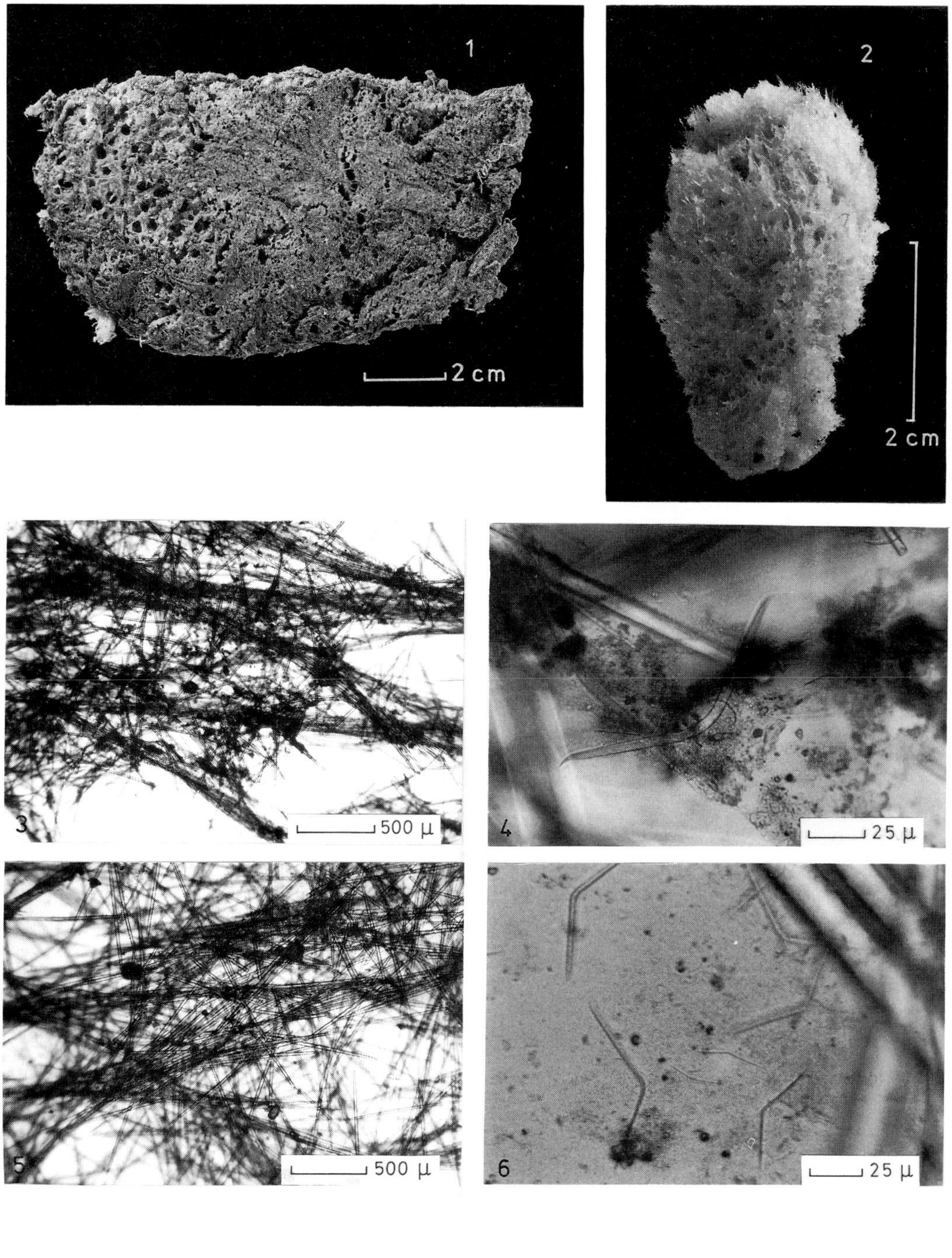
1
2 cm
2
2 cm
3
500 µ
4
25 µ
5
500 µ
6
25 µ

PARTICIPANTS IN THE N.A.T.O. WORKSHOP
(Observers*)

A. BIBILONI*
Centre d'Estudis Avançats de Blanes - Cami de Santa Barbara - Blanes (Gerona) - Spain.

N. BOURY-ESNAULT
Centre d'Océanologie de Marseille - Station Marine d'Endoume - Rue de la Batterie des Lions - 13007 Marseille - France.

J.C. BREAKMAN
Université libre de Bruxelles - Service de Chimie bio-organique - 50 Av. Franklin D. Roosevelt - 1050 Bruxelles - Belgium.

G. CORRIERO
Istituto di Zoologia - Via Archinafi 18 - 90123 Palermo - Italy.

C. DONADEY
Centre d'Océanologie de Marseille - Station Marine d'Endoume - Rue de la Batterie des Lions - 13007 Marseille - France.

D. DOUMENC
Museum National d'Histoire Naturelle - Laboratoire de Biologie des Invertébrés marins - 57 Rue Cuvier - 75231 Paris Cedex 05 - France.

R. FAUNDEZ-DESQUEYROUX
Musée d'Histoire Naturelle - Case postale n°284 - 1211 Genève 6 - Switzerland.

P. FRY
Luton college of Higher Education - Park square - Luton, Bedsforshire LU1 3JU - U.K.

M.F. GALLISSIAN*
Centre d'Océanologie de Marseille - Station Marine d'Endoume - Rue de la Batterie des Lions - 13007 Marseille - France.

J. HUYSECOM*
Laboratoire de Biologie animale et cellulaire - Université libre de Bruxelles - 50 Av. Franklin D. Roosevelt - 1050 Bruxelles - Belgium.

M. ILAN*
Department of Zoology - Tel-Aviv University - Ramat-Aviv - 69978 Tel-Aviv - Israël.

W.C. JONES
School of Animal Biology - University College of North-Wales - Bangor, Gwynedd LL57 2UW - U.K.

F. van LENT
Instituut voor Taxonomische Zoologie - Postbus 20125 - 1000HC Amsterdam - Netherlands.

C. LEVI
Museum National d'Histoire Naturelle - Laboratoire de Biologie des Invertébrés marins - 57 Rue Cuvier - 75231 Paris Cedex 05 - France.

R. MANCONI*
Istituto di Zoologia dell'Università - Via Balbi, 5 - 16126 Genova - Italy.

F. NOBBE
Haupstrasse 204 - D. 4690 Herne 2 - R.F.A.

J.A. OLIVER
Facultad de Ciencias Biologicas - Departamento de Zoologia - 50 Burjasot (Valencia) - Spain.

M. PANSINI
Istituto di Zoologia dell'Università - Via Balbi, 5 - 16126 Genova - Italy.

D. ROSSEL*
Centre d'Estudis Avançats de Blanes - Cami de Santa Barbara - Blanes (Gerona) - Spain.

K. RÜTZLER
Smithsonian Institution - Division of Echinoderms - Washington DC 20560 - U.S.A.

M. SARA
Istituto di Zoologia dell'Università - Via Balbi, 5 - 16126 Genova - Italy.

L. SCALERA-LIACI
Istituto di Zoologia dell'università degli Studi di Bari - Via Amendola, 165/A - 70126 Bari - Italy.

R. van SOEST
Institute of Taxonomic Zoology - University of Amsterdam -
Postbus 20125 - 1000HC Amsterdam - Netherlands.

A. SOLE-CAVA
Marine Biological Station - Port Erin, Isle of Man - U.K.

S. STONE
Zoology Department - British Museum (Natural History) -
Cromwell Road - London SW7 5BD - U.K.

O. TENDAL
Zoologisk Museum - Universitetsparken 15 DK - 2100 Kobenhavn -
Denmark.

M.J. URIZ
Centre d'Estudis Avançats de Blanes - Cami de Santa Barbara -
Blanes (Gerona) - Spain.

J. VACELET
Centre d'Océanologie de Marseille - Station Marine d'Endoume -
Rue de la Batterie des Lions - 13007 Marseille - France.

B. VERDENAL*
Centre d'Océanologie de Marseille - Station Marine d'Endoume -
Rue de la Batterie des Lions - 13007 Marseille - France.

E. VOULTSIADOU-KOUKOURA
University of Thessaloniki - Faculty of Sciences -
Department of Zoology - Thessaloniki - Greece.

W. de WEERDT
Instituut voor Taxonomische Zoologie - Postbus 20125 -
1000HC Amsterdam - Netherlands.

INDEX

NATO ASI Series G